D0138416

INTRODUCTION
TO DISCRETE-TIME
SIGNAL PROCESSING

INTRODUCTION TO DISCRETE-TIME SIGNAL PROCESSING

STEVEN A. TRETTER

University of Maryland, College Park

JOHN WILEY & SONS

New York • Santa Barbara • London • Sydney • Toronto

Library of Congress Cataloging in Publication Data:

Tretter, Steven A
    Introduction to discrete-time signal processing.

    Bibliography: p.
    Includes index.
    1. Discrete-time systems. 2. Signal processing.
I. Title.
QA402.T73   621.38'043'015157   76-25943
ISBN 0-471-88760-9

Printed in the United States of America

10 9 8 7 6 5 4 3 2 1

To Anne, David, and Terry

PREFACE

This book grew out of a one-semester graduate electrical engineering course I have been teaching periodically at the University of Maryland since 1969. In the course I included standard sampled-data theory, digital filtering techniques using the difference equation and fast Fourier transform (FFT) approaches, classical linear parameter estimation methods, and the Kalman filter. No single textbook was available that adequately presented all of these important and related topics.

The book emphasizes discrete-time systems and signal processing. Discrete-time systems almost always interface with continuous-time systems, and it is important to take into account the effects of these interfaces in overall system design. Therefore, methods for analyzing mixed continuous and discrete-time systems (that is, sampled-data systems) as well as strictly discrete-time systems have been included.

To thoroughly comprehend the book the reader must understand (1) the theory of complex variables up to the evaluation of contour integrals by the method of residues; (2) Fourier transforms, Laplace transforms, and linear system analysis at the senior or first-year graduate level; and (3) random processes at the senior or first-year graduate level (for example, as presented in Papoulis, *Probability, Random Variables, and Stochastic Processes*). In general, the mathematical sophistication required increases as the book progresses. Even though the book was written for a graduate course, practicing engineers and scientists with a solid undergraduate math background will be able to understand most of the material in Chapters 1 to 6 and 8 to 10. These chapters provide a good working knowledge of digital filtering. A knowledge of random processes at the introductory graduate level is required to understand most of the theoretical derivations in Chapters 7, 11, 13, and 14.

Chapter 1 is a brief introduction. Chapters 2 to 5 present what might be called the classical methods for analyzing sampled-data systems with deterministic inputs. In particular, the effects of uniform sampling and the theory of Z-transforms are developed. Chapter 6 presents the more recent state-space approach to system representation and analysis and provides background for Chapter 14. Chapter 7 extends the standard results for stationary continuous-time random processes and linear systems to sampled-data systems. Chapters 8 to 10 discuss methods for designing digital filters with prescribed frequency responses and some of the practical problems involved in implementing them by difference equations or the FFT. Chapter 11 is an introduction to the subject of power spectral density estimation. Chapters 12 to 14 form an introduction to linear parameter estimation. Chapter 12 introduces the elements of Hilbert spaces to provide the mathematical foundation for a unified approach to linear estimation; Chapter 13 discusses the estimation of a constant parameter vector from a fixed-length data record; and Chapter 14 presents the recursive estimation algorithm known as the Kalman filter.

Working problems is an important part of thoroughly learning a technical subject. Approximately 240 problems are included here. Some are simple drill problems while others extend the textual material.

The book includes more material than can be covered in a one-semester course to allow instructors a choice of topics, to provide advanced topics for motivated students, and to provide reference material for practicing engineers and scientists. My preference for an advanced graduate course is to cover, in order, these chapters and sections: Chapter 1, 2.1–2.4, 3.5–3.9, 4.1–4.8, 4.11, 4.12, 5.1–5.8, 5.12–5.14, 7.1–7.8, 8.1–8.5, 8.7, 8.8, 8.10, 9.1–9.7, Chapter 10, Chapter 12 (two 50-minute lectures), 13.5, 6.1–6.5 (one 50-minute lecture), and 14.1–14.3. Some instructors may prefer to replace Chapter 6 and Chapters 12 to 14, which form the block on linear parameter estimation, with Chapter 11, "Estimation of Power Spectral Densities."

The remarkable advances in digital computer technology have made the signal processing techniques discussed here practical to use in many situations. I think that knowledge of discrete-time signal processing techniques should now be and will shortly become a requirement for the well-educated electrical engineer.

Finally, I thank Professor K. Steiglitz of Princeton University for enthusiastically introducing me to this subject in 1963.

Steven A. Tretter

CONTENTS

6 STATE SPACE REPRESENTATION OF SAMPLED-DATA SYSTEMS   127

7 LINEAR SYSTEMS AND UNIFORMLY SAMPLED RANDOM PROCESSES   165

INTRODUCTION
TO DISCRETE-TIME
SIGNAL PROCESSING

1

INTRODUCTION

1.1 SOME COMMENTS, DEFINITIONS, AND HISTORICAL BACKGROUND

Recent advances in digital circuit technology, in particular, the development of large-scale integration (LSI) technology, have caused a revolution in system design. Many functions that only a few years ago were most practically implemented with analog circuitry can now be implemented more practically in digital form. Using the digital approach, the designer no longer has to be concerned with the realizability constraints of analog devices. As a result, significantly more sophisticated algorithms can now be chosen for problem solving.

Digital systems operate in discrete time steps. Therefore, when an analog device is replaced by a digital system, the continuous-time input signal must be converted into a sequence of numbers. This is usually done by sampling the input signal at uniformly spaced time instants. We will call the sequence of samples a *discrete-time signal*. Sampling can also arise in systems for other reasons. For example, a revolving radar antenna can see a particular target only once per revolution. A system in which continuous-time signals appear at some points and discrete-time signals appear at others is known as a *sampled-data system*. A system in which only discrete-time signals appear is known as a *discrete-time system*. For example, a computational algorithm for converting an input sequence into an output sequence would be a discrete-time system.

Intensive development of the theory of sampled-data systems began in the late 1940s. Through the 1950s, the research efforts centered on the design and

1

analysis of sampled-data control systems. These systems typically involved low-frequency signals, and it was realistic to consider building digital controllers using the elementary digital circuit technology of the times. Through the use of the Z-transform, the classical techniques for designing analog control systems were extended to sampled-data control systems. By 1958, the subject was well-formulated and the first textbooks appeared [60,114]. With the advent of the aerospace era in the late 1950s and improved digital technology, research efforts shifted to the design of control systems by the time-domain state space approach rather than the classical frequency-domain transform approach.

In this book, we are primarily concerned with *signal processing* (i.e., filtering and estimation) rather than control system design. The distinction between signal processing and control system design lies in the differences of the desired goals. In control system design, the goal is to make the outputs of a system behave in some desired manner relative to its inputs, usually through the use of appropriate feedback. The systems must operate in real time. Delays around the feedback loops must be held to a minimum to maintain stability. On the other hand, the goal of filtering (for example, in a communication system) is to modify the spectrum of a received signal in some desired manner. A moderate overall delay is usually perfectly acceptable if it is needed to achieve the desired filter frequency response. The goal of estimation is to estimate the parameters of a signal masked by noise, an autocorrelation function, a power spectral density, and so on. Signal processing is often performed off-line in nonreal time.

Before high-speed digital computers were developed, discrete-time signal processing was limited to relatively simple algorithms. Advancements in digital computer technology revealed the enormous potential of computers, and extensive research to develop sophisticated discrete-time signal processing techniques began. In 1958 Blackman and Tukey published classic articles in the *Bell System Technical Journal* describing how to estimate power spectra from a finite set of signal samples [10]. Techniques were also developed for designing discrete-time filters, or digital filters as they are commonly called, to closely approximate specified frequency responses. J. F. Kaiser of Bell Labs was one of the principal early contributors to the field of digital filter design. His Chapter 7 in Reference 81 was the first extensive description of digital filter design methods to appear in a book. The standard approach at this time was to implement the digital filters as difference equations. In 1965 Cooley and Tukey published an article describing an algorithm, now known as the fast Fourier transform (FFT), for very efficiently computing Fourier series at a set of uniformly spaced points [26]. The FFT changed the approach to digital power spectrum estimation and significantly reduced the computation time. It also made a frequency-domain approach to digital filtering competitive with the time-domain difference equation approach. The development of discrete-time parameter estimation techniques was influenced by the needs of the aerospace industry. Kalman's 1960 paper [67], which presented a recursive algorithm (now known as the Kalman filter) for estimating the state of a linear dynamical system, was a milestone in the development of estimation techniques.

1.2 OVERVIEW OF THE BOOK

Chapters 2 through 5 present what might be called the classical methods for analyzing linear sampled-data systems with deterministic inputs. In Chapter 2 the frequency-domain effects of sampling a signal at uniformly spaced time instants are examined. In general, sampling introduces an irreversible phenomenon known as aliasing. However, we will see that a signal band limited to half the sampling rate can be exactly reconstructed from its samples. This result is known as the sampling theorem and is fundamental to the theory of discrete-time signal processing. The practical technique of approximately reconstructing a signal from its samples by polynomial interpolation and extrapolation is discussed in Chapter 3. The theory of Z-transforms is developed in Chapter 4. The Z-transform plays the same role for discrete-time systems as the Laplace transform does for continuous-time systems. In Chapter 5 methods for analyzing sampled-data systems by a combination of Fourier or Laplace and Z-transform techniques are presented. We will see that the time-domain behavior of a linear time-invariant discrete-time system can be described by a linear constant-coefficient difference equation. We will also find that, like continuous-time systems, this type of system can be described in the z-domain by a transfer function that is the Z-transform of the system response to a unit pulse. We will also see that the sinusoidal steady-state frequency response of the system can be obtained by evaluating its transfer function at $z = e^{j\omega T}$ where T is the sampling period and ω is the frequency variable. Techniques for checking the stability of discrete-time systems are presented. The modified Z-transform and fast output sampling methods of determining the behavior of a sampled-data system between sampling instants are discussed. Data reduction by skip-sampling is also considered. As the book is studied, it will soon become clear that all the standard analysis techniques and results for continuous-time systems have direct parallels for discrete-time systems.

Chapter 6 introduces the state space method of representing discrete- and continuous-time systems. The state space representation is used extensively in modern control and system theory. This chapter provides background for Chapter 14.

Chapter 7 presents what might be called the classical theory of sampled-data systems with random inputs. Only linear time-invariant systems with uniform sampling and stationary random processes are considered. We will see that the power spectral density for a sampled random process is the Z-transform of its autocorrelation function. Formulas for finding correlation functions and power spectral densities of discrete-time random processes passed through linear discrete-time filters are derived and are directly analogous to those for the continuous-time case. The discrete-time version of the Wiener filter is discussed. The problem of optimum one-step linear prediction using the last N samples is solved. The linear predictor has recently found applications in speech analysis and synthesis. Methods for analyzing mixed continuous and discrete-time systems with random inputs are presented and applied to the problem of optimum signal reconstruction.

Digital filtering is discussed in Chapters 8, 9, and 10. In Chapter 8 methods for finding digital filter transfer functions that closely approximate desired frequency responses are presented, and methods for designing both recursive and nonrecursive filters are examined. The design methods in Chapter 8 are based on the assumption of perfect implementation. Chapter 9 discusses some of the practical considerations that must be taken into account when a filter is actually implemented digitally. The problems of quantization, finite-word-length arithmetic, and limit cycles are investigated. In Chapter 10 a frequency-domain approach to digital filtering through the use of the discrete Fourier transform (DFT) is presented. This approach became practical after the discovery of the FFT algorithm for efficiently computing DFTs. Prior to that time, the time-domain difference equation approach was used almost exclusively. Both approaches have advantages and disadvantages. The choice between the two depends on the problem constraints.

Estimation of autocorrelation functions and power spectral densities from a finite length sequence of uniformly spaced samples is discussed in Chapter 11. Three methods for estimating power spectral densities are presented: (1) averaging periodograms, (2) smoothing a single periodogram, and (3) the autoregressive model. The use of the FFT in efficiently performing the required computations is discussed.

Chapters 12, 13, and 14 introduce the subject of linear parameter estimation. Chapter 12 introduces the elements of Hilbert space theory to provide the mathematical background for a unified approach to linear estimation. The projection theorem, which is a powerful tool for solving linear estimation problems, is proved. Three linear parameter estimation problems are solved in Chapter 13. The first problem is strictly deterministic—the least-squares approximation of a finite-length observed data sequence by a linear combination of a set of functions (such as a set of polynomials). In the second problem the linear minimum-variance unbiased estimate of a constant but unknown parameter vector is determined from a finite-length observed data sequence that is the sum of a linear transformation of the parameter vector and noise. The third problem is that of finding the linear minimum-mean-square-error estimate of a random parameter vector from a related finite-length observed data sequence. The three problems are similar in that the observed data record length is fixed and the parameters do not vary with time. In Chapter 14 the results of Chapter 13 are extended to the case where the state of a linear dynamical system is to be estimated, and the observed data sequence is received sequentially in time. Recursive estimation algorithms, commonly called Kalman filters, are derived.

The majority of this book is devoted to discrete-time systems and signal processing techniques. However, discrete-time systems almost always interface with continuous-time systems. To properly design an overall system, it is important to take into account the effects of these interfaces. Therefore, methods for analyzing mixed continuous and discrete-time systems, that is, sampled-data systems, have been included.

2

UNIFORM SAMPLING

2.1 INTRODUCTION

In most applications signals are sampled periodically every T seconds. This is called *uniform sampling*. Uniform sampling is commonly used because it leads to the simplest system designs and computational algorithms. In addition, systems using uniform sampling can be analyzed relatively easily by transform methods.

In this chapter we will see that uniform sampling of a signal $f(t)$ can be represented as amplitude modulation of a periodic pulse train by $f(t)$. This leads to an expression relating the spectrum of the sampled signal to the spectrum of $f(t)$. This expression shows how uniform sampling introduces a phenomenon known as *aliasing*, which is related to the ambiguity resulting from sampling. In the special case of low-pass band-limited signals we will see that $f(t)$ is uniquely represented by its samples if the sampling rate is at least twice the signal bandwidth. This important result, known as the *sampling theorem*, is the theoretical basis for modern pulse-code modulation communication systems. A similar sampling theorem is derived for band-pass signals.

2.2 SAMPLING AS A MODULATION PROCESS

An ideal uniform sampler would measure the exact values of the signal $f(t)$ at the sampling instants nT to provide the sequence $\{f(nT)\}$. In practice, ideal

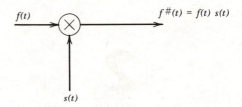

FIGURE 2.2.1.   Sampling by modulation.

sampling is not possible but can be approximated closely. One way of approaching nonideal sampling is to consider the sampled signal $f^{\#}(t)$ to be the product of $f(t)$ and a periodic sampling pulse train $s(t)$. This is shown in Figs. 2.2.1 and 2.2.2. Normally $s(t)$ would consist of a train of short pulses separated by T seconds. Ideal sampling results when $s(t)$ is a train of Dirac delta functions. In this representation the sampling process is a type of amplitude modulation. The carrier signal is $s(t)$ and the modulating signal $f(t)$.

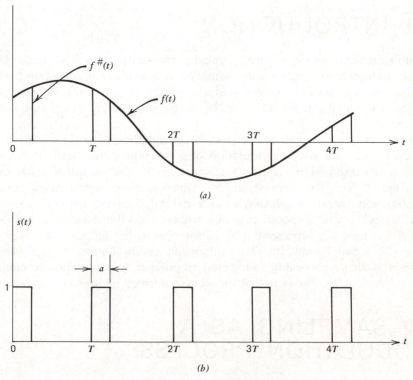

FIGURE 2.2.2.   Sampling with a periodic rectangular pulse train. (a) Original and sampled signals; (b) sampling pulse train.

Since $s(t)$ is periodic, it can be represented by the Fourier series

$$s(t) = \sum_{n=-\infty}^{\infty} c_n e^{jn\omega_s t} \tag{2.2.1}$$

where $\omega_s = 2\pi/T$ is the sampling rate in radians/second and

$$c_n = \frac{1}{T} \int_{-T/2}^{T/2} f(t) e^{-jn\omega_s t} \, dt \tag{2.2.2}$$

Therefore

$$f^{\#}(t) = f(t) \sum_{n=-\infty}^{\infty} c_n e^{jn\omega_s t} = \sum_{n=-\infty}^{\infty} c_n f(t) e^{jn\omega_s t} \tag{2.2.3}$$

The Fourier transform of the sampled signal is

$$F^{\#}(\omega) = \mathcal{F}\left\{ \sum_{n=-\infty}^{\infty} c_n f(t) e^{jn\omega_s t} \right\} = \sum_{n=-\infty}^{\infty} c_n \mathcal{F}\{f(t) e^{jn\omega_s t}\}$$

$$= \sum_{n=-\infty}^{\infty} c_n F(\omega - n\omega_s) \tag{2.2.4}$$

where $F(\omega) = \mathcal{F}\{f(t)\}$. Similarly the Laplace transform of $f^{\#}(t)$ is

$$\mathcal{F}^{\#}(s) = \sum_{n=-\infty}^{\infty} c_n \mathcal{F}(s - jn\omega_s) \tag{2.2.5}$$

Thus the sampling introduces new spectral components that are translations of the baseband transform $F(\omega)$. This is illustrated in Fig. 2.2.3. If $f(t)$ is band-limited, that is, $F(\omega) \equiv 0$ for $|\omega| > \omega_c$, and $\omega_s > 2\omega_c$ then it is clear from Fig. 2.2.3d that the translations do not overlap and that the original signal can be recovered to within a scale factor by passing $f^{\#}(t)$ through a filter with the transfer function

$$H(\omega) = \begin{cases} 1 & \text{for} & |\omega| < \omega_c \\ 0 & \text{for} & |\omega| \geq \omega_c \end{cases}$$

$H(\omega)$ is an ideal low-pass filter with cutoff frequency ω_c. It is physically unrealizable but can be approximated as closely as desired with sufficient delay. If $\omega_s < 2\omega_c$ or $f(t)$ is not band-limited, then the translations overlap and the original signal cannot be recovered by low-pass filtering $f^{\#}(t)$. This overlapping is known as *aliasing*. This is a result of the ambiguity introduced by sampling. Clearly, if the sampling pulse train $s(t)$ is identically zero over finite intervals, then $f(t)$ cannot be uniquely determined from $f^{\#}(t)$ in general.

Further insight into the phenomenon of aliasing can be gained by considering the sampling of a pure sinusoid. If $f(t) = e^{j\omega_0 t}$, then its samples are $f(nT) = e^{j\omega_0 nT}$. But

$$e^{j\omega_0 nT} = e^{j\omega_0 nT} e^{j2\pi nk} = e^{jnT[\omega_0 + (2\pi/T)k]}$$

$$= e^{jnT(\omega_0 + k\omega_s)}$$

Therefore $f_1(t) = e^{j(\omega_0 + k\omega_s)t}$ has the same samples as $f(t)$.

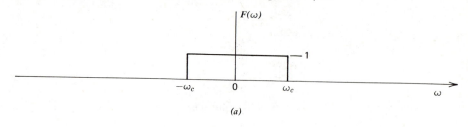

(a)

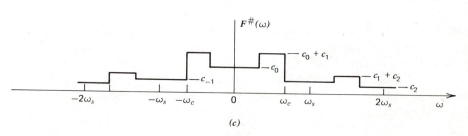

(b)

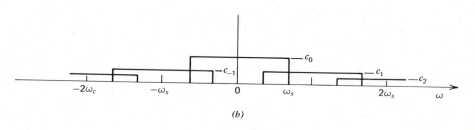

(c)

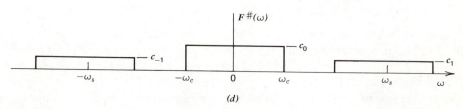

(d)

FIGURE 2.2.3. Spectrum of sampled signal. (a) Original transform; (b) Weighted baseband transform and translations for $\omega_s < 2\omega_c$; (c) $F^{\#}(\omega)$ for $\omega_s < 2\omega_c$; (d) $F^{\#}(\omega)$ for $\omega_s > 2\omega_c$.

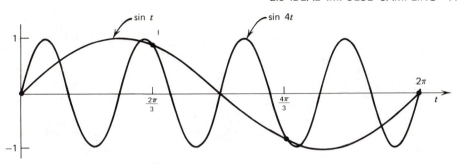

FIGURE 2.2.4.   Sinusoids with the same samples.

Example 2.2.1

Let $\omega_0 = 1$, $\omega_s = 3$, $k = 1$. Then $\text{Im}\{f(t)\} = \sin t$ and $\text{Im}\{f_1(t)\} = \sin 4t$ have the same samples. This is shown in Fig. 2.2.4   ◄

2.3 IDEAL IMPULSE SAMPLING

Ideal sampling can be represented by letting the sampling pulse train $s(t)$ be a sequence of Dirac delta or impulse functions. Then for uniform sampling

$$s(t) = \sum_{n=-\infty}^{\infty} \delta(t - nT) \qquad (2.3.1)$$

The resulting sampled signal is

$$f^*(t) = f(t) \sum_{n=-\infty}^{\infty} \delta(t - nT) = \sum_{n=-\infty}^{\infty} f(t)\,\delta(t - nT) \qquad (2.3.2)$$

The symbol * will be used to denote signals sampled by ideal impulse sampling. The impulse function $\delta(t - nT)$ can be considered to be zero, except in the vicinity of $t = nT$, and have area 1. Therefore, (2.3.2) is equivalent to

$$f^*(t) = \sum_{n=-\infty}^{\infty} f(nT)\,\delta(t - nT) \qquad (2.3.3)$$

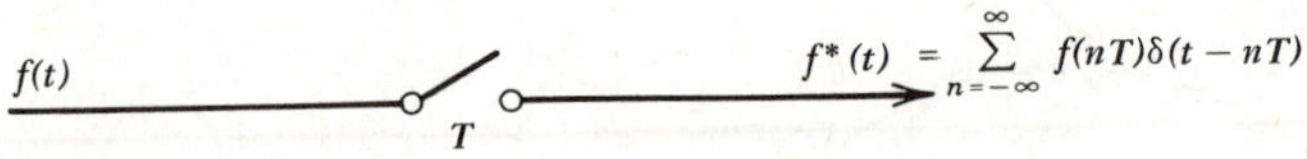

$$f(t) \qquad\qquad\qquad\qquad f^*(t) = \sum_{n=-\infty}^{\infty} f(nT)\delta(t-nT)$$

$$T$$

FIGURE 2.3.1. Ideal impulse sampler.

This can be justified more rigorously if impulses are interpreted in the distribution sense [103]. Thus the sequence of samples is now represented by a train of impulse functions whose areas correspond to the sample values.

In the remainder of this book, ideal impulse sampling is assumed. The sampler or impulse modulator will be represented schematically by a switch, as shown in Fig. 2.3.1, with the sampling period indicated below the switch.

From (2.2.1) and (2.2.2) we see that

$$c_n = \frac{1}{T} \int_{-T/2}^{T/2} \delta(t)e^{-jn\omega_s t}\, dt = \frac{1}{T} \qquad\qquad (2.3.4)$$

and

$$s(t) = \frac{1}{T} \sum_{n=-\infty}^{\infty} e^{jn\omega_s t} \qquad\qquad (2.3.5)$$

According to (2.2.4) and (2.2.5) the Fourier and Laplace transforms of $f^*(t)$ are

$$F^*(\omega) = \frac{1}{T} \sum_{n=-\infty}^{\infty} F(\omega - n\omega_s) \qquad\qquad (2.3.6)$$

and

$$\mathscr{F}^*(s) = \frac{1}{T} \sum_{n=-\infty}^{\infty} \mathscr{F}(s - jn\omega_s) \qquad\qquad (2.3.7)$$

On the other hand, the Fourier transform of (2.3.3) yields

$$F^*(\omega) = \sum_{n=-\infty}^{\infty} f(nT)e^{-j\omega nT} \qquad\qquad (2.3.8)$$

and its Laplace transform is

$$\mathscr{F}^*(s) = \sum_{n=-\infty}^{\infty} f(nT)e^{-snT} \qquad\qquad (2.3.9)$$

Thus we have two different methods for calculating the transforms of the sampled signal. Usually (2.3.8) and (2.3.9) lead most easily to a closed form solution. However, (2.3.6) allows the effects of aliasing to be visualized more clearly and leads to answers more directly when $f(t)$ is band-limited or has discrete frequency components that give impulses in $F(\omega)$.

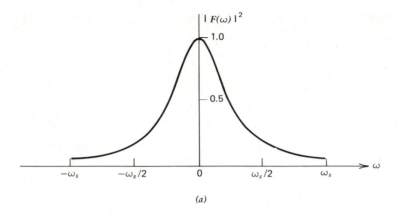

(a)

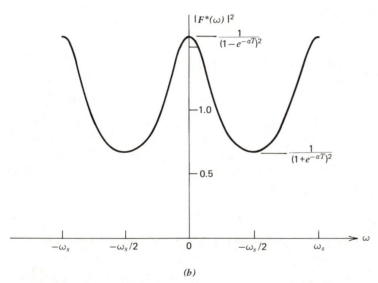

(b)

FIGURE 2.3.2. $|F(\omega)|^2$ and $|F^*(\omega)|^2$ vs. ω for $f(t) = e^{-at}u(t)$ with $a = 1$ and $\omega_s = 4$. (a) $|F(\omega)|^2$ vs. ω; (b) $|F^*(\omega)|^2$ vs. ω.

From (2.3.6) or (2.3.8) it is clear that $F^*(\omega)$ is periodic with period ω_s. Also $\mathscr{F}^*(s + jk\omega_s) = \mathscr{F}^*(s)$ so that $\mathscr{F}^*(s)$ repeats with period ω_s when evaluated along a path parallel to the $j\omega$ axis. Equation 2.3.8 is a Fourier series expansion for $F^*(\omega)$ and we see immediately that the Fourier coefficients are just the sample values $f(nT)$. Given $F^*(\omega)$, the signal samples can be calculated as

$$f(nT) = \frac{1}{\omega_s} \int_{-\omega_s/2}^{\omega_s/2} F^*(\omega)e^{j\omega nT}\, d\omega \qquad (2.3.10)$$

Equating (2.3.7) and (2.3.9) shows that

$$\frac{1}{T}\sum_{n=-\infty}^{\infty} \mathcal{F}(s-jn\omega_s) = \sum_{n=-\infty}^{\infty} f(nT)e^{-snT} \qquad (2.3.11)$$

This is known as the *Poisson sum formula*.

Example 2.3.1

Let $f(t) = e^{-at}u(t)$ with $a > 0$ and $u(t)$ the unit step function defined as

$$u(t) = \begin{cases} 0 & \text{for} & t < 0 \\ 1 & \text{for} & t \geq 0 \end{cases}$$

In this case $F(\omega) = 1/(j\omega + a)$. From (2.3.8)

$$F^*(\omega) = \sum_{n=0}^{\infty} e^{-anT}e^{-j\omega nT} = \frac{1}{1-e^{-aT}e^{-j\omega T}}$$

To see the effect of aliasing

$$|F^*(\omega)|^2 = \frac{1}{1+e^{-2aT}-2e^{-aT}\cos \omega T} \qquad \text{and} \qquad |F(\omega)|^2 = \frac{1}{a^2+\omega^2}$$

are plotted in Fig. 2.3.2 for $a = 1$ and $\omega_s = 4$. Note that even though the sampling rate is four times the 3 dB signal bandwidth, the aliasing is still significant. If T is large, or equivalently, the sampling rate is slow, then $|F^*(\omega)|^2$ becomes very flat and is not at all like $|F(\omega)|^2$. ◀

2.4 THE SAMPLING THEOREM

We have observed that a signal cannot be uniquely recovered from its samples in general. In Section 2.2 we saw that for the special case of a band-limited signal sampled at a rate greater than twice the highest frequency component in the signal, the original signal could be recovered by passing the sampled signal through an ideal low-pass filter. This important result is derived for the case of ideal sampling in the following theorem, which was popularized by Shannon [98].

THEOREM 2.4.1. The Sampling Theorem
Let $f(t)$ be a signal with a Fourier transform $F(\omega)$ such that $F(\omega) \equiv 0$ for $|\omega| \geq \omega_c$.

Then

$$f(t) = \sum_{n=-\infty}^{\infty} f(nT) \frac{\sin \omega_c (t-nT)}{\omega_c (t-nT)} \tag{2.4.1}$$

where $\omega_s = 2\omega_c$ and $T = 2\pi/\omega_s$.

Proof In the interval $|\omega| < \omega_c$, $F(\omega)$ can be represented by the Fourier series

$$F(\omega) = \sum_{n=-\infty}^{\infty} f_n e^{-jn(\pi/\omega_c)\omega} \tag{2.4.2}$$

where

$$f_n = \frac{1}{2\omega_c} \int_{-\omega_c}^{\omega_c} F(\omega) e^{jn(\pi/\omega_c)\omega} \, d\omega \tag{2.4.3}$$

and since $F(\omega) = 0$ for $|\omega| \geq \omega_c$

$$f_n = \frac{2\pi}{2\omega_c} \frac{1}{2\pi} \int_{-\infty}^{\infty} F(\omega) e^{jn(\pi/\omega_c)\omega} \, d\omega$$

$$= \frac{\pi}{\omega_c} f\left(n \frac{\pi}{\omega_c}\right) \tag{2.4.4}$$

Substituting (2.4.4) into (2.4.2) gives

$$F(\omega) = \begin{cases} \sum_{n=-\infty}^{\infty} \frac{\pi}{\omega_c} f\left(n \frac{\pi}{\omega_c}\right) e^{-jn(\pi/\omega_c)\omega} & \text{for} \quad |\omega| < \omega_c \\ 0 & \text{for} \quad |\omega| \geq \omega_c \end{cases}$$

and

$$f(t) = \frac{1}{2\pi} \int_{-\omega_c}^{\omega_c} F(\omega) e^{j\omega t} \, d\omega = \sum_{n=-\infty}^{\infty} f\left(n \frac{\pi}{\omega_c}\right) \frac{1}{2\omega_c} \int_{-\omega_c}^{\omega_c} e^{j[t-n(\pi/\omega_c)]\omega} \, d\omega$$

$$= \sum_{n=-\infty}^{\infty} f\left(n \frac{\pi}{\omega_c}\right) \frac{\sin \omega_c (t-n\pi/\omega_c)}{\omega_c (t-n\pi/\omega_c)} \tag{2.4.5}$$

Q.E.D.

The sampling theorem can also be obtained by observing from the aliasing formula (2.3.6) that $F^*(\omega) = F(\omega)/T$ for $|\omega| < \omega_c$ if $\omega_s = 2\omega_c$. If $f^*(t)$ is passed through the ideal low-pass filter

$$G(\omega) = \begin{cases} T & \text{for} \quad |\omega| < \omega_c \\ 0 & \text{for} \quad |\omega| \geq \omega_c \end{cases}$$

the output is $F^*(\omega)G(\omega) = F(\omega)$ or using (2.3.8)

$$F(\omega) = \sum_{n=-\infty}^{\infty} f(nT) G(\omega) e^{-j\omega nT} \tag{2.4.6}$$

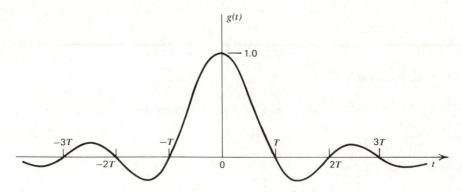

FIGURE 2.4.1. Impulse response of cardinal hold.

The inverse Fourier transform of both sides of (2.4.6) yields

$$f(t) = \sum_{n=-\infty}^{\infty} f(nT)g(t-nT) \qquad (2.4.7)$$

where

$$g(t) = \mathcal{F}^{-1}\{G(\omega)\} = \frac{\sin \omega_c t}{\omega_c t} \qquad (2.4.8)$$

The interpolation formula (2.4.1) is known as the *cardinal series* and $g(t)$ is called the *cardinal data hold*. In the data communications literature $\omega_c = \omega_s/2$ is usually called the *Nyquist frequency* and ω_s is called the *Nyquist rate*.

The cardinal hold $g(t)$ is shown in Fig. 2.4.1. Notice that the zeros of $\sin \omega_c t$ occur when $\omega_c t = k\pi$ or $t = k2\pi/\omega_s = kT$. Thus $g(kT) = 0$ for k not equal to zero. For $t = 0$, l'Hospital's rule shows that $g(0) = 1$. The sampling theorem shows how to calculate $f(t)$ from its samples. For t between sampling instants, it is not obvious that the sum converges to $f(t)$. However, at the sampling instants it is obvious since

$$f(kT) = \sum_{n=-\infty}^{\infty} f(nT)g(kT-nT)$$

and $g(kT-nT) = 0$ except when $n = k$. If the interpolation is considered to be performed by passing $f^*(t)$ through an ideal low-pass filter, then the impulse responses from samples before and after $f(nT)$ cause no interference at time nT. In synchronous pulse-amplitude modulation data communication systems, a channel impulse response with this characteristic is said to cause no "inter-symbol interference" [87].

Numerous generalizations of the sampling theorem have been derived. It has been shown that a band-limited signal can be reconstructed from samples of the signal and its first $n-1$ derivatives if these functions are sampled at the rate $2\omega_c/n$ [84]. Sampling theorems for stochastic signals have been derived [7]. The interpolation formulas have been extended to N-dimensional spaces [107]. Some results for nonuniform sampling have been derived [143] and are presented in Reference 34.

2.5 SAMPLING BAND-PASS SIGNALS

A signal $f(t)$ whose Fourier transform is identically zero except in the intervals $\omega_1 < \omega < \omega_2$ and $-\omega_2 < \omega < -\omega_1$ with $0 < \omega_1 < \omega_2 < \infty$ is called a band-pass signal with bandwidth $B = \omega_2 - \omega_1$. Clearly a band-pass signal is also a band-limited signal with cutoff frequency ω_2. According to Theorem 2.4.1 it appears that $f(t)$ must be sampled at a rate $\omega_s > 2\omega_2$ to uniquely reconstruct it from its samples. Intuitively one would expect that samples taken at the rate $2B$ should be sufficient. It will be shown in this section that, in a sense, our intuition is correct.

Hilbert transforms are a convenient tool in analyzing band-pass systems. The Hilbert transform of a signal $f(t)$ is defined to be [103]

$$\check{f}(t) = \frac{1}{\pi} \int_{-\infty}^{\infty} \frac{f(x)}{t - x} \, dx = f(t) * h(t) \tag{2.5.1}$$

where $*$ denotes convolution and $h(t) = 1/(\pi t)$. It can be shown that

$$H(\omega) = -j \operatorname{sign} \omega = \begin{cases} -j & \text{for} \quad \omega > 0 \\ 0 & \text{for} \quad \omega = 0 \\ j & \text{for} \quad \omega < 0 \end{cases} \tag{2.5.2}$$

From (2.5.1) it follows that

$$\check{F}(\omega) = F(\omega) H(\omega) = -j F(\omega) \operatorname{sign} \omega \tag{2.5.3}$$

Thus the Hilbert transform of a signal is found by passing it through an ideal 90° phase shifter.

Consider the signal

$$f(t) = a(t) \cos \omega_0 t - b(t) \sin \omega_0 t \tag{2.5.4}$$

where $a(t)$ and $b(t)$ are low-pass signals band-limited to $\omega_c = B/2$ and $\omega_0 > B/2$. Since

$$\mathcal{F}\{a(t) \cos \omega_0 t\} = \frac{A(\omega - \omega_0) + A(\omega + \omega_0)}{2}$$

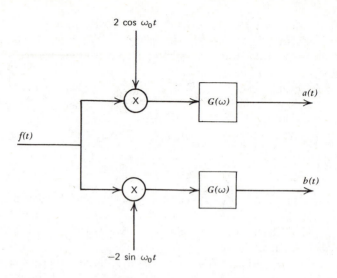

FIGURE 2.5.1. System for deriving in-phase and quadrature signal components.

and

$$\mathscr{F}\{b(t)\sin\omega_0 t\} = \frac{B(\omega-\omega_0)-B(\omega+\omega_0)}{2j}$$

it can be seen that $F(\omega)$ is identically zero except for $\omega_0-(B/2)<|\omega|<\omega_0+(B/2)$ so that $f(t)$ is a band-pass signal. The system shown in Fig. 2.5.1 can be used to derive $a(t)$ and $b(t)$ from $f(t)$. In the upper branch the input to $G(\omega)$ is

$$2f(t)\cos\omega_0 t = a(t)+a(t)\cos 2\omega_0 t - b(t)\sin 2\omega_0 t \qquad (2.5.5)$$

The second two terms on the right-hand side of (2.5.5) have transforms centered around $2\omega_0$ and do not overlap $A(\omega)$. Therefore, if $G(\omega)$ is an ideal low-pass filter with cutoff frequency $B/2$, only $a(t)$ appears at its output. For the lower branch

$$-2f(t)\sin\omega_0 t = b(t)-b(t)\cos 2\omega_0 t - a(t)\sin 2\omega_0 t \qquad (2.5.6)$$

so that only $b(t)$ appears at the output of $G(\omega)$. The signal $a(t)$ is called the *in-phase component* of $f(t)$ and $b(t)$ the *quadrature component*. It will now be shown in the following theorem that any band-pass signal can be represented in the form of (2.5.4).

THEOREM 2.5.1. Band-Pass Representation Theorem
Let $f(t)$ be a signal with $F(\omega)$ identically zero except for $\omega_1<|\omega|<\omega_2$ with

$0 < \omega_1 < \omega_2 < \infty$. Then

$$f(t) = a(t) \cos \omega_0 t - b(t) \sin \omega_0 t \qquad (2.5.7)$$

where

ω_0 is a positive constant

$$a(t) = f(t) \cos \omega_0 t + \check{f}(t) \sin \omega_0 t \qquad (2.5.8)$$
$$b(t) = \check{f}(t) \cos \omega_0 t - f(t) \sin \omega_0 t \qquad (2.5.9)$$

and

$$A(\omega) = \begin{cases} F(\omega + \omega_0) + \overline{F(\omega_0 - \omega)} & \text{for} \quad |\omega| < \max(|\omega_1 + \omega_0|, |\omega_2 - \omega_0|) \\ 0 & \text{elsewhere} \end{cases}$$
$$(2.5.10)$$

$$jB(\omega) = \begin{cases} F(\omega + \omega_0) - \overline{F(\omega_0 - \omega)} & \text{for} \quad |\omega| < \max(|\omega_1 - \omega_0|, |\omega_2 - \omega_0|) \\ 0 & \text{elsewhere} \end{cases}$$
$$(2.5.11)$$

with $\overline{F(\omega)}$ the complex conjugate of $F(\omega)$.

Proof   Let $q(t) = f(t) + j\check{f}(t)$. This is called the *analytic signal* associated with $f(t)$. Then

$$Q(\omega) = F(\omega) + jF(\omega)(-j \, \text{sign} \, \omega)$$
$$= \begin{cases} 2F(\omega) & \text{for} \quad \omega > 0 \\ F(0) & \text{for} \quad \omega = 0 \\ 0 & \text{for} \quad \omega < 0 \end{cases} \qquad (2.5.12)$$

This is illustrated in Fig. 2.5.2b. Now let

$$q_L(t) = q(t)e^{-j\omega_0 t} \qquad (2.5.13)$$
$$= [f(t) + j\check{f}(t)][\cos \omega_0 t - j \sin \omega_0 t]$$
$$= a(t) + jb(t) \qquad (2.5.14)$$

where $a(t)$ and $b(t)$ are given by (2.5.8) and (2.5.9) respectively. By taking the Fourier transform of (2.5.13) and using (2.5.12), we have

$$Q_L(\omega) = Q(\omega + \omega_0) = \begin{cases} 2F(\omega + \omega_0) & \text{for} \quad \omega_1 - \omega_0 < \omega < \omega_2 - \omega_0 \\ 0 & \text{elsewhere} \end{cases} \qquad (2.5.15)$$

If $\omega_1 < \omega_0 < \omega_2$ then $Q_L(\omega)$ is a low-pass transform as shown in Fig. 2.5.2c. Since

$$\mathcal{F}\{\overline{q_L(t)}\} = \overline{Q_L(-\omega)} \qquad (2.5.16)$$

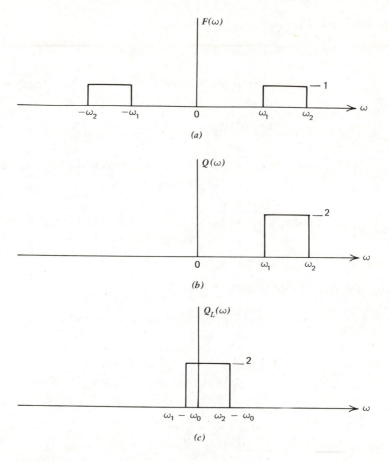

FIGURE 2.5.2. Transforms used in deriving the band-pass representation theorem. (a) Original Fourier transform; (b) Fourier transform of analytic signal; (c) Fourier transform of low-pass equivalent signal.

it follows from (2.5.14) that

$$A(\omega) + jB(\omega) = Q_L(\omega) \qquad (2.5.17)$$

and

$$A(\omega) - jB(\omega) = \overline{Q_L(-\omega)} \qquad (2.5.18)$$

Adding (2.5.17) and (2.5.18) yields

$$A(\omega) = \frac{Q_L(\omega) + \overline{Q_L(-\omega)}}{2} \qquad (2.5.19)$$

and subtracting gives

$$jB(\omega) = \frac{Q_L(\omega) - \overline{Q_L(-\omega)}}{2} \tag{2.5.20}$$

Substituting (2.5.15) into (2.5.19) and (2.5.20) yields (2.5.10) and (2.5.11). Finally

$$f(t) = \text{Re}\,\{q(t)\} = \text{Re}\,\{q_L(t)e^{j\omega_0 t}\} \tag{2.5.21}$$

$$= \text{Re}\,\{[a(t) + jb(t)][\cos \omega_0 t + j \sin \omega_0 t]\}$$

$$= a(t) \cos \omega_0 t - b(t) \sin \omega_0 t \tag{2.5.22}$$

Q.E.D.

We have just seen that a band-pass signal is uniquely determined by its in-phase and quadrature components. If ω_0 is chosen in the pass-band, then $a(t)$ and $b(t)$ are low-pass signals and can be represented by the cardinal series expansion of Theorem 2.4.1. These observations lead to the following theorem.

THEOREM 2.5.2. Band-Pass Sampling Theorem
Let $f(t)$ be a band-pass signal with $F(\omega)$ identically zero except for $\omega_1 < |\omega| < \omega_2$ with $0 < \omega_1 < \omega_2 < \infty$. Let $\omega_0 = (\omega_1 + \omega_2)/2$ be the center frequency of the pass-band and $B = \omega_2 - \omega_1$ be the signal bandwidth. Then

$$f(t) = \sum_{n=-\infty}^{\infty} [a(nT) \cos \omega_0 t - b(nT) \sin \omega_0 t] \frac{\sin \omega_c(t - nT)}{\omega_c(t - nT)} \tag{2.5.23}$$

where $a(t)$ and $b(t)$ are given by (2.5.8) and (2.5.9) respectively, $\omega_s = 2\omega_c = B$ and $T = 2\pi/\omega_s$.

Proof With these assumptions, $|\omega_1 - \omega_0| = |\omega_2 - \omega_0| = B/2$ so that $q_L(t)$, $a(t)$, and $b(t)$ are low-pass signals band-limited to $\omega_c = B/2$ according to (2.5.10) and (2.5.11). Therefore

$$a(t) = \sum_{n=-\infty}^{\infty} a(nT) \frac{\sin \omega_c(t - nT)}{\omega_c(t - nT)} \tag{2.5.24}$$

and

$$b(t) = \sum_{n=-\infty}^{\infty} b(nT) \frac{\sin \omega_c(t - nT)}{\omega_c(t - nT)} \tag{2.5.25}$$

where $\omega_s = 2\omega_c = B$ and $T = 2\pi/\omega_s$. Substituting (2.5.24) and (2.5.25) into (2.5.7) completes the proof.

Q.E.D.

Equation 2.5.7 can also be written as

$$f(t) = c(t) \cos [\omega_0 t + \theta(t)] \qquad (2.5.26)$$

where

$$c(t) = [a^2(t) + b^2(t)]^{1/2} \qquad (2.5.27)$$

and

$$\theta(t) = \arctan b(t)/a(t) \qquad (2.5.28)$$

The function $c(t)$ is the envelope of $f(t)$ and $\theta(t)$ is its instantaneous phase. Notice that $c(t) = |q_L(t)|$ and $\theta(t) = \arg q_L(t)$. The band-pass sampling interpolation formula (2.5.23) can also be written as

$$f(t) = \sum_{n=-\infty}^{\infty} c(nT) \cos [\omega_0 t + \theta(nT)] \frac{\sin \omega_c(t - nT)}{\omega_c(t - nT)} \qquad (2.5.29)$$

Thus a band-pass signal can be reconstructed from samples of its in-phase and quadrature components or from samples of its envelope and instantaneous phase. In both cases two signals must be sampled at the rate $\omega_s = B$ so that the effective sampling rate is $2\omega_s = 2B$.

3

DATA RECONSTRUCTION BY POLYNOMIAL INTERPOLATION AND EXTRAPOLATION

3.1 INTRODUCTION

In Chapter 2 we saw that a band-limited signal can be reconstructed exactly from its samples with a cardinal hold if the sampling rate is at least twice the signal bandwidth. The cardinal hold is not a physically realizable filter but can be approximated arbitrarily closely if sufficient delay is allowed. In communication systems and off-line processing a reasonable delay is often acceptable. On the other hand, significant delays are not usually tolerable in feedback control systems. In addition, the signals involved may not be highly band-limited. In these cases it is usually necessary to sample at a faster rate to separate the translated spectra $F(\omega - n\omega_s)$ from the baseband spectrum $F(\omega)$. This reduces aliasing and allows the use of a simpler filter with smaller delay for approximate signal reconstruction. These filters will be called *reconstruction filters* or *data holds*. In this chapter "reconstructing a signal" means "approximately reconstructing a signal."

In many situations it is known a priori that the signal spectra fall off sufficiently rapidly with frequency to insure that the signals are smooth and well-behaved. These types of signals can be closely approximated over finite intervals by polynomials. Since polynomials are easily generated, they are a

23

natural choice for simple data holds. In addition, polynomial fits can be used to reconstruct nonuniformly sampled signals.

3.2 POLYNOMIAL INTERPOLATION AND EXTRAPOLATION

Only one mth degree polynomial passes through a set of $m+1$ samples $f(t_0)$, $f(t_1), \ldots, f(t_m)$ with $t_0 < t_1 < \cdots < t_m$. To see this let

$$a(t) = \sum_{k=0}^{m} a_k t^k \qquad (3.2.1)$$

Then we would like to choose the coefficients $a_0, \ldots, a_m$ so that

$$
\begin{bmatrix} f(t_0) \\ f(t_1) \\ \cdot \\ \cdot \\ \cdot \\ f(t_m) \end{bmatrix}
=
\begin{bmatrix}
1 & t_0 & t_0^2 \cdots t_0^m \\
1 & t_1 & t_1^2 \cdots t_1^m \\
\cdot \\
\cdot \\
\cdot \\
1 & t_m & t_m^2 \cdots t_m^m
\end{bmatrix}
\begin{bmatrix} a_0 \\ a_1 \\ \cdot \\ \cdot \\ \cdot \\ a_m \end{bmatrix}
\qquad (3.2.2)
$$

or

$$\mathbf{F} = \mathbf{T}_m \mathbf{A} \qquad (3.2.3)$$

The matrix $\mathbf{T}_m$ is a Vandermonde matrix and it can be shown that

$$\det(\mathbf{T}_m) = \prod_{i>j} (t_i - t_j) \qquad (3.2.4)$$

Since the sampling instants are all distinct, $\det(\mathbf{T}_m)$ is nonzero, $\mathbf{T}_m$ is nonsingular, and the set of coefficients is uniquely given by

$$\mathbf{A} = \mathbf{T}_m^{-1} \mathbf{F} \qquad (3.2.5)$$

If the sampling instants are sufficiently close together, $a(t)$ will closely approximate the original signal in the interval $(t_0 - d, t_m + d)$ for some positive d. If $a(t)$ is used to approximate $f(t)$ for $t \in (t_0, t_m)$ we say that we are *interpolating* between samples and if $t \notin (t_0, t_m)$ we say that we are *extrapolating* from the samples. The extrapolation error $f(t) - a(t)$ will normally become large as t gets far from (t_0, t_m).

In real-time applications the samples usually arrive in a continuing stream. There are a variety of ways in which blocks of $m+1$ samples can be used for signal reconstruction. For example, blocks of $m+1$ samples with the first sample of the successive blocks equal to the last sample of the previous block can be used to interpolate the data over the disjoint intervals (t_0, t_m), (t_m, t_{2m}), ... This would require a delay of m samples. A little thought will show that blocks of $m + 1$ samples overlapping by more than one sample can be used in various ways for signal reconstruction.

In most real-time applications signals are reconstructed over single sampling intervals rather than over blocks of sampling intervals. This is because of the intuitive feeling that the most recent data should be used in reconstruction so that the approximation polynomial should be changed with each newly received sample. In this chapter we only consider reconstruction of the signal for the time interval (t_{n-r}, t_{n-r+1}) from samples at times $t_{n-m}, t_{n-m+1}, \ldots, t_n$ where t_n is the latest sampling instant and $0 \le r \le m$. If $r = 0$ this corresponds to extrapolating into the future until the next sample is received at time t_{n+1}. If $0 < r \le m$ this corresponds to interpolation. Reconstruction of this type introduces a delay of r samples.

3.3 LAGRANGE INTERPOLATION FORMULA

The mth degree polynomial $a(t)$ passing through the $m + 1$ samples $f(t_0), \ldots,$ $f(t_m)$ can be calculated from (3.2.1) and (3.2.5). This involves inverting the matrix $\mathbf{T}_m$. However, $a(t)$ can be found without a matrix inversion by using the Lagrange interpolation formula.

Consider the set of $m + 1$ polynomials

$$q_{m,i}(t) = \prod_{\substack{k=0 \\ k \ne i}}^{m} \frac{t - t_k}{t_i - t_k} \qquad \text{for} \qquad i = 0, \ldots, m \qquad (3.3.1)$$

These are known as *Lagrange polynomials*. Each polynomial in the set has degree m. Also

$$q_{m,i}(t_r) = \delta_{ri} = \begin{cases} 0 & \text{for} \quad r \ne i \\ 1 & \text{for} \quad r = i \end{cases} \qquad (3.3.2)$$

with $0 \le r \le m$. Let the mth degree polynomial $b(t)$ be defined as

$$b(t) = \sum_{i=0}^{m} f(t_i) q_{m,i}(t) \qquad (3.3.3)$$

Then

$$b(t_r) = \sum_{i=0}^{m} f(t_i) \delta_{ri} = f(t_r) = a(t_r) \qquad \text{for} \qquad r = 0, \ldots, m$$

Since $a(t)$ and $b(t)$ are both polynomials with degree m passing through the same $m+1$ points, they must be identical. Equation 3.3.3 is known as the *Lagrange interpolation formula*.

Example 3.3.1

Let $m = 2$, $t_0 = 0$, $t_1 = 3$, and $t_2 = 4$. Then

$$q_{2,0}(t) = \frac{(t-3)(t-4)}{12}, \qquad q_{2,1}(t) = -\frac{t(t-4)}{3}, \qquad q_{2,2}(t) = \frac{t(t-3)}{4}$$

and

$$a(t) = f(0)q_{2,0}(t) + f(3)q_{2,1}(t) + f(4)q_{2,2}(t) \qquad \blacktriangleleft$$

3.4 RECONSTRUCTION FILTERS FOR UNIFORM SAMPLING

The results up to this point in Chapter 3 can be used for both uniform and nonuniform sampling. To obtain linear, time-invariant reconstruction filters, the results will now be specialized to the important case of uniform sampling. For uniform sampling the sampling instants are $t_n = nT$.

The Lagrange polynomials for the sampling instants $0, T, \ldots, mT$ are

$$q_{m,i}(t) = \prod_{\substack{k=0 \\ k \neq i}}^{m} \frac{t - kT}{iT - kT} \qquad \text{for} \qquad i = 0, \ldots, m \qquad (3.4.1)$$

The function

$$a(t; n) = \sum_{i=0}^{m} f[(i + n - m)T]q_{m,i}[t - (n - m)T] \qquad (3.4.2)$$

with $q_{m,i}(t)$ defined by (3.4.1) is a polynomial of degree m passing through the $m+1$ samples $f[(n-m)T], \ldots, f(nT)$. If the output of the reconstruction filter is

$$y(t) = a(t - rT; n) \qquad \text{for} \qquad nT \leq t < (n+1)T, \qquad 0 \leq r \leq m \qquad (3.4.3)$$

where r is a fixed integer independent of n, then $y(t)$ is an estimate of $f(t - rT)$. From (3.4.2) and (3.4.3) it follows that

$$y(t) = \sum_{i=0}^{m} f[(i + n - m)T]q_{m,i}[t - nT - (r - m)T] \qquad \text{for} \qquad nT \leq t < (n+1)T$$

$$(3.4.4)$$

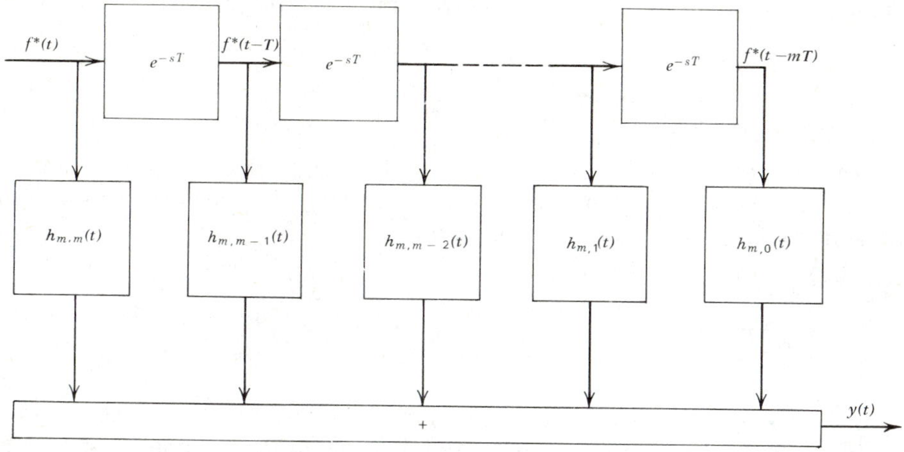

FIGURE 3.4.1.   Polynomial reconstruction filter.

If the $m+1$ functions $h_{m,i}(t)$ for $i=0,\ldots,m$ are defined as

$$h_{m,i}(i) = \begin{cases} q_{m,i}[t-(r-m)T] & \text{for} \quad 0\le t<T \\ 0 & \text{elsewhere} \end{cases} \tag{3.4.5}$$

then

$$y(t) = \sum_{i=0}^{m} f[(i+n-m)T]h_{m,i}(t-nT) \qquad \text{for} \qquad nT\le t<(n+1)T \tag{3.4.6}$$

The block diagram of a system for generating $y(t)$ based on (3.4.6) is shown in Fig. 3.4.1. It contains a set of m delays and a set of $m+1$ filters with impulse responses $h_{m,0}(t),\ldots,h_{m,m}(t)$. Fig. 3.4.1 is intended to provide only a pictorial representation of (3.4.6) and not necessarily the structure for a hardware implementation. In fact, it is unlikely that the hardware implementation would have this structure. For an interesting discussion of the implementation of polynomial reconstructors using a chain of integrators in a feedback loop see Section 3.7 of Reference 114. This technique was suggested by Porter and Stoneman [108] in connection with problems of tracking radar targets.

From Fig. 3.4.1 it is relatively easy to see that the impulse response of any filter realizing the polynomial reconstruction rule (3.4.6) is

$$h(t) = \sum_{k=0}^{m} h_{m,m-k}(t-kT) \tag{3.4.7}$$

Since $h_{m,i}(t)$ for each i is identically zero except for $0\le t<T$, $h(t)$ is identically zero except for $0\le t<(m+1)T$. In other words, the duration of the impulse response of an mth degree polynomial reconstruction filter is $(m+1)T$ seconds independent of whether it is an interpolator or extrapolator. The transfer function of the filter is

$$\mathcal{H}(s) = \sum_{k=0}^{m} \mathcal{H}_{m,m-k}(s)e^{-ksT} \qquad (3.4.8)$$

These concepts are illustrated in the following sections, where the most commonly used polynomial reconstruction filters, the zero-order hold, the first-order hold, and the linear point connector, are analyzed in detail.

3.5 THE ZERO-ORDER HOLD

The most common reconstruction filter is the zero-order hold. It is also known as a box-car circuit, sample-and-hold, and data clamp. The impulse response of the zero-order hold is

$$h(t) = \begin{cases} 1 & \text{for} \quad 0 \le t < T \\ 0 & \text{elsewhere} \end{cases} \qquad (3.5.1)$$

This corresponds to the special case $m = 0$ in Section 3.4. The impulse response $h(t)$ is sketched in Fig. 3.5.1 and a reconstructed signal is shown in Fig. 3.5.2. The output of the zero-order hold between sampling instants just remains constant at the last sample value. If T is sufficiently small, then the reconstructed signal closely approximates the original signal.

The transfer function of the zero-order hold is

$$\mathcal{H}(s) = \frac{1 - e^{-sT}}{s} \qquad (3.5.2)$$

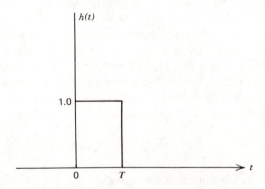

FIGURE 3.5.1. Impulse response of zero-order hold.

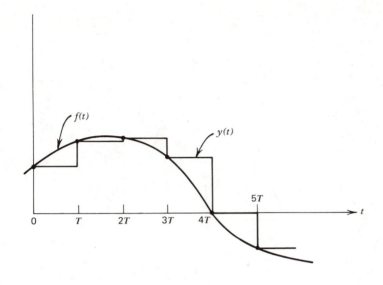

FIGURE 3.5.2.   Reconstruction with a zero-order hold.

and its frequency response is

$$\mathscr{H}(j\omega) = \frac{1 - e^{-j\omega T}}{j\omega} = Te^{-j\omega T/2} \frac{\sin \omega T/2}{\omega T/2} \tag{3.5.3}$$

This is shown in Fig. 3.5.3. The zero-order hold roughly approximates the cardinal hold with a delay of $T/2$ seconds. In practice, the zero-order hold is frequently followed by a simple low-pass filter to attenuate the undesired high-frequency components and provide a smooth output signal.

3.6 THE FIRST-ORDER HOLD

The case $m = 1$ and $r = 0$ is known as a first-order hold. This corresponds to extrapolating into the future with a straight line whose backward extension passes through $f(nT)$ and $f[(n-1)T]$. From (3.4.1) the Lagrange polynomials for $m = 1$ are

$$q_{1,0}(t) = -\frac{t - T}{T} \quad \text{and} \quad q_{1,1}(t) = \frac{t}{T} \tag{3.6.1}$$

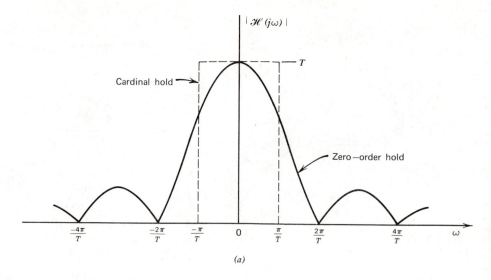

(a)

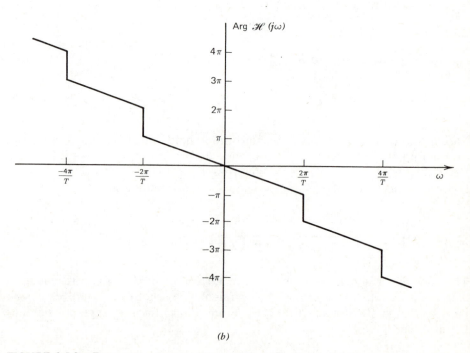

(b)

FIGURE 3.5.3. Frequency response of the zero-order hold. (*a*) Amplitude response; (*b*) phase response.

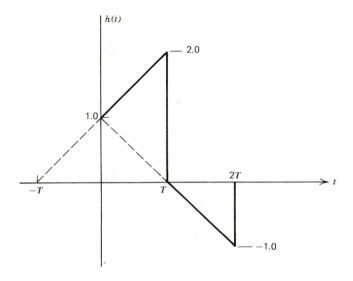

FIGURE 3.6.1.   Impulse response of the first-order hold.

According to (3.4.4)

$$y(t) = f(nT) + \frac{f(nT) - f[(n-1)T]}{T}(t - nT) \qquad \text{for} \qquad nT \le t < (n+1)T$$

$$(3.6.2)$$

The impulse response of the first-order hold calculated from (3.4.7) is

$$h(t) = \begin{cases} \dfrac{t+T}{T} & \text{for} \qquad 0 \le t < T \\[2mm] -\dfrac{t-T}{T} & \text{for} \qquad T \le t < 2T \\[2mm] 0 & \text{elsewhere} \end{cases} \qquad (3.6.3)$$

This is sketched in Fig. 3.6.1 and a reconstructed signal is shown in Fig. 3.6.2. The impulse response $h(t)$ can also be expressed as

$$h(t) = u(t) + \frac{t}{T}u(t) - 2u(t-T) - 2\frac{t-T}{T}u(t-T) + u(t-2T) + \frac{t-2T}{T}u(t-2T)$$

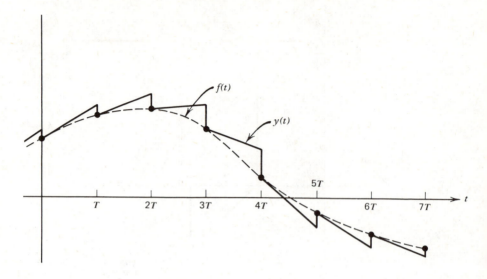

FIGURE 3.6.2.   Reconstruction with a first-order hold.

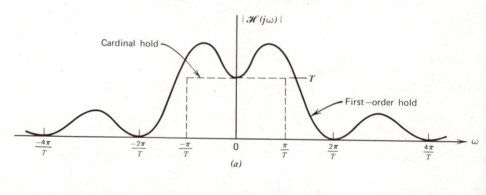

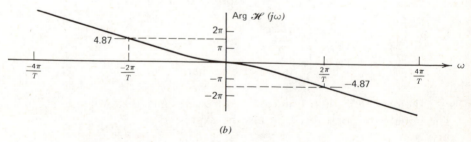

FIGURE 3.6.3.   Frequency response of a first-order hold. (*a*) Amplitude response; (*b*) phase response.

where $u(t)$ is the unit step function. The corresponding transfer function is

$$\mathcal{H}(s) = \frac{1}{s} + \frac{1}{Ts^2} - \frac{2}{s}e^{-sT} - \frac{2}{Ts^2} + \frac{e^{-2sT}}{s} + \frac{e^{-2sT}}{Ts^2}$$

$$= T(1+sT)\left[\frac{1-e^{-sT}}{sT}\right]^2 \tag{3.6.4}$$

and the frequency response can be put in the form

$$\mathcal{H}(j\omega) = T\sqrt{1+(\omega T)^2}\left[\frac{\sin \omega T/2}{\omega T/2}\right]^2 e^{-j(\omega T - \arctan \omega T)} \tag{3.6.5}$$

This is sketched in Fig. 3.6.3. It is apparent in Fig. 3.6.3 that the first-order hold significantly emphasizes some signal components in the pass-band from 0 to $2\pi/T$ radians/second. The zero-order hold attenuates most components in this band. The phase lag of the first-order hold is greater than that of the zero-order hold over most of the pass-band.

3.7 FIRST-ORDER HOLD WITH PARTIAL VELOCITY CORRECTION

If $f(t)$ is essentially linear, the first-order hold will reconstruct the signal almost perfectly. However, if the spectrum of $f(t)$ is flat in the pass-band, the reconstructed signal will be distorted and have significant ripple because of the peak in the frequency response of the first-order hold. In this case the first-order hold can be improved by using an extrapolation rule of the form

$$y(t) = f(nT) + a\frac{f(nT) - f[(n-1)T]}{T}(t-nT) \qquad \text{for} \qquad nT \leq t < (n+1)T \tag{3.7.1}$$

where $0 \leq a \leq 1$. For $a = 0$ this is a zero-order hold and for $a = 1$ it is a first-order hold. The impulse response of the filter for generating $y(t)$ is

$$h(t) = \begin{cases} 1 + at/T & \text{for} \quad 0 \leq t < T \\ a(1 - t/T) & \text{for} \quad T \leq t < 2T \\ 0 & \text{elsewhere} \end{cases} \tag{3.7.2}$$

This is sketched in Fig. 3.7.1 and is known as a first-order hold with *partial velocity correction*. Its transfer function is

$$\mathcal{H}(s) = \frac{1}{s}(1-e^{-sT})\left[1 - ae^{-sT} + \frac{a}{sT}(1-e^{-sT})\right] \tag{3.7.3}$$

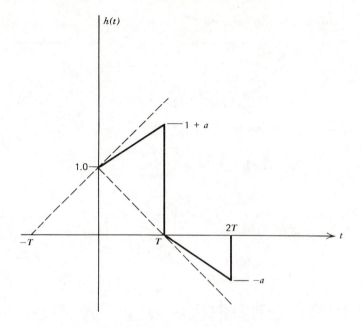

FIGURE 3.7.1. Impulse response of a first-order hold with partial velocity correction.

and the frequency response can be put in the form

$$\mathcal{H}(j\omega) = T\frac{\sin \omega T/2}{\omega T/2}\left[1 - a + a(1 + j\omega T)\frac{\sin \omega T/2}{\omega T/2}e^{-j\omega T/2}\right] \qquad (3.7.4)$$

This frequency response is plotted in Fig. 3.7.2 for several values of a. For $a = 0.3$ the amplitude response is relatively flat over a wide band. The optimum choice of the parameter a depends on the specific situation.

3.8 THE LINEAR POINT CONNECTOR

If a delay of one sample is allowable, a good reconstruction can be obtained simply by connecting the samples with straight lines. This corresponds to the case $m = 1$ and $r = 1$ with the interpolation rule

$$y(t) = f(nT)\frac{t - nT}{T} + f[(n-1)T]\frac{(n+1)T - t}{T} \qquad \text{for} \qquad nT \le t < (n+1)T$$

$$(3.8.1)$$

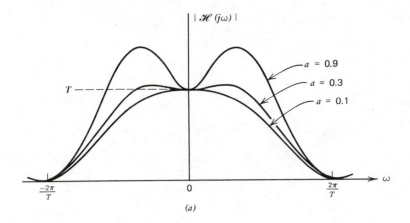

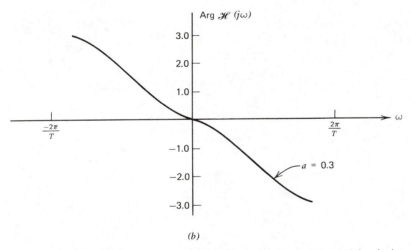

FIGURE 3.7.2. Frequency response of a first-order hold with partial velocity correction. (*a*) Amplitude response; (*b*) phase response.

According to (3.4.7) the impulse response of the *linear point connector* is

$$h(t) = \begin{cases} t/T & \text{for} \quad 0 \le t < T \\ 2 - t/T & \text{for} \quad T \le t < 2T \\ 0 & \text{elsewhere} \end{cases} \qquad (3.8.2)$$

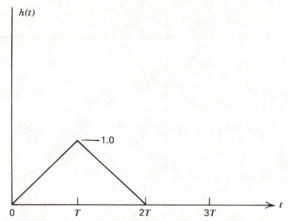

FIGURE 3.8.1.   Impulse response of a linear point connector.

This impulse response is plotted in Fig. 3.8.1 and a reconstructed signal is shown in Fig. 3.8.2.

The impulse response $h(t)$ can also be expressed as

$$h(t) = \frac{t}{T} u(t) - 2\frac{t-T}{T} u(t-T) + \frac{t-2T}{T} u(t-2T)$$

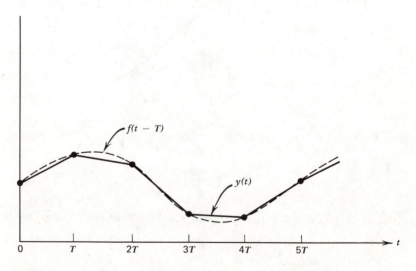

FIGURE 3.8.2.   Reconstruction with the linear point connector.

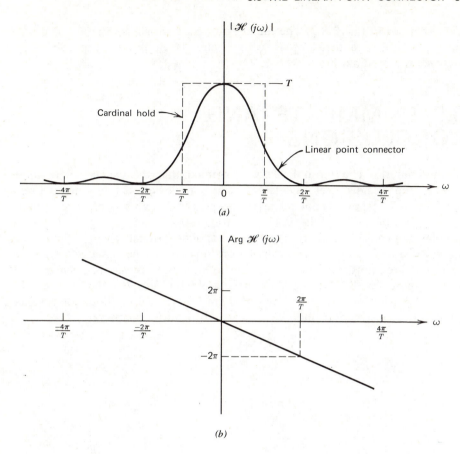

FIGURE 3.8.3. Frequency response of a linear point connector. (*a*) Amplitude response; (*b*) phase response.

so that the transfer function of the linear point connector is

$$\mathcal{H}(s) = \frac{1}{Ts^2} - \frac{2}{Ts^2} e^{-sT} + \frac{1}{Ts^2} e^{-2sT}$$

$$= \frac{(1 - e^{-sT})^2}{Ts^2} \tag{3.8.3}$$

Its frequency response is

$$\mathcal{H}(j\omega) = Te^{-j\omega T} \left[\frac{\sin \omega T/2}{\omega T/2} \right]^2 \tag{3.8.4}$$

This is shown in Fig. 3.8.3. The linear point connector has exactly linear phase corresponding to a delay of T seconds and its amplitude response roughly matches the cardinal hold.

3.9 COMMENTS AND CONCLUSIONS

Throughout this chapter it has been assumed that the exact signal samples uncorrupted by noise were available. If the samples are actually noisy, forcing the mth degree polynomial to pass through these points is not the optimum strategy. Particularly in the case of extrapolation, small errors in the signal samples can cause large errors in the extrapolated signal. Even if the samples are noiseless, large errors in extrapolation can occur if the signal is actually not an mth degree polynomial. For this reason low-order extrapolators are usually used in practice. By far the reconstruction filter most frequently used is a zero-order hold followed by a low-pass filter. Statistical methods for designing optimum reconstruction filters are discussed in future chapters.

4

THE Z-TRANSFORM

4.1 INTRODUCTION

The Z-transform is an important tool for analyzing linear, time-invariant, discrete-time systems. In this chapter we see that the Z-transform plays the same role for discrete-time systems that the Laplace transform plays for continuous-time systems. In fact, the Z-transform provides a bridge between continuous- and discrete-time signal processing because the Laplace transform $\mathscr{F}^*(s)$ of an ideal impulse sampled signal $f^*(t)$ is related to the Z-transform $F(z)$ of the discrete-time signal $f(nT)$ by the transformation $z = e^{sT}$. This transformation maps the left half plane in the complex s-plane into the unit circle in the complex z-plane. We will see that the interior of the unit circle, the unit circle, and the exterior of the unit circle in the z-plane have similar meaning for discrete-time signals as the left half s-plane, $j\omega$ axis, and right half s-plane for continuous-time signals. All the familiar properties of Laplace transforms used in analyzing continuous-time systems will be shown to have analogs for discrete-time systems in terms of Z-transforms. In particular, the notions of convolution, transfer functions, and filtering carry over to the discrete-time case as a result of the Discrete-Time Convolution Theorem.

4.2 THE Z-TRANSFORM

We define the Z-transform of a continuous-time signal $f(t)$ or a discrete-time signal $f(nT)$ to be

$$F(z) = \sum_{n=-\infty}^{\infty} f(nT)z^{-n} \tag{4.2.1}$$

39

The notation $\mathscr{Z}\{f(t)\}$ or $\mathscr{Z}\{f(nT)\}$ denotes the Z-transform $F(z)$. If this series converges to an analytic function $F(z)$ in the annular domain $R_1 < |z| < R_2$ in the complex z-plane, then it is the *Laurent series* expansion for $F(z)$ about $z = 0$ [71]. The function $F(z)$ is also called the *generating function* for this sequence of samples in the mathematical literature. As with Laplace transforms, we extend $F(z)$ to the entire z-plane by analytic continuation. The Z-transform provides a compact representation of the sequence of samples if the series converges. The sample $f(nT)$ is just the coefficient of z^{-n} in the Laurent series expansion of $F(z)$.

In previous chapters capital letters have been used to denote Fourier transforms. For example, $F(\omega)$ denotes the Fourier transform of $f(t)$. It is customary to use capital letters for Z-transforms also and we continue to use this convention. Clearly, $F(\omega)$ and $F(z)$ are defined to be different functions. Usually, it will be obvious from the context whether the capital letter refers to the Fourier or Z-transform.

Some elementary Z-transforms are derived in the following examples. A more extensive list of useful transforms is given in Appendix B. Additional transforms can be found in References 34, 79, 114, and 131, and other sampled-data books.

Example 4.2.1

If $u(t)$ is the unit step function defined as

$$u(t) = \begin{cases} 0 & \text{for} & t < 0 \\ 1 & \text{for} & t \geq 0 \end{cases}$$

then

$$U(z) = \sum_{n=0}^{\infty} z^{-n}$$

This is a geometric series and converges for $|z| > 1$ to

$$U(z) = 1/(1 - z^{-1}) \qquad \blacktriangleleft$$

Example 4.2.2

If $f(t) = e^{-bt}u(t)$, then

$$F(z) = \sum_{n=0}^{\infty} e^{-bnT} z^{-n} = \sum_{n=0}^{\infty} (e^{-bT} z^{-1})^n$$

$$= 1/(1 - e^{-bT} z^{-1}) \qquad \text{for} \qquad |z| > |e^{-bT}| \qquad \blacktriangleleft$$

Example 4.2.3

If $f(t) = e^{-b|t|}$ with $b > 0$, then

$$F(z) = \sum_{n=-\infty}^{\infty} e^{-b|nT|} z^{-n} = \sum_{n=-\infty}^{0} (e^{-bT}z)^{-n} + \sum_{n=0}^{\infty} (e^{-bT}z^{-1})^n - 1$$

The term -1 appears on the right-hand side since both summations include a term for $n = 0$. The first summation on the right-hand side is a geometric series and converges to $1/(1 - e^{-bT}z)$ for $|z| < e^{bT}$. From Example 4.2.2 we see that the second summation converges to $1/(1 - e^{-bT}z^{-1})$ for $|z| > e^{-bT}$. These regions of convergence overlap. Substituting the closed forms for the summations and combining terms yields

$$F(z) = \frac{1 - e^{-2bT}}{(1 - e^{-bT}z)(1 - e^{-bT}z^{-1})} \qquad \text{for} \qquad e^{-bT} < |z| < e^{bT} \qquad \blacktriangleleft$$

Example 4.2.4

If $f(t) = e^{-j\omega_0 t} u(t)$ with ω_0 real, then

$$F(z) = \sum_{n=0}^{\infty} e^{-j\omega_0 nT} z^{-n} = \sum_{n=0}^{\infty} z^{-n} \cos \omega_0 nT - j \sum_{n=0}^{\infty} z^{-n} \sin \omega_0 nT$$

$$= \mathscr{L}\{u(t) \cos \omega_0 t\} - j\mathscr{L}\{u(t) \sin \omega_0 t\}$$

From Example 4.2.2

$$F(z) = \frac{1}{1 - e^{-j\omega_0 T}z^{-1}} = \frac{1 - z^{-1} \cos \omega_0 T - jz^{-1} \sin \omega_0 T}{1 - 2z^{-1} \cos \omega_0 T + z^{-2}}$$

for $|z| > 1$. On comparing these two forms for $F(z)$ we see that

$$\mathscr{L}\{u(t) \cos \omega_0 t\} = \frac{1 - z^{-1} \cos \omega_0 T}{1 - 2z^{-1} \cos \omega_0 T + z^{-2}} \qquad \text{for} \qquad |z| > 1$$

and

$$\mathscr{L}\{u(t) \sin \omega_0 t\} = \frac{z^{-1} \sin \omega_0 T}{1 - 2z^{-1} \cos \omega_0 T + z^{-2}} \qquad \text{for} \qquad |z| > 1 \qquad \blacktriangleleft$$

Like Fourier and Laplace transforms, the Z-transform is linear. This means that for constants c_1 and c_2 and signals $f_1(t)$ and $f_2(t)$

$$\mathscr{L}\{c_1 f_1(t) + c_2 f_2(t)\} = c_1 \mathscr{L}\{f_1(t)\} + c_2 \mathscr{L}\{f_2(t)\} \qquad (4.2.2)$$

The linearity property follows directly from the properties of summations. The final results in Example 4.2.4 were derived by observing this linearity property.

A signal that is identically zero for negative time is called a *causal* signal or a signal that is *one-sided* for positive time. The Z-transform of a causal signal

$f(t)$ reduces to

$$F(z) = \sum_{n=0}^{\infty} f(nT)z^{-n} \qquad (4.2.3)$$

It is conventional to define the *one-sided Z-transform* of a signal $f(t)$ that is not necessarily causal by (4.2.3). The notation $\mathscr{Z}_+\{f(t)\}$ will be used to denote the one-sided Z-transform of $f(t)$. Comparing (4.2.1) and (4.2.3) we see that

$$\mathscr{Z}_+\{f(t)\} = \mathscr{Z}\{f(t)u(t)\} \qquad (4.2.4)$$

The one-sided Z-transform is also linear.

The values of z for which the series (4.2.3) converges can be determined by using the ratio test or the root test [71]. The ratio test states that a series of complex numbers

$$\sum_{n=0}^{\infty} a_n$$

with

$$\lim_{n \to \infty} \left| \frac{a_{n+1}}{a_n} \right| = L \qquad (4.2.5)$$

converges absolutely if $L < 1$ and diverges if $L > 1$. For $L = 1$ the series may converge or diverge. The root test states that if

$$\lim_{n \to \infty} \sqrt[n]{|a_n|} = L \qquad (4.2.6)$$

then the series converges absolutely if $L < 1$, diverges if $L > 1$, and may converge or diverge if $L = 1$. More generally, the series converges absolutely if

$$\overline{\lim_{n \to \infty}} \sqrt[n]{|a_n|} < 1 \qquad (4.2.7)$$

where $\overline{\lim}$ denotes the upper limit, and diverges if

$$\overline{\lim_{n \to \infty}} \sqrt[n]{|a_n|} > 1 \qquad (4.2.8)$$

Applying the root test to the series (4.2.3) we see that it will converge if

$$\overline{\lim_{n \to \infty}} \sqrt[n]{|f(nT)z^{-n}|} < 1$$

or

$$|z| > \overline{\lim_{n \to \infty}} \sqrt[n]{|f(nT)|} = R_1 \qquad (4.2.9)$$

The number R_1 is called the *radius of convergence* for the series. The root test implies that $f(t)$ must be of exponential order for the series (4.2.3) to converge. This means that

$$|f(t)| \le ce^{-\sigma_1 t} \qquad \text{for} \qquad t \ge 0 \qquad (4.2.10)$$

where c is finite. From (4.2.10) and Example 4.2.2 we can see that the series (4.2.3) converges for $|z| > e^{-\sigma_1 T}$. The number $e^{-\sigma_1 T}$ that gives the tightest bound is simply the radius of convergence R_1.

If $f(t)$ is one-sided for negative time, that is, $f(t) \equiv 0$ for $t > 0$, then

$$F(z) = \sum_{n=-\infty}^{0} f(nT)z^{-n} = \sum_{n=0}^{\infty} f(-nT)z^n \qquad (4.2.11)$$

According to the root test, this series will converge if

$$|z| < 1/\varlimsup_{n \to \infty} \sqrt[n]{|f(-nT)|} = R_2 \qquad (4.2.12)$$

Once again $f(t)$ must be of exponential order for the series (4.2.11) to converge, so that

$$|f(t)| \le ce^{\sigma_2 t} \qquad \text{for} \qquad t \le 0 \qquad (4.2.13)$$

and the series converges for $|z| < e^{\sigma_2 T}$. The value for $e^{\sigma_2 T}$ giving the tightest bound is R_2.

If a signal is not identically zero for positive or negative time, it is called a *two-sided* signal. The Z-transform defined by (4.2.1) is frequently called a *two-sided* or *bilateral* transform. In this book two-sided transforms are assumed unless stated otherwise. If a signal is causal, its two-sided and one-sided transforms will automatically be identical. The two-sided Z-transform can be written as

$$F(z) = F_1(z) + F_2(z) \qquad (4.2.14)$$

where

$$F_1(z) = \mathscr{Z}_+\{f(t)\} = \sum_{n=0}^{\infty} f(nT)z^{-n} \qquad (4.2.15)$$

and

$$F_2(z) = \sum_{n=-\infty}^{-1} f(nT)z^{-n} \qquad (4.2.16)$$

The series (4.2.15) will converge for $|z| > R_1$ according to (4.2.9) and the series (4.2.16) will converge for $|z| < R_2$ according to (4.2.12). If $R_1 < R_2$ these regions overlap. In this case the two-sided Z-transform exists and is given by (4.2.14) with the region of convergence

$$R_1 < |z| < R_2 \qquad (4.2.17)$$

The knowledge that a function $F(z)$ is a two-sided Z-transform is not enough to uniquely specify the transform. The region of convergence must also be known. For example, consider the two functions

$$f(t) = e^{-at}u(t)$$

and

$$g(t) = \begin{cases} -e^{-at} & \text{for} & t < 0 \\ 0 & \text{for} & t \geq 0 \end{cases}$$

According to Example 4.2.2

$$F(z) = 1/(1 - e^{-aT}z^{-1}) \qquad \text{for} \qquad |z| > |e^{-aT}|$$

But

$$G(z) = \sum_{n=-\infty}^{-1} -e^{-anT}z^{-n} = -e^{-aT}z \sum_{n=0}^{\infty} (e^{aT}z)^n$$

$$= \frac{-e^{aT}z}{1 - e^{aT}z} \qquad \text{for} \qquad |z| < |e^{-aT}| \qquad (4.2.18)$$

Multiplying both the numerator and denominator of (4.2.18) by $-e^{-aT}z^{-1}$ we see that

$$G(z) = 1/(1 - e^{-aT}z^{-1}) \qquad \text{for} \qquad |z| < |e^{-aT}|$$

Thus $F(z)$ and $G(z)$ are identical except for the different regions of convergence. In general, the same function of z can correspond to the Z-transform of a signal that is one-sided for positive time, one-sided for negative time, or two-sided. The different cases are distinguished by different regions of convergence. This point is discussed further when the Z-transform inversion formula is presented.

4.3 THE RELATIONSHIP BETWEEN $F(z)$, $\mathscr{F}^*(s)$, AND $F^*(\omega)$

In Chapter 2 we saw that the samples of a signal $f(t)$ could be represented by the amplitude modulated impulse train

$$f^*(t) = \sum_{n=-\infty}^{\infty} f(nT)\, \delta(t - nT) \qquad (4.3.1)$$

with the corresponding Laplace transform

$$\mathscr{F}^*(s) = \sum_{n=-\infty}^{\infty} f(nT)e^{-snT} \qquad (4.3.2)$$

and Fourier transform

$$F^*(\omega) = \sum_{n=-\infty}^{\infty} f(nT)e^{-j\omega nT} \tag{4.3.3}$$

If the substitution $z = e^{sT}$ is made in the Z-transform definition (4.2.1) we see that

$$\mathscr{F}^*(s) = F(z)|_{z=e^{sT}} \tag{4.3.4}$$

If the region of convergence for $F(z)$ includes the unit circle, $|z| = 1$, then

$$F^*(\omega) = F(z)|_{z=e^{j\omega T}} \tag{4.3.5}$$

From (4.3.2) we can see that

$$\mathscr{F}^*(s + j\omega_s) = \sum_{n=-\infty}^{\infty} f(nT)e^{-snT}e^{-j\omega_s nT} = \sum_{n=-\infty}^{\infty} f(nT)e^{-snT}$$

$$= \mathscr{F}^*(s) \tag{4.3.6}$$

Therefore, $\mathscr{F}^*(s)$ repeats periodically with period ω_s when evaluated along a line parallel to the $j\omega$ axis in the complex s-plane. $\mathscr{F}^*(s)$ is highly redundant since knowledge of $\mathscr{F}^*(s)$ in the strip $-\omega_s/2 < \omega \le \omega_s/2$ determines $\mathscr{F}^*(s)$ for all s. The transformation

$$z = e^{sT}$$

maps this strip uniquely onto the complex z-plane so that $F(z)$ contains all the information in $\mathscr{F}^*(s)$ without the redundancy. To see the nature of this mapping let $s = \sigma + j\omega$. Then

$$z = e^{\sigma T}e^{j\omega T} \tag{4.3.8}$$

Since $|z| = e^{\sigma T}$, it is clear that

$$|z| \begin{cases} <1 & \text{for} \quad \sigma < 0 \\ =1 & \text{for} \quad \sigma = 0 \\ >1 & \text{for} \quad \sigma > 0 \end{cases} \tag{4.3.9}$$

Thus points in the left half s-plane get mapped inside the unit circle in the z-plane, points on the $j\omega$ axis get mapped onto the unit circle, and points in the right half s-plane get mapped outside the unit circle. Lines parallel to the $j\omega$ axis get mapped into circles of the form $|z| = e^{\sigma T}$ and lines parallel to the σ axis get mapped into rays of the form $\arg z = \omega T$ radiating from $z = 0$. Figure 4.3.1 illustrates these concepts. The origin of the s-plane corresponds to $z = 1$ and the σ axis corresponds to the positive $u = \text{Re } z$ axis. As ω varies between $-\omega_s/2$ and $\omega_s/2$, $\arg z = \omega T$ varies between $-\pi$ and π radians. Successive strips of the form $-\omega_s/2 + k\omega_s < \omega \le \omega_s/2 + k\omega_s$ in the s-plane get mapped on top of each other in the z-plane.

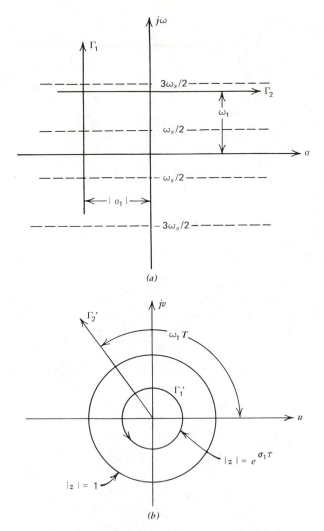

FIGURE 4.3.1. Mapping induced by $z = e^{sT}$. (a) s-plane; (b) z-plane.

In working with continuous-time, linear, time-invariant systems and Laplace transforms, we are accustomed to visualizing various system properties from plots of poles and zeros in the s-plane. It will become evident that the z-plane plays a role analogous to the s-plane for discrete-time, linear, time-invariant systems with the interior of the unit circle corresponding to the left half s-plane, the exterior of the unit circle corresponding to the right half s-plane, and the unit circle corresponding to the $j\omega$ axis.

4.4 FINDING THE Z-TRANSFORM OF A CAUSAL SIGNAL DIRECTLY FROM ITS LAPLACE TRANSFORM

In this section we will see how to find the Z-transform of a causal signal from its Laplace transform without going through the time domain. This technique is useful in linear system analysis where signals are often specified by their Laplace transforms. It provides a systematic method for finding a closed form for the series (4.2.1) defining the Z-transform.

This method is based on the fact that the Laplace transform of the product of two signals is the convolution of their Laplace transforms. If $f(t)$ and $g(t)$ are causal signals with Laplace transforms $\mathcal{F}(s)$ and $\mathcal{G}(s)$ which converge absolutely for $\mathrm{Re}\, s > \sigma_f$ and $\mathrm{Re}\, s > \sigma_g$ respectively, then

$$\mathcal{L}\{f(t)g(t)\} = \frac{1}{2\pi j} \int_{c-j\infty}^{c+j\infty} \mathcal{F}(p)\mathcal{G}(s-p)\, dp \qquad (4.4.1)$$

The integral is taken along a path parallel to the imaginary axis in the complex p-plane with

$$\sigma = \mathrm{Re}\, s > \sigma_f + \sigma_g \qquad \text{and} \qquad \sigma_f < c < \sigma - \sigma_g \qquad (4.4.2)$$

With this choice the poles of $\mathcal{G}(s-p)$ lie to the right of the path of integration.

When $f(t)$ is causal, $f^*(t)$ can be written as

$$f^*(t) = f(t) \sum_{n=0}^{\infty} \delta(t-nT) \qquad (4.4.3)$$

If

$$g(t) = \sum_{n=0}^{\infty} \delta(t-nT) \qquad (4.4.4)$$

then

$$\mathcal{G}(s) = \sum_{n=0}^{\infty} e^{-nTs} = 1/(1-e^{-sT}) \qquad \text{for} \qquad \mathrm{Re}\, s > 0 \qquad (4.4.5)$$

and according to (4.4.1)

$$\mathcal{F}^*(s) = \frac{1}{2\pi j} \int_{c-j\infty}^{c+j\infty} \frac{\mathcal{F}(p)}{1-e^{-(s-p)T}}\, dp \qquad (4.4.6)$$

for $\sigma > \sigma_f$ and $\sigma_f < c < \sigma$. This integral can be evaluated by using Cauchy's Residue Theorem. A brief summary of the theory of residues is presented in Appendix A. Consider the closed contour $\Gamma_1\Gamma_2$ shown in Fig. 4.4.1 where along Γ_2

$$p = c + Re^{j\theta} \qquad \text{and} \qquad \pi/2 \leq \theta \leq 3\pi/2 \qquad (4.4.7)$$

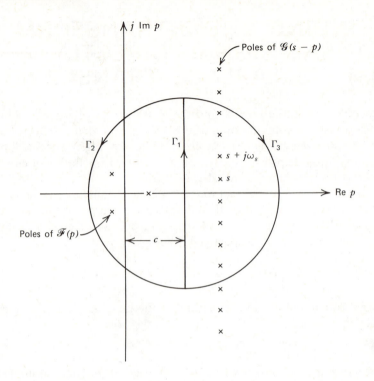

FIGURE 4.4.1. Contours for finding $\mathscr{F}^*(s)$.

If $\mathscr{F}(p)$ is analytic for $|p|$ greater than some finite number R_0 and has a zero at $p = \infty$, then in the limit as R approaches infinity the integral along Γ_1 converges to $\mathscr{F}^*(s)$ and the contour $\Gamma_1 \Gamma_2$ encloses all the poles of $\mathscr{F}(p)$. According to Cauchy's Residue Theorem

$$\mathscr{F}^*(s) = \sum_k \mathrm{Res}\left[\frac{\mathscr{F}(p)}{1 - e^{pT}e^{-sT}}, p_k\right] - \lim_{R \to \infty} \frac{1}{2\pi j} \int_{\Gamma_2} \frac{\mathscr{F}(p)}{1 - e^{pT}e^{-sT}} \, dp \qquad (4.4.8)$$

where $\{p_k\}$ are the poles of $\mathscr{F}(p)$. As a result of the assumptions about $\mathscr{F}(p)$, it must have a Laurent series expansion of the form

$$\mathscr{F}(p) = \frac{a_{-1}}{p} + \frac{a_{-2}}{p^2} + \cdots$$

$$= \frac{a_{-1}}{p} + \frac{Q(p)}{p^2} \qquad \text{for} \qquad |p| > R_0 \qquad (4.4.9)$$

$Q(p)$ must be analytic in this domain so that

$$|Q(p)| < M < \infty \qquad \text{for} \qquad |p| > R_0 \tag{4.4.10}$$

From (4.4.9) it follows that

$$a_{-1} = \lim_{p \to \infty} p \mathscr{F}(p) \tag{4.4.11}$$

In complex variable theory $-a_{-1}$ is known as the residue of $\mathscr{F}(p)$ at ∞. From the Initial Value Theorem [72] we also can see that

$$a_{-1} = f(0^+) \tag{4.4.12}$$

By substituting (4.4.12) and (4.4.9) into the integral on the right-hand side of (4.4.8) we get

$$\lim_{R \to \infty} \frac{1}{2\pi j} \int_{\Gamma_2} \frac{\mathscr{F}(p)}{1 - e^{pT} e^{-sT}} \, dp = \lim_{R \to \infty} \frac{1}{2\pi j} \int_{\Gamma_2} \frac{f(0^+)}{p(1 - e^{pT} e^{-sT})} \, dp$$

$$+ \lim_{R \to \infty} \frac{1}{2\pi j} \int_{\Gamma_2} \frac{Q(p)}{p^2(1 - e^{pT} e^{-sT})} \, dp \tag{4.4.13}$$

Since the poles of $\mathscr{G}(s - p)$ all lie to the right of Γ_1

$$|\mathscr{G}(s - p)| < L < \infty \qquad \text{along } \Gamma_2 \tag{4.4.14}$$

The magnitude of an integral along a contour is no greater than the length of the contour times the maximum magnitude of the integrand along the contour so that

$$\left| \lim_{R \to \infty} \int_{\Gamma_2} \frac{Q(p)}{p^2(1 - e^{pT} e^{-sT})} \, dp \right| \leq \lim_{R \to \infty} \frac{ML}{(|c| - R)^2} (\pi R) = 0 \tag{4.4.15}$$

Making the change of variables $p = c + Re^{j\theta}$ in the first integral on the right of (4.4.13) gives

$$\lim_{R \to \infty} \frac{1}{2\pi} \int_{\pi/2}^{3\pi/2} \frac{f(0^+) Re^{j\theta}}{(c + Re^{j\theta})(1 - e^{-sT} e^{(c + R\cos\theta + jR\sin\theta)T})} \, d\theta$$

$$= \frac{f(0^+)}{2\pi} \int_{\pi/2}^{3\pi/2} \lim_{R \to \infty} \frac{Re^{j\theta}}{(c + Re^{j\theta})(1 - e^{-sT} e^{(c + R\cos\theta + jR\sin\theta)T})} \, d\theta \tag{4.4.16}$$

$$= \frac{f(0^+)}{2\pi} \int_{\pi/2}^{3\pi/2} d\theta = f(0^+)/2 \tag{4.4.17}$$

The limit of the integrand in (4.4.16) is 1 since the numerator and first factor in the denominator cancel for large R and the second factor in the denominator becomes 1 because $\cos \theta < 0$ for $\pi/2 < \theta < 3\pi/2$. Thus (4.4.8) reduces to

$$\mathscr{F}^*(s) = \sum_k \text{Res}\left[\frac{\mathscr{F}(p)}{1 - e^{pT}e^{-sT}}, p_k\right] - f(0^+)/2 \qquad (4.4.18)$$

where $\{p_k\}$ are the poles of $\mathscr{F}(p)$ and $\sigma = \text{Re } s > \sigma_f$. Letting $z = e^{sT}$ this becomes

$$F(z) = \sum_k \text{Res}\left[\frac{\mathscr{F}(p)}{1 - e^{pT}z^{-1}}, p_k\right] - f(0^+)/2 \qquad (4.4.19)$$

for $|z| > e^{\sigma_f T}$.

Example 4.4.1

Let $f(t) = tu(t)$ so that $\mathscr{F}(s) = 1/s^2$ for $\text{Re } s > 0$. Then from (4.4.19)

$$F(z) = \text{Res}\left[\frac{1}{p^2(1 - e^{pT}z^{-1})}, 0\right]$$

$$= \frac{d}{dp}\frac{1}{1 - e^{pT}z^{-1}}\bigg|_{p=0} = \frac{Tz^{-1}}{(1 - z^{-1})^2} \qquad \blacktriangleleft$$

Example 4.4.2

Let $f(t) = e^{-bt}u(t)$. Then $\mathscr{F}(s) = 1/(s+b)$ for $\text{Re } s > -\text{Re } b$ or $|z| > |e^{-bT}|$. From (4.4.19)

$$F(z) = \text{Res}\left[\frac{1}{(p+b)(1 - e^{pT}z^{-1})}, -b\right] - \frac{1}{2}$$

$$= \frac{1}{1 - e^{-bT}z^{-1}} - \frac{1}{2} \qquad \blacktriangleleft$$

If the results in Examples 4.4.2 and 4.2.2 are compared, it seems that there is a discrepancy in the answers. From Example 4.2.2 we can see that $F(z)$ for Example 4.4.2 can be expanded in the series

$$F(z) = \sum_{n=0}^{\infty} e^{-bnT}z^{-n} - \frac{1}{2} \qquad (4.4.20)$$

so that for Example 4.4.2

$$f(nT) = \begin{cases} 0 & \text{for} & n < 0 \\ 1/2 & \text{for} & n = 0 \\ e^{-bnT} & \text{for} & n > 0 \end{cases} \qquad (4.4.21)$$

This discrepancy is a consequence of the discontinuity at $t = 0$. In Example 4.2.2 the value $f(0) = 1$ was used. If $f(t)$ is found from $\mathscr{F}(s)$ by using the Laplace transform integral inversion formula, then the value calculated for $f(t)$

at a step discontinuity is halfway between the values immediately to the left
and right of the discontinuity. Thus to make a causal signal $f(t)$ consistent with
$\mathscr{F}(s)$ and the inversion formula, $f(0)$ should be assigned the value $f(0^+)/2$. This
accounts for the discrepancy between Examples 4.2.2 and 4.4.2.

It is conventional in calculating the Z-transforms of causal signals to assign
the value $f(0^+)$ to $f(0)$. With this convention the formula for calculating $F(z)$
from $\mathscr{F}(s)$ reduces to

$$F(z) = \sum_k \text{Res}\left[\frac{\mathscr{F}(p)}{1 - e^{pT}z^{-1}}, p_k\right] \tag{4.4.22}$$

for $|z| > e^{\sigma_f T}$ with $\{p_k\}$ the set of poles of $\mathscr{F}(p)$. This convention is used in the
rest of the book.

Equation 4.4.22 can also be derived by using the artifice of advancing $f(t)$ by
an infinitesimal amount ϵ in time and sampling the signal $f_1(t) = f(t + \epsilon)$. Since
$\mathscr{F}_1(p) = \mathscr{F}(p)e^{p\epsilon}$, it can be shown that the integral along Γ_2 becomes zero.

If $\mathscr{F}(s)$ has only simple poles, it can be partial fractioned and expressed as

$$\mathscr{F}(s) = \sum_k a_k/(s - p_k) \tag{4.4.23}$$

where a_k is the residue of $\mathscr{F}(s)$ at the pole p_k. Since

$$\mathscr{L}\{e^{p_k t}u(t)\} = 1/(s - p_k)$$

and

$$\mathscr{Z}\{e^{p_k t}u(t)\} = 1/(1 - e^{p_k T}z^{-1})$$

it follows that

$$F(z) = \sum_k a_k/(1 - e^{p_k T}z^{-1}) \tag{4.4.24}$$

This is exactly equivalent to the calculations that would be performed in using
(4.4.22). Equation 4.4.22 generalizes the method of partial fractions to the case
where the poles of $\mathscr{F}(s)$ can have order greater than one.

An alternate form for $\mathscr{F}^*(s)$ can be derived if the contour is closed along the
path Γ_3 shown in Fig. 4.4.1. Along Γ_3

$$p = c + Re^{j\theta} \qquad \text{with} \qquad -\pi/2 < \theta < \pi/2 \tag{4.4.25}$$

The poles of $\mathscr{G}(s - p) = 1/(1 - e^{pT}e^{-sT})$ are located at the points

$$q_k = s - jk\omega_s \qquad \text{for integral } k \tag{4.4.26}$$

Applying Cauchy's Residue Theorem we see that

$$\mathscr{F}^*(s) = -\sum_{k=-\infty}^{\infty} \text{Res}\left[\frac{\mathscr{F}(p)}{1 - e^{pT}e^{-sT}}, s - jk\omega_s\right]$$

$$-\lim_{R\to\infty} \frac{1}{2\pi j}\int_{\Gamma_3} \frac{\mathscr{F}(p)}{1 - e^{pT}e^{-sT}}\,dp \tag{4.4.27}$$

Since it has been assumed that $\mathcal{F}(p)$ is analytic for $|p| > R_0$ and $|1 - e^{pT}e^{-sT}|$ grows exponentially with R for $-\pi/2 < \theta < \pi/2$, it can be shown by an argument similar to that used in deriving (4.4.15) that the second term on the right-hand side of (4.4.27) converges to 0. The residues in the summation are

$$\text{Res}\left[\frac{\mathcal{F}(p)}{1 - e^{pT}e^{-sT}}, q_k\right] = \frac{\mathcal{F}(q_k)}{\dfrac{d}{dp}(1 - e^{pT}e^{-sT})|_{q_k}}$$

$$= \frac{\mathcal{F}(s - jk\omega_s)}{-T} \tag{4.4.28}$$

so that

$$\mathcal{F}^*(s) = \frac{1}{T}\sum_{k=-\infty}^{\infty}\mathcal{F}(s - jk\omega_s) \tag{4.4.29}$$

This is the same result that was obtained in Chapter 2 by a different method.

Equating (4.4.29) and (4.4.18) we see that

$$\sum_k \text{Res}\left[\frac{\mathcal{F}(p)}{1 - e^{pT}e^{-sT}}, p_k\right] = \frac{1}{T}\sum_{k=-\infty}^{\infty}\mathcal{F}(s - jk\omega_s) + f(0^+)/2 \tag{4.4.30}$$

If the convention of assigning the value $f(0^+)$ to $f(0)$ is used in calculating $F(z)$, then according to (4.4.22) and (4.4.30)

$$F(z)|_{z=e^{sT}} = \frac{1}{T}\sum_{k=-\infty}^{\infty}\mathcal{F}(s - jk\omega_s) + f(0^+)/2 \tag{4.4.31}$$

4.5 THE INVERSE Z-TRANSFORM

In general, only the sequence of sample values $\{f(nT)\}$ can be found from the Z-transform $F(z)$. The discrete-time signal $f(nT)$ is called the *inverse Z-transform* of $F(z)$. The inverse Z-transform operation will be denoted as

$$f(nT) = \mathcal{Z}^{-1}\{F(z)\}$$

If it is known that a rational function $F(z)$ corresponds to a causal discrete-time signal, then this signal can be found by dividing the denominator into the numerator to generate a power series in z^{-1} and recognizing that $f(nT)$ is the coefficient of z^{-n}. Similarly, if it is known that $f(nT)$ is identically zero for positive n, the samples can be found by dividing the denominator into the numerator to generate a power series in z. Normally this method is only useful for finding sample values for n close to zero. However, sometimes it is possible to recognize the general term.

Example 4.5.1

Let $F(z) = (1 + 2z^{-1})/(1 - 2z^{-1} + z^{-2})$ correspond to a causal sequence. Then

$$
\begin{array}{r}
1 + 4z^{-1} + 7z^{-2} + \cdots \\
1 - 2z^{-1} + z^{-2}\overline{)1 + 2z^{-1}\phantom{+z^{-2}}} \\
\underline{1 - 2z^{-1} + z^{-2}} \\
4z^{-1} - z^{-2} \\
\underline{4z^{-1} - 8z^{-2} + 4z^{-3}} \\
7z^{-2} - 4z^{-3} \\
\cdots
\end{array}
$$

If the division is continued, it will be recognized that

$$
f(nT) = \begin{cases} 0 & \text{for} & n < 0 \\ 3n + 1 & \text{for} & n \geq 0 \end{cases}
$$

On the other hand, if $f(nT)$ is known to be zero for positive n, the division should be carried out as follows:

$$
\begin{array}{r}
2z + 5z^2 \\
z^{-2} - 2z^{-1} + 1\overline{)2z^{-1} + 1} \\
\underline{2z^{-1} - 4 + 2z} \\
5 - 2z \\
\underline{5 - 10z + 5z^2} \\
8z - 5z^2 \\
\cdots
\end{array}
$$

If the division is continued, it will be recognized that

$$
f(nT) = \begin{cases} -3n - 1 & \text{for} & n < 0 \\ 0 & \text{for} & n \geq 0 \end{cases} \qquad \blacktriangleleft
$$

As with Laplace transforms, inverse Z-transforms can often be found algebraically by using the method of *partial fractions* [72]. This technique results in an expression for the general term in the sequence of samples. To illustrate the method of partial fractions let us assume that $f(t)$ is a causal signal with a Z-transform of the form

$$
F(z) = \frac{a_0 + a_1 z^{-1} + \cdots + a_{N-1} z^{-(N-1)}}{\prod\limits_{k=i}^{N} (1 - e^{p_k T} z^{-1})} \tag{4.5.1}
$$

The poles of $F(z)$ are $e^{p_1 T}, \ldots, e^{p_N T}$. By inspection, the region of convergence must be $|z| > \max_k |e^{P_k T}|$. If all the poles are distinct, $F(z)$ can be expanded into

$$F(z) = \sum_{k=1}^{N} \frac{A_k}{1 - e^{P_k T} z^{-1}} \tag{4.5.2}$$

where

$$A_k = \lim_{z \to p_k} (1 - e^{P_k T} z^{-1}) F(z) \tag{4.5.3}$$

We have already observed that

$$\mathcal{Z}\{e^{P_k n T} u(nT)\} = 1/(1 - e^{P_k T} z^{-1}) \tag{4.5.4}$$

so that as a result of the linearity property of Z-transforms

$$f(nT) = \sum_{k=1}^{N} A_k e^{P_k n T} u(nT) \tag{4.5.5}$$

This method can be extended for higher order poles. However, we will not pursue the method of partial fractions further because the computation becomes tedious and it is much easier to find the samples by applying the Z-transform integral inversion formula, which will be derived shortly. The partial fraction method can also be used for discrete-time signals that are zero except for negative n and for two-sided signals by making the proper correspondences.

Example 4.5.2

Let

$$F(z) = \frac{1}{(1 - z^{-1})(1 - e^T z^{-1})} \qquad \text{for} \qquad |z| > 1$$

then

$$F(z) = \frac{A_1}{1 - z^{-1}} + \frac{A_2}{1 - e^{-T} z^{-1}}$$

where

$$A_1 = 1/(1 - e^{-T}) \qquad \text{and} \qquad A_2 = 1/(1 - e^T)$$

so that

$$f(nT) = \frac{1}{1 - e^{-T}} u(nT) + \frac{e^{-nT}}{1 - e^T} u(nT) \qquad \blacktriangleleft$$

The most general method for inverting a Z-transform $F(z)$ is based on the fact that $F(z)$ is defined as a Laurent series. Therefore, the Laurent integral inversion formula can be used to find the coefficients in the series [72]. This formula is conventionally called the Z-transform integral inversion formula in the system theory literature. It is presented in the following theorem.

THEOREM 4.5.1.   The Z-Transform Integral Inversion Formula

If

$$F(z) = \sum_{m=-\infty}^{\infty} f(mT)z^{-m} \tag{4.5.6}$$

converges to an analytic function in the annular domain $R_1 < |z| < R_2$, then

$$f(nT) = \frac{1}{2\pi j} \oint_C F(z) z^n \frac{dz}{z} \tag{4.5.7}$$

where C is any simple closed curve separating $|z| = R_1$ from $|z| = R_2$.

Proof   Multiplying (4.5.6) by z^{n-1} and integrating both sides around the contour C, we see that when the integration and summation are interchanged

$$\frac{1}{2\pi j} \oint_C F(z) z^n \frac{dz}{z} = \sum_{m=-\infty}^{\infty} f(mT) \frac{1}{2\pi j} \oint_C z^{n-m} \frac{dz}{z} \tag{4.5.8}$$

Interchanging the integration and summation is justified because the series (4.5.6) can be shown to converge uniformly along C. Making the change of variables $z = Re^{j\theta}$ with $R_1 < R < R_2$ it follows that

$$\frac{1}{2\pi j} \oint_C z^k \frac{dz}{z} = \frac{1}{2\pi} \int_0^{2\pi} e^{jk\theta}\, d\theta$$

$$= \begin{cases} 1 & \text{for} \quad k = 0 \\ 0 & \text{elsewhere} \end{cases} \tag{4.5.9}$$

Therefore, the summation on the right-hand side of (4.5.8) reduces to $f(nT)$.

Q.E.D.

If the region of convergence includes the unit circle, the contour C can be chosen as the unit circle. Making the change of variables $z = e^{j\omega T}$, (4.5.7) becomes

$$f(nT) = \frac{1}{\omega_s} \int_{-\omega_s/2}^{\omega_s/2} F^*(\omega) e^{j\omega nT}\, d\omega \tag{4.5.10}$$

which is identical to (2.3.10).

The inversion integral can be easily evaluated by the method of residues if the singularities of $F(z)$ are poles of finite order. Let $\{a_k\}$ be the set of poles of $F(z)z^{n-1}$ inside the contour C and $\{b_k\}$ be the set of poles of $F(z)z^{n-1}$ outside C in the finite z-plane. According to Cauchy's Residue Theorem

$$f(nT) = \sum_{\iota} \text{Res}\,[F(z)z^{n-1}, a_k] \tag{4.5.11}$$

or

$$f(nT) = -\sum_k \text{Res}\,[F(z)z^{n-1}, b_k] - \text{Res}\,[F(z)z^{n-1}, \infty] \qquad (4.5.12)$$

Usually either (4.5.11) or (4.5.12) will be easier to evaluate for a particular value of n. The computation involved in evaluating $f(nT)$ by the partial fraction method is essentially the same as that for evaluating (4.5.11) and (4.5.12).

Example 4.5.3

Let

$$F(z) = \frac{1}{(1 - z^{-1})(1 - e^{-T}z^{-1})} \qquad \text{for} \qquad |z| > 1.$$

The poles of $F(z)$ are at $z = 1$ and $z = e^{-T}$. The contour C can be chosen to be any simple closed path in the domain $|z| > 1$ that encloses the poles of $F(z)$. For simplicity C is shown as a circle in Fig. 4.5.1. Now

$$F(z)z^{n-1} = \frac{z^{n+1}}{(z - 1)(z - e^{-T})}$$

For $n \geq -1$ only the poles of $F(z)z^{n-1}$ at 1 and e^{-T} are enclosed by C and it is easiest to evaluate $f(nT)$ by (4.5.11). This gives

$$f(nT) = \text{Res}\,[F(z)z^{n-1}, 1] + \text{Res}\,[F(z)z^{n-1}, e^{-T}]$$

$$= \frac{1}{1 - e^{-T}} + \frac{e^{-(n+1)T}}{e^{-T} - 1} \qquad \text{for} \qquad n \geq -1$$

$F(z)z^{n-1}$ has no poles outside C and at least a zero of order two at ∞ for $n \leq -1$. Equation 4.5.12 is easiest to evaluate in this case. For $n \leq -1$

$$\text{Res}\,[F(z)z^{n-1}, \infty] = -\lim_{z \to \infty} zF(z)z^{n-1} = 0$$

so that

$$f(nT) = 0 \qquad \text{for} \qquad n \leq -1$$

Compare these results with Example 4.5.2. ◄

The contour C encloses all the poles of a Z-transform $F(z)$ if the region of convergence has the form $|z| > R$. $F(z)z^{n-1}$ will have a zero of order two or greater at ∞ for $n < 0$ so that $\text{Res}\,[F(z)z^{n-1}, \infty] = 0$ in this case. Therefore, $f(nT) \equiv 0$ for $n < 0$ according to (4.5.12). This shows that $F(z)$ corresponds to a causal signal if the region of convergence has the form $|z| > R$. On the other

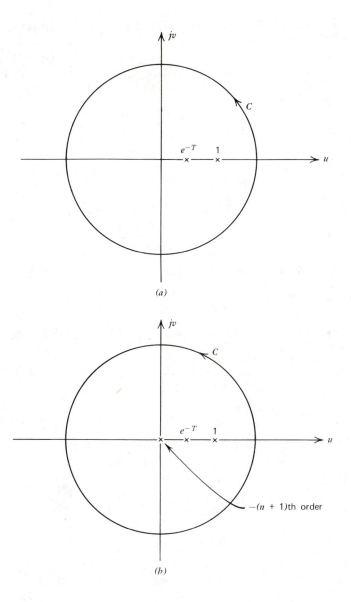

(a)

(b)

FIGURE 4.5.1. Pole locations of $F(z)z^{n-1}$ for Example 4.5.3. (a) Poles of $F(z)z^{n-1}$ in the finite z-plane for $n \geq -1$; (b) poles of $F(z)z^{n-1}$ for $n < -1$.

hand, all the poles of $F(z)$ are outside C if the region of convergence has the form $|z| < R$. $F(z)z^{n-1}$ will have no poles inside C for $n > 0$ so that $f(nT) \equiv 0$ according to (4.5.11) in this case. If the region of convergence has the form $0 < R_1 < |z| < R_2 < \infty$, then $f(nT)$ will not be identically zero for either positive or negative n.

Example 4.5.4

Let

$$F(z) = \frac{1 - a^2}{(1 - az)(1 - az^{-1})} \qquad \text{for} \qquad a < |z| < a^{-1}$$

with $0 < a < 1$. The poles of $F(z)$ are at a and a^{-1}. Now

$$F(z)z^{n-1} = \frac{(1 - a^2)z^n}{(1 - az)(z - a)}$$

For $n \geq 0$ the contour C encloses only the pole at $z = a$ so that using (4.5.11) we see that

$$f(nT) = \text{Res}\,[F(z)z^{n-1}, a] = a^n$$

For $n < 0$ only one pole at $z = a^{-1}$ is outside C and $F(z)z^{n-1}$ has a zero of order greater than two at ∞. Therefore, the residue at ∞ is zero and according to (4.5.12)

$$f(nT) = -\text{Res}\,[F(z)z^{n-1}, a^{-1}] = a^{-n}$$

Compare these results with Example 4.2.3.   ◄

The behavior of $f(nT)$ depends on the location of the poles of $F(z)$. If $F(z)$ corresponds to a causal signal, $f(nT)$ is most easily calculated from (4.5.11) for $n \geq 0$. A pole of order L at z_0 contributes to $f(nT)$ a term of the form

$$\text{Res}\,[F(z)z^{n-1}, z_0] = \lim_{z \to z_0} \frac{1}{(L-1)!} \frac{d^{L-1}}{dz^{L-1}} [(z - z_0)^L F(z)z^{n-1}]$$

$$= g(n)z_0^n \tag{4.5.13}$$

where $g(n)$ is a polynomial of degree $L-1$ in n. If $|z_0| < 1$ this term decreases to zero with n, if $|z_0| > 1$ it grows exponentially in magnitude with n, and if $z_0 = 1$ it maintains constant magnitude for $L = 1$ and grows algebraically in magnitude with n for $L > 1$. If $f(nT)$ is a real signal and z_0 is complex, then $F(z)$ will also contain a pole of order L at $\bar{z}_0$, the complex conjugate of z_0. If $z_0 = e^{(\sigma_0 + j\omega_0)T}$, then when the terms for z_0 and $\bar{z}_0$ are combined a component that looks like the samples of a damped oscillatory signal with frequency ω_0 will result. Once again we see the correspondence between the left half plane, the $j\omega$ axis, and the right half plane for Laplace transforms and the interior of the unit circle, the unit circle, and the exterior of the unit circle for Z-transforms.

4.6 USEFUL TRANSFORM RELATIONSHIPS

In this section a group of relatively simple but useful Z-transform relationships is presented.

1. *Multiplication by e^{-bt}*
   If $\mathscr{L}\{f(t)\} = F(z)$ for $R_1 < |z| < R_2$, then

   $$\mathscr{L}\{e^{-bt}f(t)\} = F(e^{bT}z) \qquad \text{for} \qquad |e^{-bT}|\,R_1 < |z| < |e^{-bT}|\,R_2$$

   $$(4.6.1)$$

   Proof

   $$\mathscr{L}\{e^{-bt}f(t)\} = \sum_{n=-\infty}^{\infty} f(nT)(e^{bT}z)^{-n}$$

   $$= F(e^{bT}z) \qquad \text{for} \qquad R_1 < |e^{bT}z| < R_2$$

   Q.E.D.

2. *Frequency Translation*
   If the region of convergence for $F(z)$ includes the unit circle and $g(t) = e^{j\omega_0 t}f(t)$, then

   $$G^*(\omega) = F^*(\omega - \omega_0) \qquad (4.6.2)$$

   Proof From (4.6.1) $G(z) = F(e^{-j\omega_0 T}z)$ and has the same region of convergence as $F(z)$. Therefore

   $$G^*(\omega) = G(z)|_{z=e^{j\omega T}} = F[e^{j(\omega-\omega_0)T}]$$

   $$= F^*(\omega - \omega_0)$$

   Q.E.D.

3. *Time Reversal*
   If $\mathscr{L}\{f(t)\} = F(z)$ for $R_1 < |z| < R_2$, then

   $$\mathscr{L}\{f(-t)\} = F(z^{-1}) \qquad \text{for} \qquad 1/R_2 < |z| < 1/R_1 \qquad (4.6.3)$$

   Proof

   $$\mathscr{L}\{f(-t)\} = \sum_{n=-\infty}^{\infty} f(-nT)z^{-n} = \sum_{n=-\infty}^{\infty} f(nT)z^{n}$$

   $$= F(z^{-1}) \qquad \text{for} \qquad R_1 < |z^{-1}| < R_2$$

   Q.E.D.

4.   *Multiplication by* t

If $\mathscr{L}\{f(t)\} = F(z)$ for $R_1 < |z| < R_2$, then

$$\mathscr{L}\{tf(t)\} = -zT\frac{d}{dz}F(z) \qquad \text{for} \qquad R_1 < |z| < R_2 \qquad (4.6.4)$$

Proof   A Laurent series can be differentiated term by term in its region of convergence and the resulting series has the same region of convergence. Therefore

$$\frac{d}{dz}F(z) = \frac{d}{dz}\sum_{n=-\infty}^{\infty}f(nT)z^{-n}$$

$$= \sum_{n=-\infty}^{\infty}-nf(nT)z^{-n-1} \qquad \text{for} \qquad R_1 < |z| < R_2 \qquad (4.6.5)$$

Multiplying both sides of (4.6.5) by $-zT$ yields

$$-zT\frac{d}{dz}F(z) = \sum_{n=-\infty}^{\infty}(nT)f(nT)z^{-n} \qquad \text{for} \qquad R_1 < |z| < R_2$$

Q.E.D.

Example 4.6.1

Let $f(t) = u(t)$ so that $F(z) = 1/(1 - z^{-1})$ for $|z| > 1$. Then from (4.6.4)

$$\mathscr{L}\{tu(t)\} = -zT\frac{d}{dz}F(z) = Tz^{-1}/(1 - z^{-1})^2$$

Compare this example with Example 4.4.1.   ◄

5.   *Division by* t

If $f(nT) = 0$ for $n < 2$, $f(t + T)/t$ is defined to be 0 for $t = 0$, and

$$\mathscr{L}\{f(t)\} = F(z) \qquad \text{for} \qquad |z| > R$$

then

$$\mathscr{L}\{f(t + T)/t\} = \frac{1}{T}\int_z^{\infty}F(x)\,dx \qquad \text{for} \qquad |z| > R \qquad (4.6.6)$$

Proof   By definition

$$F(z) = \sum_{n=2}^{\infty}f(nT)z^{-n} \qquad \text{for} \qquad |z| > R \qquad (4.6.7)$$

This series converges uniformly for $|z| > R$ so that it can be integrated term by term [71]. Integrating both sides of (4.6.7) from z to ∞ gives

$$\int_z^{\infty}F(x)\,dx = \sum_{n=2}^{\infty}f(nT)z^{-n+1}/(n-1) \qquad \text{for} \qquad |z| > R \qquad (4.6.8)$$

Making the change of variables $m = n - 1$, the series in (4.6.8) becomes

$$\sum_{m=1}^{\infty} \frac{f[(m+1)T]}{m} z^{-m} = T \sum_{m=1}^{\infty} \frac{f(mT+T)}{mT} z^{-m}$$

Q.E.D.

Example 4.6.2

Let $f(t) = u(t - 2T)$. Then $F(z) = z^{-2}/(1 - z^{-1})$ for $|z| > 1$ and according to (4.6.6)

$$\mathscr{L}\{u(t-T)/t\} = \frac{1}{T} \int_z^{\infty} \frac{x^{-2}}{1-x^{-1}} \, dx$$

$$= -\frac{1}{T} \log_e (1 - z^{-1}) \qquad \text{for} \qquad |z| > 1 \qquad \blacktriangleleft$$

6. *Delay Theorem*
    If k is an integer, then
    $$\mathscr{L}\{f(t - kT)\} = z^{-k} F(z) \tag{4.6.9}$$

Proof

$$\mathscr{L}\{f(t - kT)\} = \sum_{n=-\infty}^{\infty} f(nT - kT) z^{-n} = z^{-k} \sum_{n=-\infty}^{\infty} f(mT) z^{-m}$$

The last step results from making the change of variables $m = n - k$.

Q.E.D.

7. *One-Sided Advance Theorem*
    If $f(t)$ is a causal signal with Z-transform $F(z)$, then for $k \geq 0$ the one-sided Z-transform of $f(t + kT)$ is

    $$\mathscr{L}_+\{f(t + kT)\} = z^k F(z) - \sum_{n=0}^{k-1} f(nT) z^{k-n} \tag{4.6.10}$$

Proof By (4.6.9) the two-sided Z-transform of $f(t + kT)$ is

$$\mathscr{L}\{f(t + kT)\} = z^k F(z) = \sum_{n=0}^{\infty} f(nT) z^{k-n}$$

$$= \sum_{n=0}^{k-1} f(nT) z^{k-n} + \sum_{n=k}^{\infty} f(nT) z^{k-n} \tag{4.6.11}$$

The one-sided Z-transform can be obtained from the two-sided Z-transform of a signal by eliminating the part of the series with positive powers of z. Therefore, the sum on the far right-hand side of (4.6.11) is $\mathscr{L}_+\{f(t + kT)\}$.

Q.E.D.

The One-Sided Advance Theorem will be used later to find the solutions to Nth order, constant coefficient, linear difference equations when the initial conditions are specified. This theorem is analogous to the Differentiation Theorem for one-sided Laplace transforms which is used to solve differential equations with specified initial conditions.

8.   *Z-Transform of Complex Conjugate Signal*
   If $f(t)$ is a complex signal with Z-transform $F(z)$ which converges for $R_1 < |z| < R_2$, then

$$\mathscr{Z}\{\overline{f(t)}\} = \overline{F(\bar{z})} \qquad \text{for} \qquad R_1 < |z| < R_2 \qquad (4.6.12)$$

Proof   By definition

$$F(z) = \sum_{n=-\infty}^{\infty} f(nT) z^{-n} \qquad (4.6.13)$$

Replacing z by $\bar{z}$ and taking the conjugate of both sides of (4.6.13) we see that

$$\overline{F(\bar{z})} = \sum_{n=-\infty}^{\infty} \overline{f(nT)} z^{-n} \qquad \text{for} \qquad R_1 < |z| < R_2$$

Q.E.D.

9.   *Symmetry of $F^*(\omega)$ for Real Signals*
   If $f(t)$ is a real signal and the region of convergence for $F(z)$ includes the unit circle, then

$$F^*(-\omega) = \overline{F^*(\omega)} \qquad (4.6.14)$$

so that

$$|F^*(\omega)| \text{ is even} \qquad (4.6.15)$$

and

$$\arg F^*(\omega) \text{ is odd} \qquad (4.6.16)$$

Proof   When $f(t)$ is real, (4.6.12) is equivalent to

$$F(z) = \overline{F(\bar{z})} \qquad \text{or} \qquad \overline{F(z)} = F(\bar{z})$$

Making the substitution $z = e^{j\omega T}$ gives (4.6.14). Equations 4.6.15 and 4.6.16 follow directly from (4.6.14).

Q.E.D.

10.   *Initial Value Theorem*
   If $f(t)$ is a causal signal with Z-transform $F(z)$, then

$$f(0) = \lim_{z \to \infty} F(z) \qquad (4.6.17)$$

Proof

$$\lim_{z \to \infty} F(z) = \lim_{z \to \infty} \sum_{n=0}^{\infty} f(nT) z^{-n} = \sum_{n=0}^{\infty} f(nT) \lim_{z \to \infty} z^{-n} = f(0)$$

Interchanging the summation and limit is justified because the series converges uniformly in its region of convergence.

Q.E.D.

11. *Final Value Theorem*

If $f(t)$ is a causal signal with Z-transform $F(z)$ that has all its poles inside the unit circle except for possibly a first-order pole at $z = 1$, then

$$\lim_{n \to \infty} f(nT) = \lim_{z \to 1} (1 - z^{-1})F(z) \qquad (4.6.18)$$

Proof According to (4.5.11) and (4.5.13) the contributions to $f(nT)$ from poles inside the unit circle decay to zero with n. If there is a pole of order $L > 1$ at $z = 1$, its contribution becomes infinite as n becomes large. Finally, if there is a simple pole at $z = 1$, its contribution is

$$\mathrm{Res}\,[F(z)z^{n-1}, 1] = \lim_{z \to 1} (z - 1)F(z)z^{-1}z^n$$

$$= \lim_{z \to 1} (1 - z^{-1})F(z) \qquad \text{for} \qquad n \geq 0$$

Q.E.D.

Example 4.6.3

Let $F(z) = \dfrac{1}{(1 - z^{-1})(1 - e^{-T}z^{-1})}$ for $|z| > 1$. Then from (4.6.17)

$$f(0) = \lim_{z \to \infty} F(z) = 1$$

and from (4.6.18)

$$\lim_{n \to \infty} f(nT) = \lim_{z \to 1} (1 - z^{-1})F(z) = 1/(1 - e^{-T})$$

Compare these results with Example 4.5.2. ◀

12. *One-Sided Periodic Repetition*

If $f_1(t)$ is identically zero except for $0 \leq t < NT$ and we define the one-sided periodic repetition of $f_1(t)$ to be

$$f(t) = \sum_{r=0}^{\infty} f_1(t - rNT) \qquad (4.6.19)$$

then

$$F(z) = F_1(z)/(1 - z^{-N}) \qquad \text{for} \qquad |z| > 1 \qquad (4.6.20)$$

Proof Taking the Z-transform of (4.6.19) yields

$$F(z) = F_1(z) \sum_{r=0}^{\infty} z^{-rN}$$

The geometric series converges for $|z| > 1$.

Q.E.D.

4.7 Z-TRANSFORM OF PRODUCTS

The Laplace or Fourier transform of the product of two continuous-time signals is the frequency domain convolution of their transforms. A similar result is true for uniformly sampled signals and is presented in the following theorem.

THEOREM 4.7.1. *Z-Transform of Products*
Let
$$h(t) = f(t)g(t) \tag{4.7.1}$$
with
$$\mathscr{Z}\{f(t)\} = F(z) \quad \text{for} \quad R_{1f} < |z| < R_{2f} \tag{4.7.2}$$
and
$$\mathscr{Z}\{g(t)\} = G(z) \quad \text{for} \quad R_{1g} < |z| < R_{2g} \tag{4.7.3}$$
Then
$$\mathscr{Z}\{f(t)g(t)\} = H(z) = \sum_{n=-\infty}^{\infty} f(nT)g(nT)z^{-n}$$
$$= \frac{1}{2\pi j} \oint_C F(x)G\left(\frac{z}{x}\right)\frac{dx}{x} \tag{4.7.4}$$
for
$$R_{1f}R_{1g} < |z| < R_{2f}R_{2g} \tag{4.7.5}$$
where C is any simple closed curve encircling the origin with
$$\max(R_{1f}, |z|/R_{2g}) < |x| < \min(R_{2f}, |z|/R_{1g}) \tag{4.7.6}$$

Proof The series in (4.7.4) will converge to an analytic function $H(z)$ for $R_{1h} < |z| < R_{2h}$ where from (4.2.9)
$$R_{1h} = \overline{\lim_{n\to\infty}} \sqrt[n]{|f(nT)g(nT)|}$$
$$\leq \overline{\lim_{n\to\infty}} \sqrt[n]{|f(nT)|}\; \overline{\lim_{n\to\infty}} \sqrt[n]{|g(nT)|}$$
$$= R_{1f}R_{1g} \tag{4.7.7}$$
and from (4.2.12)
$$R_{2h} = 1/\overline{\lim_{n\to\infty}} \sqrt[n]{|f(-nT)g(-nT)|}$$
$$\geq R_{2f}R_{2g} \tag{4.7.8}$$
Replacing $f(nT)$ in (4.7.4) by the inversion formula (4.5.7) we see that
$$H(z) = \sum_{n=-\infty}^{\infty} \frac{1}{2\pi j} \oint_C F(x)x^n \frac{dx}{x} g(nT)z^{-n} \tag{4.7.9}$$

Interchanging the summation and integration, (4.7.9) can be written as

$$H(z) = \frac{1}{2\pi j} \oint_C F(x) \sum_{n=-\infty}^{\infty} g(nT) \left(\frac{z}{x}\right)^{-n} \frac{dx}{x} \qquad (4.7.10)$$

This step is justified if the integrand in (4.7.10) converges uniformly for some choice of C and z. For the inversion formula to be correct, C must be chosen so that

$$R_{1f} < |x| < R_{2f} \qquad (4.7.11)$$

If

$$R_{1g} < \left|\frac{z}{x}\right| < R_{2g}$$

or equivalently

$$|z|/R_{2g} < |x| < |z|/R_{1g} \qquad (4.7.12)$$

the series in the integrand of (4.7.10) will converge uniformly to $G(z/x)$. Otherwise it will diverge. The boundaries of the domains specified by (4.7.11) and (4.7.12) are shown in Fig. 4.7.1. These domains will intersect if

$$|z|/R_{2g} < R_{2f} \qquad \text{or} \qquad |z| < R_{2f}R_{2g}$$

and

$$|z|/R_{1g} > R_{1f} \qquad \text{or} \qquad |z| > R_{1f}R_{1g}$$

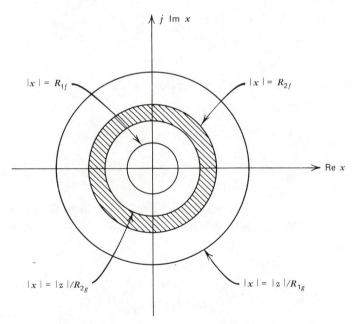

FIGURE 4.7.1. Regions of convergence for $F(x)$ and $G(z/x)$.

or, equivalent, if

$$R_{1f}R_{1g} < |z| < R_{2f}R_{2g} \tag{4.7.13}$$

When z satisfies (4.7.13), the intersection of the domains (4.7.11) and (4.7.12) is

$$(R_{1f} < |x| < R_{2f}) \cap (|z|/R_{2g} < |x| < |z|/R_{1g})$$
$$= \max (R_{1f}, |z|/R_{2g}) < |x| < \min (R_{2f}, |z|/R_{1g}) \tag{4.7.14}$$

C must be chosen in this intersection as specified in the theorem statement.

Q.E.D.

In the special case of causal signals, $R_{2f} = R_{2g} = \infty$ and the conditions on $|z|$ and C reduce to

$$R_{1f}R_{1g} < |z| \tag{4.7.15}$$

and

$$R_{1f} < |x| < |z|/R_{1g} \tag{4.7.16}$$

With these choices all the poles of $F(x)$ lie inside C and all the poles of $G(z/x)$ lie outside C.

Example 4.7.1

Let $f(t) = u(t)$ and $g(t) = e^{-|t|}$. Then $F(z) = 1/(1 - z^{-1})$ for $|z| > 1$ and

$$G(z) = \frac{1 - e^{-2T}}{(1 - e^{-T}z^{-1})(1 - e^{-T}z)} \quad \text{for} \quad e^{-T} < |z| < e^{T}$$

To evaluate $\mathcal{L}\{f(t)g(t)\}$ using Theorem 4.7.1, z must be chosen so that

$$e^{-T} < |z| < \infty$$

according to (4.7.5), and C must be chosen so that

$$\max (1, |z| \, e^{-T}) < |x| < |z| \, e^{T}$$

according to (4.7.6). A choice with $|z| > e^{T}$ is shown in Fig. 4.7.2. According to (4.7.4)

$$\mathcal{L}\{f(t)g(t)\} = \frac{1}{2\pi j} \oint_{C} \frac{1}{1 - x^{-1}} \frac{1 - e^{-2T}}{(1 - e^{-T}x/z)(1 - e^{-T}z/x)} \frac{dx}{x}$$

The poles of $G(z/x)$ are at $x = ze^{-T}$ and $x = ze^{T}$. Therefore C encloses the pole of $F(x)$ at $x = 1$ and the pole of $G(z/x)$ at $x = ze^{-T}$. According to Cauchy's Residue Theorem

$$\mathcal{L}\{f(t)g(t)\} = \text{Res}\,[F(x)G(z/x)x^{-1}, x = 1]$$
$$+ \text{Res}\,[F(x)G(z/x)x^{-1}, x = ze^{-T}]$$
$$= 1/(1 - e^{-T}z^{-1}) \quad \text{for} \quad |z| > e^{-T}$$

as expected since $f(t)g(t) = e^{-t}u(t)$. ◀

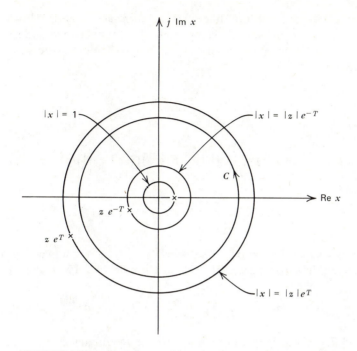

FIGURE 4.7.2. Choice of C and z for Example 4.7.1.

Example 4.7.2

Let $a(t) = u(t)$ so that $A(z) = 1/(1-z^{-1})$ for $|z| > 1$ and let $b(t)$ be a causal signal with Z-transform $B(z)$ for $|z| > R^2$. Then $F(z) = A(z^2)$ for $|z| > 1$ corresponds to a signal $f(t)$ whose samples are zero for n odd and $G(z) = z^{-1}B(z^2)$ for $|z| > R$ corresponds to signal $g(t)$ whose samples are zero for n even. This is easily seen by observing the coefficients in the series defining $F(z)$ and $G(z)$. For this contrived example

$$h(nT) = f(nT)g(nT) \equiv 0$$

so that

$$R_{1h} = \overline{\lim_{n \to \infty}} \sqrt[n]{|f(nT)g(nT)|} = 0 < R_{1f}R_{1g} = R$$

Using Theorem 4.7.1 to calculate $H(z)$, z must be chosen so that $|z| > R$ and C must be chosen so that $1 < |x| < |z|/R$. With this choice C contains only the poles of $F(x)$ at $x = 1$ and $x = -1$, so that

$$H(z) = \text{Res}\,[F(x)G(z/x)x^{-1}, 1] + \text{Res}\,[F(x)G(z/x)x^{-1}, -1]$$

On evaluating the residues, we find that $H(z) = 0$ as it must ◀

Equation 4.7.4 does not look like a convolution as it stands. However, if we let $z = e^{sT}$ where $s = \sigma + j\omega$ is chosen so that (4.7.5) is satisfied and let C be the circle $x = e^{pT}$ where $p = \gamma + j\lambda$ with $-\omega_s/2 < \lambda \leq \omega_s/2$ and γ fixed so that (4.7.6) is satisfied, then

$$\frac{dx}{x} = T\,dp$$

and (4.7.4) becomes

$$H(e^{sT}) = \frac{1}{j\omega_s} \int_{\gamma - j\omega_s/2}^{\gamma + j\omega_s/2} F(e^{pT}) G[e^{(s-p)T}]\,dp$$

or

$$\mathcal{H}^*(s) = \frac{1}{j\omega_s} \int_{\gamma - j\omega_s/2}^{\gamma + j\omega_s/2} \mathcal{F}^*(p)\mathcal{G}^*(s-p)\,dp \tag{4.7.17}$$

If $|z| = 1$ satisfies (4.7.5) and $|x| = 1$ satisfies (4.7.6), then σ and γ can be chosen as 0. Then, making the change of variables $p = j\lambda$, (4.7.17) becomes

$$H^*(\omega) = \frac{1}{\omega_s} \int_{-\omega_s/2}^{\omega_s/2} F^*(\lambda) G^*(\omega - \lambda)\,d\lambda \tag{4.7.18}$$

Equations 4.7.17 and 4.7.18 are similar to the frequency domain convolution formulas for the continuous-time case.

4.8 PARSEVAL'S THEOREM

In analyzing systems, the sum of the squares of a sequence of samples or the sum of the products of two sequences of samples is sometimes needed. Parseval's Theorem relates the time-domain sum to a z-domain integral.

THEOREM 4.8.1.   Parseval's Theorem
Let

$$\mathcal{Z}\{f(t)\} = F(z) \qquad \text{for} \qquad R_{1f} < |z| < R_{2f} \tag{4.8.1}$$

and

$$\mathcal{Z}\{g(t)\} = G(z) \qquad \text{for} \qquad R_{1g} < |z| < R_{2g} \tag{4.8.2}$$

with

$$R_{1f}R_{1g} < 1 < R_{2f}R_{2g} \tag{4.8.3}$$

Then

$$\sum_{n=-\infty}^{\infty} f(nT)g(nT) = \frac{1}{2\pi j} \oint_C F(z)G(z^{-1})\frac{dz}{z} \tag{4.8.4}$$

where C is any simple closed curve encircling the origin with

$$\max (R_{1f}, 1/R_{2g}) < |z| < \min (R_{2f}, 1/R_{1g}) \tag{4.8.5}$$

Proof If (4.8.3) is true, then z can be chosen as 1 in (4.7.4) and

$$\sum_{n=-\infty}^{\infty} f(nT)g(nT) = \frac{1}{2\pi j} \oint_C F(x)G(x^{-1}) \frac{dx}{x} \tag{4.8.6}$$

where C satisfies (4.7.6). Replacing x by z in (4.8.6) completes the proof.

Q.E.D.

In dealing with complex signals, a slight modification of the sum in (4.8.4) is frequently used. From (4.8.4) and (4.6.12) we can see that

$$\sum_{n=-\infty}^{\infty} f(nT)\overline{g(nT)} = \frac{1}{2\pi j} \oint_C F(z)\overline{G(\bar{z}^{-1})} \frac{dz}{z} \tag{4.8.7}$$

If $|z| = 1$ satisfies (4.8.5), then C can be chosen as the unit circle. Making the change of variables $z = e^{j\omega T}$, (4.8.7) becomes

$$\sum_{n=-\infty}^{\infty} f(nT)\overline{g(nT)} = \frac{1}{\omega_s} \int_{-\omega_s/2}^{\omega_s/2} F(e^{j\omega T})\overline{G(e^{j\omega T})} \, d\omega$$

$$= \frac{1}{\omega_s} \int_{-\omega_s/2}^{\omega_s/2} F^*(\omega)\overline{G^*(\omega)} \, d\omega \tag{4.8.8}$$

In the special case $f(t) = g(t)$, (4.8.3) becomes $R_{1f}^2 < 1 < R_{2f}^2$. This can be true only if the region of convergence for $F(z)$ includes the unit circle. Under these conditions $|z| = 1$ satisfies (4.8.5), and (4.8.8) becomes

$$\sum_{n=-\infty}^{\infty} |f(nT)|^2 = \frac{1}{\omega_s} \int_{-\omega_s/2}^{\omega_s/2} |F^*(\omega)|^2 \, d\omega \tag{4.8.9}$$

Example 4.8.1

Let $f(t) = e^{-t}u(t)$. Then $F(z) = 1/(1 - e^{-T}z^{-1})$ for $|z| > e^{-T}$. From (4.8.4)

$$S = \sum_{n=-\infty}^{\infty} f(nT)^2 = \frac{1}{2\pi j} \oint_C \frac{1}{1 - e^{-T}z^{-1}} \frac{1}{1 - e^{-T}z} \frac{dz}{z}$$

C encloses only the pole at $z = e^{-T}$ so that

$$S = \text{Res} \left[\frac{1}{(z - e^{-T})(1 - e^{-T}z)}, e^{-T} \right]$$

$$= 1/(1 - e^{-2T})$$

◀

4.9 SOME PROPERTIES OF THE REAL AND IMAGINARY PARTS OF $F^*(\omega)$

A signal $g(t)$ is called *Hermitian* if $g(-t) = \overline{g(t)}$ and *anti-Hermitian* if $g(-t) = -g(t)$. Any signal $f(t)$ can be expressed as the sum of a Hermitian and an anti-Hermitian part. To see this let us assume that

$$f(t) = f_1(t) + f_2(t) \tag{4.9.1}$$

where $f_1(t)$ is Hermitian and $f_2(t)$ is anti-Hermitian. Then

$$\overline{f(-t)} = \overline{f_1(-t)} + \overline{f_2(-t)}$$

or

$$\overline{f(-t)} = f_1(t) - f_2(t) \tag{4.9.2}$$

Adding (4.9.2) and (4.9.1) we see that

$$f_1(t) = \frac{f(t) + \overline{f(-t)}}{2} \tag{4.9.3}$$

Subtracting (4.9.2) from (4.9.1) we see that

$$f_2(t) = \frac{f(t) - \overline{f(-t)}}{2} \tag{4.9.4}$$

Notice that

$$f_1(0) = \mathrm{Re}\, f(0) \tag{4.9.5}$$

and

$$f_2(0) = j\, \mathrm{Im}\, f(0) \tag{4.9.6}$$

If the region of convergence for $F(z)$ includes the unit circle, then so does the region of convergence for $F_1(z)$ and $F_2(z)$. In this case

$$F_1^*(\omega) = \sum_{n=-\infty}^{\infty} f_1(nT) e^{-j\omega nT}$$

$$= f_1(0) + \sum_{n=1}^{\infty} [f_1(nT) e^{-j\omega nT} + f_1(-nT) e^{j\omega nT}]$$

and since $f_1(t)$ is Hermitian

$$F_1^*(\omega) = \mathrm{Re}\, f(0) + 2 \sum_{n=1}^{\infty} \mathrm{Re}\, [f_1(nT) e^{-j\omega nT}] \tag{4.9.7}$$

Similarly

$$F_2^*(\omega) = \sum_{n=-\infty}^{\infty} f_2(nT)e^{-j\omega nT}$$

$$= f_2(0) + \sum_{n=1}^{\infty} [f_2(nT)e^{-j\omega nT} + f_2(-nT)e^{j\omega nT}]$$

and since $f_2(nT)$ is anti-Hermitian

$$F_2^*(\omega) = j \operatorname{Im} f(0) + 2j \sum_{n=1}^{\infty} \operatorname{Im}[f_2(nT)e^{-j\omega nT}] \qquad (4.9.8)$$

According to (4.9.7) and (4.9.8), $F_1^*(\omega)$ is real and $F_2^*(\omega)$ is imaginary. Since $F^*(\omega) = F_1^*(\omega) + F_2^*(\omega)$, we see that

$$\operatorname{Re} F^*(\omega) = F_1^*(\omega) \qquad (4.9.9)$$

and

$$j \operatorname{Im} F^*(\omega) = F_2^*(\omega) \qquad (4.9.10)$$

If $f(t)$ is real, then $f_1(t)$ and $f_2(t)$ are the even and odd parts of $f(t)$. Equations 4.9.7 and 4.9.8 become

$$F_1^*(\omega) = f(0) + 2 \sum_{n=1}^{\infty} f_1(nT) \cos \omega nT \qquad (4.9.11)$$

and

$$F_2^*(\omega) = -2j \sum_{n=1}^{\infty} f_2(nT) \sin \omega nT \qquad (4.9.12)$$

Thus for real signals, $\operatorname{Re} F^*(\omega)$ is even and corresponds to the even part of $f(t)$ and $j \operatorname{Im} F^*(\omega)$ is odd and corresponds to the odd part of $f(t)$.

4.10 z-DOMAIN ANALOG OF THE HILBERT TRANSFORM

The real and imaginary parts of the Fourier transform of a causal signal are related by the Hilbert transform [103]. As expected, an analogous relationship relating the real and imaginary parts of the z-transform of a causal signal on the unit circle exists. This relationship will be called the z-domain *Hilbert transform* and is presented in the following theorem.

THEOREM 4.10.1.   z-Domain Hilbert Transform

Let $f(t)$ be a causal signal with z-transform $F(z)$ for $|z| > R$ with $R < 1$. Let the real and imaginary parts of $F(z)$ on the unit circle be

$$A^*(\omega) = A(e^{j\omega T}) = \operatorname{Re} F(e^{j\omega T}) \qquad (4.10.1)$$

and

$$B^*(\omega) = B(e^{j\omega T}) = \mathrm{Im}\, F(e^{j\omega T}) \tag{4.10.2}$$

so that

$$F^*(\omega) = F(e^{j\omega T}) = A^*(\omega) + jB^*(\omega) \tag{4.10.3}$$

Then

$$B^*(\omega) = -\frac{1}{\omega_s}(P_\omega)\int_{-\omega_s/2}^{\omega_s/2} A^*(y)\cot\frac{(\omega-y)T}{2}\,dy + \mathrm{Im}\, f(0) \tag{4.10.4}$$

and

$$A^*(\omega) = \frac{1}{\omega_s}(P_\omega)\int_{-\omega_s/2}^{\omega_s/2} B^*(y)\cot\frac{(\omega-y)T}{2}\,dy + \mathrm{Re}\, f(0) \tag{4.10.5}$$

where

$$(P_\omega)\int_{-\omega_s/2}^{\omega_s/2} Q(y)\,dy = \lim_{\substack{\epsilon\to 0 \\ \epsilon > 0}}\left[\int_{-\omega_s/2}^{\omega-\epsilon} Q(y)\,dy + \int_{\omega+\epsilon}^{\omega_s/2} Q(y)\,dy\right] \tag{4.10.6}$$

is the principal value of the integral when $Q(y)$ has a discontinuity at ω.

Proof Consider the integral

$$I = \oint_C F(x)\frac{x+e^{j\omega T}}{x-e^{j\omega T}}\frac{dx}{x} \tag{4.10.7}$$

where the contour C is shown in Fig. 4.10.1. The function

$$G(x) = \frac{x+e^{j\omega T}}{x-e^{j\omega T}} \tag{4.10.8}$$

has a pole at $x = e^{j\omega T}$. The path Γ_2 is chosen to loop around this pole. The paths Γ_4 and Γ_6 are separated by an infinitesimal distance. The integrand is analytic for $|x| > 1$ and along C so that $I = 0$ according to Cauchy's Residue Theorem. Therefore

$$\lim_{R_1\to 0}\left[\int_{\Gamma_1} + \int_{\Gamma_3}\right] + \lim_{R_1\to 0}\int_{\Gamma_2} + \lim_{R_2\to\infty}\int_{\Gamma_5} + \lim_{R_2\to\infty}\left[\int_{\Gamma_4} + \int_{\Gamma_6}\right] = 0 \tag{4.10.9}$$

If $x = e^{jyT}$ then $G(x)$ has the form

$$G(e^{jyT}) = \frac{e^{jyT}+e^{j\omega T}}{e^{jyT}-e^{j\omega T}} \tag{4.10.10}$$

Multiplying the numerator and denominator of the right-hand side of (4.10.10) by $e^{-j(\omega+y)T/2}$, we see that

$$G(e^{jyT}) = \frac{e^{-j(\omega-y)T/2}+e^{j(\omega-y)T/2}}{e^{-j(\omega-y)T/2}-e^{j(\omega-y)T/2}}$$

$$= \frac{2\cos\dfrac{(\omega-y)T}{2}}{-2j\sin\dfrac{(\omega-y)T}{2}} = j\cot\frac{(\omega-y)T}{2} \tag{4.10.11}$$

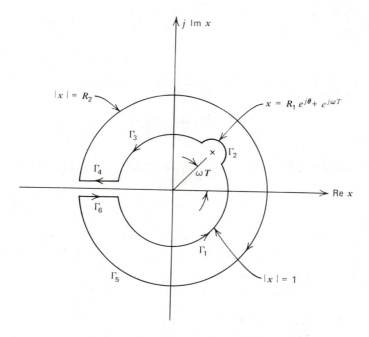

FIGURE 4.10.1. Contours for deriving the z-domain Hilbert transform.

Making the change of variables $x = e^{jyT}$, the first term on the left-hand side of (4.10.9) becomes

$$\lim_{\substack{\varepsilon \to 0 \\ \varepsilon \to 0}} \left[-T \int_{-\omega_s/2}^{\omega - \varepsilon} F^*(y) \cot \frac{(\omega - y)T}{2} \, dy - T \int_{\omega + \varepsilon}^{\omega_s/2} F^*(y) \cot \frac{(\omega - y)T}{2} \, dy \right]$$

$$= -T(P_\omega) \int_{-\omega_s/2}^{\omega_s/2} F^*(y) \cot \frac{(\omega - y)T}{2} \, dy \quad (4.10.12)$$

Along the path Γ_2 let $x = R_1 e^{j\theta} + e^{j\omega T}$. If Γ_2 meets Γ_1 for $\theta = \theta_1$ and meets Γ_3 for $\theta = \theta_2$, then

$$\lim_{R_1 \to 0} (\theta_2 - \theta_1) = \pi$$

The integral along Γ_2 becomes

$$\lim_{R_1 \to 0} j \int_{\theta_1}^{\theta_2} F(x) \frac{x + e^{j\omega T}}{x} \, d\theta = j2\pi F^*(\omega) \quad (4.10.13)$$

since

$$\lim_{R_1 \to 0} F(x) \frac{x + e^{j\omega T}}{x} = 2F(e^{j\omega T})$$

Because $F(x)G(x)$ is analytic for $|x| > 1$, it must have a Laurent series expansion of the form

$$F(x)G(x) = a_0 + a_1/x + a_2/x^2 + \cdots$$
$$= a_0 + Q(x)/x \qquad (4.10.14)$$

in this domain. $Q(x)$ must be an analytic function and is therefore bounded for $|x| > 1$. Thus

$$a_0 = \lim_{x \to \infty} F(x)G(x) = \lim_{x \to \infty} F(x) = f(0) \qquad (4.10.15)$$

The last step is just an application of the Initial Value Theorem. Making the change of variables $x = R_2 e^{j\theta}$ and using (4.10.14) and (4.10.15), it follows that

$$\lim_{R_2 \to \infty} \int_{\Gamma_5} F(x)G(x) \frac{dx}{x} = -j2\pi f(0) \qquad (4.10.16)$$

The integrals along Γ_4 and Γ_6 cancel. Substituting (4.10.12), (4.10.13), and (4.10.16) into (4.10.9), we see that

$$-T(P_\omega) \int_{-\omega_s/2}^{\omega_2/2} F^*(y) \cot \frac{(\omega - y)T}{2} \, dy + j2\pi F^*(\omega) - j2\pi f(0) = 0 \quad (4.10.17)$$

Equating the real and imaginary parts of the left-hand side of (4.10.17) to zero, we obtain the two equations

$$-T(P_\omega) \int_{-\omega_s/2}^{\omega_s/2} A^*(y) \cot \frac{(\omega - y)T}{2} \, dy - 2\pi B^*(\omega) + 2\pi \, \mathrm{Im} \, f(0) = 0 \quad (4.10.18)$$

and

$$-T(P_\omega) \int_{-\omega_s/2}^{\omega_s/2} B^*(y) \cot \frac{(\omega - y)T}{2} \, dy + 2\pi A^*(\omega) - 2\pi \, \mathrm{Re} \, f(0) = 0 \quad (4.10.19)$$

Rearranging (4.10.18) and (4.10.19) gives (4.10.4) and (4.10.5).

Q.E.D.

The z-domain Hilbert transform relations can also be derived heuristically in an interesting way. According to (4.9.9) and (4.9.10), $a(nT)$ is the Hermitian part and $jb(nT)$ is the anti-Hermitian part of $f(nT)$. Therefore, from (4.9.3) and (4.9.4) we see that

$$a(nT) = \frac{f(nT) + \overline{f(-nT)}}{2} \qquad (4.10.20)$$

and

$$jb(nT) = \frac{f(nT) - \overline{f(-nT)}}{2} \qquad (4.10.21)$$

Since $f(t)$ is causal

$$jb(nT) = \begin{cases} a(nT) & \text{for} & n > 0 \\ j \operatorname{Im} f(0) & \text{for} & n = 0 \\ -a(nT) & \text{for} & n < 0 \end{cases} \qquad (4.10.22)$$

If we define $y(nT)$ as

$$y(nT) = \begin{cases} 1 & \text{for} & n > 0 \\ 0 & \text{for} & n = 0 \\ -1 & \text{for} & n < 0 \end{cases} \qquad (4.10.23)$$

then (4.10.22) can be concisely written as

$$jb(nT) = a(nT)y(nT) + j \operatorname{Im} f(0) \, \delta_{n0} \qquad (4.10.24)$$

Similarly we can show that

$$a(nT) = jb(nT)y(nT) + \operatorname{Re} f(0) \, \delta_{n0} \qquad (4.10.25)$$

To complete the derivation we will have to know $Y^*(\omega)$. This does not exist in the ordinary sense but does exist in the distribution sense. Formally we can write that

$$
\begin{aligned}
Y^*(\omega) &= \lim_{N \to \infty} \left[\sum_{n=0}^{N-1} e^{-j\omega nT} - \sum_{n=0}^{N-1} e^{j\omega nT} \right] \\
&= \lim_{N \to \infty} \left[\frac{1 - e^{-j\omega NT}}{1 - e^{-j\omega T}} - \frac{1 - e^{j\omega NT}}{1 - e^{j\omega T}} \right] \\
&= \frac{1}{1 - e^{-j\omega T}} - \frac{1}{1 - e^{j\omega T}} + \lim_{N \to \infty} \left[\frac{e^{j\omega NT}}{1 - e^{j\omega T}} - \frac{e^{-j\omega NT}}{1 - e^{-j\omega T}} \right] \quad (4.10.26)
\end{aligned}
$$

The last term on the right-hand side of (4.10.26) is zero in the distribution sense as a result of the Riemann-Lebesgue Lemma [103]. By combining the remaining two terms, we have

$$Y^*(\omega) = -j \cot \frac{\omega T}{2} \qquad (4.10.27)$$

Ignoring the fact that $Y^*(\omega)$ has a singularity on the unit circle and evaluating the Z-transforms of (4.10.24) and (4.10.25) on the unit circle using the frequency domain convolution formula (4.7.18), we obtain the z-domain Hilbert transform formulas (4.10.4) and (4.10.5) without the indication that the principal values of the integrals should be used.

It is often difficult to evaluate the z-domain Hilbert transform integrals directly. In solving problems it is usually easier to find $B^*(\omega)$ from $A^*(\omega)$ or vice-versa by observing from (4.10.20) and (4.10.21) that

$$a(nT) = jb(nT) = f(nT)/2 \qquad \text{for} \qquad n > 0 \qquad (4.10.28)$$

Therefore, given $A^*(\omega)$ or $B^*(\omega)$ and $f(0)$ we can calculate $f(nT)$ first and then $F(z)$ and the unknown part of $F^*(\omega)$.

Example 4.10.1

Let $f(t)$ be real and

$$A^*(\omega) = \frac{1 - r \cos \omega T}{1 + r^2 - 2r \cos \omega T}$$

where r is real and $0 < |r| < 1$. Making the substitution $z = e^{j\omega T}$, we can see that

$$A(z) = \frac{1 - r(z + z^{-1})/2}{(1 - rz^{-1})(1 - rz)}$$

Therefore

$$a(nT) = \text{Res}\,[A(z)z^{n-1}, r]$$

$$= \begin{cases} 1 & \text{for} \quad n = 0 \\ r^n/2 & \text{for} \quad n > 0 \end{cases}$$

so that according to (4.10.28) and the fact that $a(0) = f(0)$

$$f(nT) = r^n u(nT)$$

and

$$F(z) = 1/(1 - rz^{-1}) \qquad \blacktriangleleft$$

If the Z-transform $F(z)$ of a causal signal $f(t)$ has all its poles and zeros inside the unit circle, then $\log |F^*(\omega)|$ and $\arg F^*(\omega)$ are related by the z-domain Hilbert transform. To see this, let

$$G(z) = \log F(z) = \log |F(z)| + j \arg F(z) \qquad (4.10.29)$$

The signal $g(nT)$ is known as the *complex cepstrum* for $f(nT)$ [101,39]. The singularities of $G(z)$ occur at the poles and zeros of $F(z)$ and so must be inside the unit circle. Therefore, $g(nT)$ is a causal signal and $\text{Re}\,G^*(\omega)$ and $\text{Im}\,G^*(\omega)$ are related by the z-domain Hilbert transform. Consequently

$$\arg F^*(\omega) = -\frac{1}{\omega_s}(P_\omega)\int_{-\omega_s/2}^{\omega_s/2} \log |F^*(y)| \cot \frac{(\omega - y)T}{2}\, dy + \text{Im}\, g(0) \quad (4.10.30)$$

If, in addition, $f(nT)$ is a real signal, then it is called a minimum phase sequence. This is because, of all the real causal signals with amplitude response $|F^*(\omega)|$, the minimum phase sequence has the least phase lag, $-\arg F^*(\omega)$, for

$\omega > 0$. When $f(nT)$ is real, $|F^*(\omega)|$ is even and arg $F^*(\omega)$ is odd. Thus Re $G^*(\omega)$ is even and Im $G^*(\omega)$ is odd so that $g(nT)$ must be real and the last term, Im $g(0)$, in (4.10.30) is zero.

4.11 SOLUTION OF CONSTANT COEFFICIENT, LINEAR, DIFFERENCE EQUATIONS

An equation of the form

$$a_0 y(n+N) + a_1 y(n+N-1) + \cdots + a_N y(n) = f(n) \qquad \text{for} \qquad n \in \mathcal{N} \quad (4.11.1)$$

where a_0 and a_N are not zero, $\mathcal{N}$ is a set of consecutive integers, and $f(n)$ is known for $n \in \mathcal{N}$, is called an *Nth order, linear, constant coefficient, difference equation*. The signal $f(n)$ is called the *input* or *driving function* and the signal $y(n)$ is called the *solution*. A difference equation may describe a system that is actually discrete in nature or may be an approximation to the differential equation describing a continuous-time system. Several mathematical treatises on difference equations have been written. For example, see References 59 and 93.

If $\mathcal{N} = [0, \infty)$ and the N initial conditions $y(0)$, $y(1), \ldots, y(N-1)$ are specified, then $y(N)$, $y(N+1), \ldots$ can be calculated recursively from (4.11.1). It may or may not be easy to recognize the general term in the solution. An expression for the general term, if it exists, can be found easily by Z-transform methods. The one-sided Z-transform $Y(z)$ can be found by taking the one-sided Z-transform of each side of the difference equation term by term. Clearly, T should be assumed to be 1 for this application. The initial conditions get automatically included by the One-Sided Advance Theorem (4.6.10). The one-sided Z-transform of (4.11.1) has the form

$$(a_0 z^N + a_1 z^{N-1} + \cdots + a_N) Y(z) - Q(z) = F(z) \qquad (4.11.2)$$

where $Q(z)$ includes all the terms introduced by the initial conditions. Therefore

$$Y(z) = \frac{F(z)}{A(z)} + \frac{Q(z)}{A(z)} \qquad (4.11.3)$$

where

$$A(z) = a_0 z^N + a_1 z^{N-1} + \cdots + a_N \qquad (4.11.4)$$

We can see from (4.11.3) that the solution consists of two components. The component corresponding to $F(z)/A(z)$ is the solution when all the initial conditions are zero and is called the *forced solution*. The component corresponding to $Q(z)/A(z)$ is the solution when the input signal $f(n)$ is identically zero and is called the *autonomous solution*. A difference equation with zero input is called an *autonomous* or *homogeneous* equation.

Example 4.11.1

Consider the difference equation

$$y(n+2)+2y(n+1)+y(n)=1 \qquad \text{for} \qquad n \geq 0$$

with $y(0)=1$ and $y(1)=0$. Taking one-sided Z-transforms gives

$$[z^2 Y(z) - z^2] + 2[zY(z) - z] + Y(z) = 1/(1 - z^{-1})$$

or

$$Y(z) = \frac{1}{(z+1)^2(1-z^{-1})} + \frac{z^2 + 2z}{(z+1)^2}$$

Inverting $Y(z)$ shows that

$$y(n) = (-1)^n[(3-2n)/4] + 1/4 \qquad \text{for} \qquad n \geq 0$$

Calculate a few values of $y(n)$ recursively from the difference equation and compare them with the answer obtained here by Z-transforms. ◀

The solution to a linear, constant coefficient, difference equation can also be found by a method analogous to the classical method for solving linear, constant coefficient, differential equations. The classical method involves three steps. In one step a parametric form is assumed for the solution without regard to boundary conditions and substituted into the difference equation. If the parameters can be chosen to make the left-hand side equal to the driving function, this solution is called a *particular* solution $y_p(n)$.

In another step, solutions to the homogeneous equation are found. Suppose we assume that $y(n) = x^n$ for some nonzero x. Substituting this assumed solution into the homogeneous difference equation gives

$$x^n(a_0 x^N + a_1 x^{N-1} + \cdots + a_N) = 0 \qquad (4.11.5)$$

Since x^n is not zero, the term in parentheses must be zero so that x must satisfy the equation

$$a_0 x^N + a_1 x^{N-1} + \cdots + a_N = 0 \qquad (4.11.6)$$

This is called the *characteristic equation* and the left-hand side is called the *characteristic polynomial* for the difference equation. Notice that the characteristic polynomial is just $A(x)$ defined by (4.11.4). If $A(x)$ has N distinct roots $x_1, \ldots, x_N$, then

$$x_k^n \qquad \text{for} \qquad k = 1, \ldots, N \qquad (4.11.7)$$

are N distinct solutions to the homogeneous equation. On the other hand, suppose $A(x)$ has a root x_1 of order M. Then

$$\frac{d^r}{dx^r}[x^n A(x)]|_{x_1} = \sum_{k=0}^{N} a_k \frac{(n+N-k)!}{(n+N-k-r)!} x_1^{n+N-k-r}$$

$$= 0 \quad \text{for} \quad r = 0, 1, \ldots, M-1 \quad (4.11.8)$$

By substituting into the difference equation and from (4.11.8) we can see that

$$n(n-1)\cdots(n-r+1)x_1^{n-r} = \frac{n!}{(n-r)!} x_1^{n-r} \quad \text{for} \quad r = 0, \ldots, M-1 \quad (4.11.9)$$

are M distinct solutions to the homogeneous equation. Thus, in general, we can find N distinct solutions to the homogeneous equation. It can be shown that these N solutions are linearly independent. If these solutions are denoted by $g_1(n), \ldots, g_N(n)$, then any solution to the homogeneous equation must have the form

$$y_c(n) = \sum_{k=1}^{N} d_k g_k(n) \quad (4.11.10)$$

The third step in the classical method is to assume that the general solution is the sum of the particular solution and a solution to the homogeneous equation, that is

$$y(n) = y_p(n) + y_c(n)$$

The N unknown coefficients in $y_c(n)$ are chosen so that $y(n)$ satisfies the N boundary conditions. The resulting component $y_c(n)$ is called the *complementary solution*.

Example 4.11.2

Consider the difference equation in Example 4.11.1. If we assume that the particular solution has the form $y_p(n) = K$ and substitute into the difference equation, we get

$$K + 2K + K = 1$$

or

$$K = \tfrac{1}{4}$$

The characteristic polynomial is

$$x^2 + 2x + 1 = (x+1)^2$$

which has a double root at $x = -1$. Thus the complementary solution must have the form

$$y_c(n) = d_1(-1)^n + d_2 n(-1)^n$$

and the total solution must have the form

$$y(n) = \tfrac{1}{4} + d_1(-1)^n + d_2 n(-1)^n$$

Since we have specified that $y(0) = 1$ and $y(1) = 0$, it readily follows that $d_1 = \tfrac{3}{4}$ and $d_2 = -\tfrac{1}{2}$ so that

$$y(n) = (-1)^n[(3 - 2n)/4] + \tfrac{1}{4}$$

as before.   ◀

4.12 DISCRETE-TIME CONVOLUTION

The *discrete-time convolution* of the signals $f(nT)$ and $g(nT)$ is defined to be the signal

$$h(nT) = \sum_{k=-\infty}^{\infty} f(kT)g[(n-k)T] \qquad \text{for} \qquad -\infty < n < \infty \qquad (4.12.1)$$

The notation $h(nT) = f(nT) * g(nT)$ is sometimes used to denote the convolution operation. The convolution sum is analogous to the convolution integral for continuous-time signals and we might expect that $H(z) = F(z)G(z)$. We see that this is indeed true in the following theorem.

THEOREM 4.12.1.   Discrete-Time Convolution Theorem
Let

$$\mathscr{L}\{f(nT)\} = F(z) \qquad \text{for} \qquad R_{1f} < |z| < R_{2f} \qquad (4.12.2)$$

and

$$\mathscr{L}\{g(nT)\} = G(z) \qquad \text{for} \qquad R_{1g} < |z| < R_{2g} \qquad (4.12.3)$$

Then if $h(nT)$ is defined by (4.12.1)

$$\mathscr{L}\{h(nT)\} = H(z) = F(z)G(z) \qquad (4.12.4)$$

for

$$\max{(R_{1f}, R_{1g})} < |z| < \min{(R_{2f}, R_{2g})} \qquad (4.12.5)$$

Proof   The domain specified by (4.12.5) is simply the intersection of the regions of convergence for $F(z)$ and $G(z)$. In this intersection the series expansions for $F(z)$ and $G(z)$ converge absolutely and these series can be multiplied together term by term. The resulting series can be rearranged in any order. Therefore

$$F(z)G(z) = \sum_{k=-\infty}^{\infty} f(kT)z^{-k} \sum_{m=-\infty}^{\infty} g(mT)z^{-m}$$

$$= \sum_{k=-\infty}^{\infty} \sum_{m=-\infty}^{\infty} f(kT)g(mT)z^{-(k+m)} \qquad (4.12.6)$$

Letting $m = n - k$, (4.12.6) becomes

$$F(z)G(z) = \sum_{n=-\infty}^{\infty} \sum_{k=-\infty}^{\infty} f(kT)g[(n-k)T]z^{-n}$$

$$= \sum_{n=-\infty}^{\infty} h(nT)z^{-n} = H(z) \qquad (4.12.7)$$

Q.E.D.

The Discrete-Time Convolution Theorem provides the basis for filtering signals with digital computers. If the discrete-time signal $f(nT)$ is the sequence of samples of a continuous-time signal $f(t)$ which is band-limited to $|\omega| < \omega_s/2$, we have already observed in Chapter 2 that

$$F(\omega) = TF^*(\omega) = TF(z)|_{z=e^{j\omega T}} \qquad \text{for} \qquad |\omega| < \omega_s/2$$

Similarly, we can think of the signal $h(nT)$ as the sequence of samples of a band-limited signal $h(t)$ with

$$H(\omega) = TH^*(\omega) = TF^*(\omega)G^*(\omega) \qquad \text{for} \qquad |\omega| < \omega_s/2 \qquad (4.12.8)$$

Thus the spectrum of $f(nT)$ can be modified by convolving it with an appropriate signal $g(nT)$. In analogy with the continuous-time case, we say that $h(nT)$ is the output of a filter with transfer function $G(z)$ or frequency response $G^*(\omega)$ and input $f(nT)$. If $f(nT) = \delta_{n0}$, then $h(nT) = g(nT)$. The signal $g(nT)$ will be called the *pulse response* of the filter rather than its impulse response. The term "impulse response" will be reserved for the response of a system to a Dirac delta function. The design of discrete-time filters will be discussed in detail in Chapters 8, 9, and 10.

Example 4.12.1

Let $f(nT) = u(nT)$ and $g(nT) = e^{-nT}u(nT)$. Then

$$F(z) = 1/(1 - z^{-1}) \qquad \text{for} \qquad |z| > 1$$
$$G(z) = 1/(1 - e^{-T}z^{-1}) \qquad \text{for} \qquad |z| > e^{-T}$$

and

$$H(z) = \frac{1}{(1 - z^{-1})(1 - e^{-T}z^{-1})} \qquad for \qquad |z| > 1$$

Calculating the inverse Z-transform of $H(z)$, we see that

$$h(nT) = f(nT)*g(nT) = \frac{1 - e^{-(n+1)T}}{1 - e^{-T}} u(nT)$$

Obtain this result by calculating the convolution sum directly. ◀

5

ANALYSIS OF SAMPLED-DATA SYSTEMS BY TRANSFORM METHODS

5.1 INTRODUCTION

In this chapter we see how to analyze sampled-data systems by a combination of Fourier or Laplace and Z-transform methods. Only systems with uniform sampling and linear, time-invariant blocks are considered. The analysis techniques are formulated in terms of single input and output systems. However, they can also be applied to multi-input and output systems. (Multi-input and output systems are discussed in Chapter 6.)

The concept of the *pulse transfer function* is introduced first. We see that when a continuous-time input signal is impulse sampled and applied to a linear, time-invariant, continuous-time system, the Z-transform of the system output is the product of the Z-transform of the input signal and the Z-transform of the impulse response of the system. The Z-transform of the impulse response of the continuous-time system is called its pulse transfer function. Thus both strictly discrete-time systems and sampled-data systems containing continuous-time components can be represented by a pulse transfer function when only the input and output samples are of interest. We see that the sinusoidal steady-state frequency response of a discrete-time system can be determined by evaluating its pulse transfer function on the unit circle. Continuous-time systems with rational transfer functions correspond to constant coefficient, linear differential equations. We will see that discrete-time systems with rational pulse transfer functions correspond to constant coefficient, linear difference equations and that a given pulse transfer function can be realized by coupled sets of difference equations corresponding to a variety of structures.

Sampled-data feedback systems are discussed briefly. If all the blocks in the system block diagram are separated by samplers and only the values of signals at the sampling instants are of interest, then an equivalent discrete-time system can be substituted for the system by simply replacing each block by its pulse transfer function. Then the overall pulse transfer function can be found by the standard techniques for reducing block diagrams. If all the continuous-time blocks are not separated by samplers, this will not be true.

Next the stability of causal open and closed-loop discrete-time systems is examined. We see that a necessary and sufficient condition for stability is that all the poles of the pulse transfer function must lie inside the unit circle. Several tests for stability are presented. First, a test known as *Jury's test* or the *modified Schur-Cohn criterion* will be described. Like the Routh-Hurwitz criterion for continuous-time systems this is a strictly algebraic test. We show that the Nyquist criterion can be easily modified so that it applies to discrete-time systems and that the root locus techniques can be used without modification (except that pole locations must be interpreted with respect to the unit circle). Finally, we see that all the methods used for analyzing continuous-time systems can be used for discrete-time systems by mapping the interior of the unit circle onto the left half plane with a bilinear transformation.

In many cases the behavior of a sampled-data system between sampling instants is important. We will see that this behavior can be determined by using what is called the *advanced or modified Z-transform*. A practical approach that is sometimes implemented is to observe the desired signal with a fast rate sampler. This results in a multirate system. When the fast sampling rate is an integral multiple of the regular sampling rate, transform methods involving the regular rate and a fast rate Z-transform can be used.

To reduce data rate or storage requirements, the output of a sampled-data system is sometimes sampled at a rate that is slower by an integral factor N than the internal sampling rate. If the output happens to be a discrete-time signal, then this slow rate sampling corresponds to retaining only every Nth sample and is called *skip sampling*. Methods for analyzing the effects of slow rate and skip sampling are derived.

In Chapter 4 care was taken to specify the regions of convergence for Z-transforms. In the remainder of this book these regions are usually not specified explicitly because they will be obvious from the context of the problem being discussed.

5.2 PULSE TRANSFER FUNCTIONS

A typical block found in sampled-data systems is shown in Fig. 5.2.1. The signal $f(t)$ is a continuous-time input signal. The ideal input sampler and filter

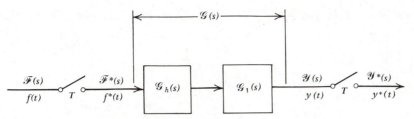

FIGURE 5.2.1.   Typical block in a sampled-data system.

$\mathscr{G}_h(s)$ represent a sample and hold circuit. $\mathscr{G}_1(s)$ could represent a continuous-time system to be controlled or a filter designed to process the reconstructed signal in some desired manner. We will assume that the two samplers are exactly synchronized. To simplify the analysis, we will lump the two transfer functions together and let $\mathscr{G}(s) = \mathscr{G}_h(s)\mathscr{G}_1(s)$ and $g(t) = \mathscr{L}^{-1}\{\mathscr{G}(s)\}$.

The input to $g(t)$ is

$$f^*(t) = \sum_{k=-\infty}^{\infty} f(kT)\,\delta(t-kT) \qquad (5.2.1)$$

By superposition we see immediately that the continuous-time output signal is

$$y(t) = \sum_{k=-\infty}^{\infty} f(kT)g(t-kT) \qquad (5.2.2)$$

The output samples are

$$y(nT) = \sum_{k=-\infty}^{\infty} f(kT)g(nT-kT) \qquad (5.2.3)$$

or, equivalently

$$y(nT) = \sum_{k=-\infty}^{\infty} g(kT)f(nT-kT) \qquad (5.2.4)$$

Thus the sequence of output samples is simply the discrete-time convolution of the sequence of input samples with the sequence of samples of the impulse response $g(t)$. According to Theorem 4.12.1

$$Y(z) = F(z)G(z) \qquad (5.2.5)$$

The Z-transform $G(z)$ is called the *pulse transfer function* for the block.

This same result can be derived by a frequency domain approach. It is customary to define the frequency domain operation [ ]* by the equation

$$[\mathscr{Y}(s)]^* = \mathscr{L}\{y^*(t)\} = \mathscr{Y}^*(s) = Y(z)\big|_{z=e^{sT}} \qquad (5.2.6)$$

This will be called the *star operator*. Also, as a matter of notational convenience in working with mixed discrete and continuous-time systems, we will frequently substitute z for e^{sT} and $Y(z)$ for $\mathcal{Y}^*(s)$. From Fig. 5.2.1 we can see that

$$\mathcal{Y}(s) = \mathcal{F}^*(s)\mathcal{G}(s) \tag{5.2.7}$$

Therefore

$$[\mathcal{Y}(s)]^* = [\mathcal{F}^*(s)\mathcal{G}(s)]^*$$

$$= \frac{1}{T}\sum_{k=-\infty}^{\infty}\mathcal{F}^*(s-jk\omega_s)\mathcal{G}(s-jk\omega_s) \tag{5.2.8}$$

We have already observed that

$$\mathcal{F}^*(s-jk\omega_s) = \mathcal{F}^*(s)$$

so that

$$\mathcal{Y}^*(s) = \mathcal{F}^*(s)\frac{1}{T}\sum_{k=-\infty}^{\infty}\mathcal{G}(s-jk\omega_s) = \mathcal{F}^*(s)\mathcal{G}^*(s) \tag{5.2.9}$$

This is equivalent to (5.2.5). In general, any factor $A(s)$ of the argument of the star operator which has the property that $A(s+j\omega_s) = A(s)$ for all s corresponds to a sampled signal and can be taken outside of the star operator.

In some applications only the impulse sampled output signal is of interest. In this case the block diagram in Fig. 5.2.1 can be replaced by the one shown in Fig. 5.2.2. Figure 5.2.2 could also represent a system in which a digital computer is used to generate a discrete-time output signal $y(nT)$ related to the input signal $f(nT)$ by the pulse transfer function $G(z)$. Conceptually, $y(nT)$ could be calculated by evaluating the convolution summation (5.2.3). In the next section we see that when $G(z)$ is a rational function of z, (5.2.3) is equivalent to a finite-order difference equation so that an infinite summation is not required.

Notice that no output sampler is present in Fig. 5.2.2 because the output transform $Y(z) = F(z)G(z)$ implicitly corresponds to an ideal impulse sampled signal. In fact, if an output sampler were present, the resulting overall system would not correspond to any real physical system. In that case, the output would have the form

$$y^{**}(t) = \sum_{n=-\infty}^{\infty} y(nT)\,\delta^2(t-nT)$$

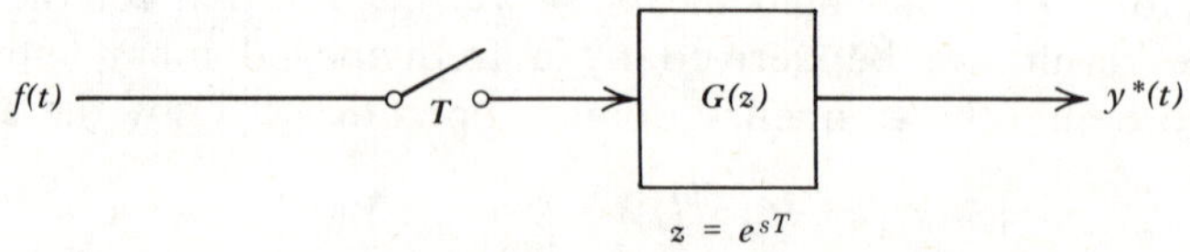

$$z = e^{sT}$$

FIGURE 5.2.2. Equivalent block diagram for Fig. 5.2.1 when only the output samples are of interest.

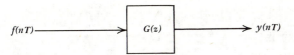

FIGURE 5.2.3. Block diagram for a discrete-time system with pulse transfer function $G(z)$.

It can be argued that $\mathcal{L}\{\delta^2(t)\}$ is infinite so that if $y^{**}(t)$ were applied to another linear system, the resulting output would also be undefined. From another point of view, we can also see from the aliasing formula (2.3.7) and periodicity of $\mathcal{Y}^*(s)$ that $[\mathcal{Y}^*(s)]^*$ is infinite except when s is a zero of $\mathcal{Y}^*(s)$. These considerations show that if $g(t)$ is the impulse response of a system separating two ideal samplers, then $g(t)$ must contain no impulses if the overall system is to be meaningful.

If all the signals of interest in a system are sampled signals, then the system can be represented by an equivalent discrete-time model. In a block diagram for a discrete-time system no samplers are shown. Each signal is a sequence of numbers and the Z-transform of the output of a block is the product of the pulse transfer function in the block and the Z-transform of the input to the block. The block diagram of the discrete-time system corresponding to the system in Fig. 5.2.1 is shown in Fig. 5.2.3.

Example 5.2.1

In Fig. 5.2.1 let $f(t) = tu(t)$, $\mathcal{G}_h(s) = (1 - z^{-1})/s$, and $\mathcal{G}_1(s) = 1/(s+1)$. Then

$$Y(z) = F(z)G(z)$$

$$= \frac{Tz^{-1}}{(1 - z^{-1})^2}\left[\frac{1 - z^{-1}}{s(s+1)}\right]^*$$

$$= \frac{Tz^{-1}}{1 - z^{-1}}\left[\frac{1}{s(s+1)}\right]^*$$

$$= \frac{T(1 - e^{-T})z^{-2}}{(1 - z^{-1})^2(1 - e^{-T}z^{-1})}$$

and

$$y(nT) = nTu(nT) - \frac{T}{1 - e^{-T}}(1 - e^{-nT})u(nT)$$

For additional insight into the behavior of the system, sketch the output of $\mathcal{G}_h(s)$ and calculate $y(t)$ between the sampling instants. ◀

5.3 RECURSIVE AND NONRECURSIVE DISCRETE-TIME SYSTEMS

A discrete-time system is called a *nonrecursive* system if its present output is calculated from past, present, and, in the nonrealizable case, future inputs. A causal discrete-time system is called a *recursive* system if its present output is calculated from past and present inputs and past outputs. A recursive or nonrecursive, causal, linear, time-invariant system with input $f(nT)$ and output $y(nT)$ can be represented by a difference equation of the form

$$y(nT) = \sum_{k=0}^{M} a_k f(nT - kT) - \sum_{k=1}^{N} b_k y(nT - kT) \qquad (5.3.1)$$

Conceptually, M and/or N can be finite or infinite. If $b_k = 0$ for $k = 1, \ldots, N$ the system is nonrecursive. This nonrecursive system is also called a *moving average filter*, a *transversal filter with $M+1$ taps*, or a *finite-duration impulse response filter* (FIR filter) if M is finite. If $N \geq 1$ and b_N is not zero, the system is often called an *Nth order recursive filter*.

Taking the Z-transform of both sides of (5.3.1) we obtain

$$Y(z) = F(z) \sum_{k=0}^{M} a_k z^{-k} - Y(z) \sum_{k=1}^{N} b_k z^{-k}$$

so that

$$G(z) = \frac{Y(z)}{F(z)} = \frac{\displaystyle\sum_{k=0}^{M} a_k z^{-k}}{1 + \displaystyle\sum_{k=1}^{N} b_k z^{-k}} \qquad (5.3.2)$$

Therefore, the discrete-time system described by (5.3.1) has the pulse transfer function $G(z)$. Conversely, any discrete-time system with a pulse transfer function of the form (5.3.2) has an input-output relationship given by a difference equation of the form (5.3.1).

If $G(z)$ cannot be reduced to a finite-degree polynomial in z^{-1}, then $g(nT)$ has infinite duration and the system is sometimes called an *infinite-duration impulse response* (IIR) filter.

Example 5.3.1

Consider a causal system with $T = 1$ and pulse transfer function

$$G(z) = 1/(1 - z^{-1})$$

so that

$$g(n) = u(n)$$

According to (5.3.1) and (5.3.2), the input and output of this system are related by the difference equation

$$y(n) = f(n) + y(n-1) \qquad (5.3.3)$$

According to the Convolution Theorem

$$y(n) = \sum_{k=-\infty}^{\infty} f(k)u(n-k) = \sum_{k=-\infty}^{n} f(k) \qquad (5.3.4)$$

It is easy to see that (5.3.3) and (5.3.4) are equivalent. A discrete-time system with this pulse transfer function is called an *accumulator*. It is the discrete-time equivalent of an integrator. ◄

Example 5.3.2

Consider a causal system with $T = 1$ and the pulse transfer function

$$G(z) = \frac{e^{-1}z^{-1}}{(1 - e^{-1}z^{-1})^2} = \frac{e^{-1}z^{-1}}{1 - 2e^{-1}z^{-1} + e^{-2}z^{-2}}$$

Using the integral inversion formula we can show that the pulse response of the system is

$$g(n) = ne^{-n}u(n)$$

According to (5.3.1) and (5.3.2), the output and input of this system are related by the difference equation

$$y(n) = e^{-1}f(n-1) + 2e^{-1}y(n-1) - e^{-2}y(n-2) \qquad (5.3.5)$$

The pulse response of this system can also be generated recursively with (5.3.5) by assuming zero initial conditions, that is, $y(n) = 0$ for $n < 0$, and letting $f(n) = \delta_{n0}$. Carrying out the recursion we see that $y(0) = 0$, $y(1) = e^{-1}$, $y(2) = 2e^{-2}$, $y(3) = 4e^{-3} - e^{-3} = 3e^{-3}, \ldots$ ◄

In theory, any pulse transfer function $G(z)$ can be realized in nonrecursive form by discrete-time convolution. If $g(nT)$ has finite duration, the convolution sum can be easily calculated in practice. Generally, $g(nT)$ has infinite duration. However, if $G(z)$ is rational and the region of convergence includes the unit circle, $g(nT)$ will decay to zero exponentially as n becomes infinite and the convolution sum can be approximated to any desired accuracy by a truncated summation. Thus, in practice, a stable recursive system can also be realized arbitrarily closely as a nonrecursive system. However, in most cases the recursive implementation is most efficient.

5.4 SINUSOIDAL STEADY-STATE FREQUENCY RESPONSE

Just as in the continuous-time case, the sinusoidal steady-state frequency response of a discrete-time system is a useful analytical tool. Consider a system with the pulse response $g(nT)$ and pulse transfer function $G(z)$ where the region of convergence includes the unit circle. If the input to the system is

$$f(nT) = e^{j\omega_0 nT} \qquad (5.4.1)$$

then its output is

$$y(nT) = \sum_{k=-\infty}^{\infty} g(kT) e^{j\omega_0(nT-kT)}$$

$$= e^{j\omega_0 nT} \sum_{k=-\infty}^{\infty} g(kT) e^{-j\omega_0 kT}$$

$$= e^{j\omega_0 nT} G^*(\omega_0) \qquad (5.4.2)$$

Thus when a sinusoid of frequency ω_0 is applied to the system, the steady-state output is also a sinusoid at the same frequency but scaled in amplitude by the magnitude of $G^*(\omega_0)$ and shifted in phase by $\arg G^*(\omega_0)$. For this reason

$$G^*(\omega) = G(z)\big|_{z=e^{j\omega T}}$$

is called the *frequency response* of the discrete-time system. $|G^*(\omega)|$ is called the *amplitude* response and $\arg G^*(\omega)$ is called the *phase* response of the system.

The frequency response of a discrete-time system with a rational pulse transfer function $G(z)$ can be visualized graphically from a plot of the locations of the poles and zeros of $G(z)$. Let us assume that the system is causal and $G(z)$ has the form

$$G(z) = Kz^{N-M} \prod_{k=1}^{M} (z - z_k) \bigg/ \prod_{k=1}^{N} (z - p_k) \qquad (5.4.3)$$

Thus $G(z)$ has zeros at $z_1, \ldots, z_M$, poles at $p_1, \ldots, p_N$, and possibly a pole or zero at $z = 0$ depending on the relationship between M and N. We will indicate the locations of the poles in the z-plane with x's and the locations of the zeros with o's. Analytically we can see that

$$|G^*(\omega)/K| = \prod_{k=1}^{M} |e^{j\omega T} - z_k| \bigg/ \prod_{k=1}^{N} |e^{j\omega T} - p_k| \qquad (5.4.4)$$

and

$$\arg[G^*(\omega)/K] = (N-M)\omega T + \sum_{k=1}^{M} \arg(e^{j\omega T} - z_k)$$

$$- \sum_{k=1}^{N} \arg(e^{j\omega T} - p_k) \qquad (5.4.5)$$

Geometrically, factors of $G^*(\omega)$ with the forms $e^{j\omega T}$, $e^{j\omega T} - z_k$, and $e^{j\omega T} - p_k$ can be represented by vectors drawn from the points 0, z_k, and p_k to the point on the unit circle with argument ωT. Therefore, the magnitudes in (5.4.4) and angles in (5.4.5) can be found graphically. With a little practice, the general nature of a frequency response can be quickly determined from the pole-zero plot.

Example 5.4.1

Let the pulse transfer function of a causal system be

$$G(z) = \frac{1 - z^{-1}}{1 + 0.81 z^{-2}}$$

This system can be realized by the difference equation

$$y(nT) = f(nT) - f(nT - T) - 0.81 y(nT - 2T)$$

The pulse transfer function can also be written in the form

$$G(z) = \frac{z(z-1)}{(z - 0.9j)(z + 0.9j)}$$

The poles and zeros of $G(z)$ are shown in Fig. 5.4.1. From this figure we can

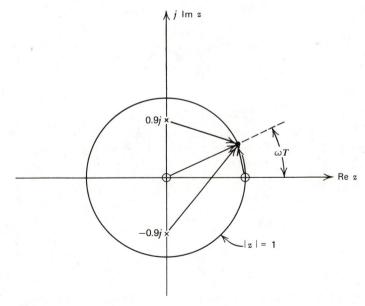

FIGURE 5.4.1. Pole-zero plot for $G(z)$ in Example 5.4.1.

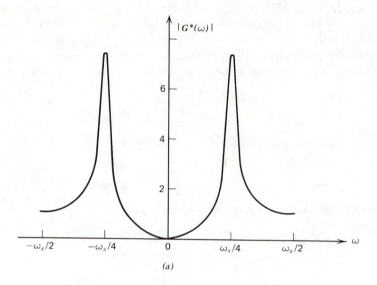

(a)

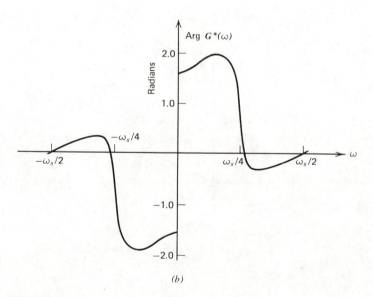

(b)

FIGURE 5.4.2.   Frequency response for Example 5.4.1. (a) Amplitude response; (b) phase response.

92

immediately see that $G^*(0)=0$ because of the zero at $z=1$ and that $|G^*(\omega)|$ peaks up for ωT in the vicinity of $\pm\pi/2$ because of the poles at $\pm 0.9j$. This system is an elementary band-pass filter. The complete frequency response is shown in Fig. 5.4.2. ◀

5.5 EQUIVALENT STRUCTURES FOR REALIZING RATIONAL PULSE TRANSFER FUNCTIONS

In this section we look at several structures that are equivalent in the sense that they all have the pulse transfer function

$$G(z) = \frac{\displaystyle\sum_{k=0}^{M} a_k z^{-k}}{1 + \displaystyle\sum_{k=1}^{N} b_k z^{-k}} \tag{5.5.1}$$

The block diagram that is shown for each structure is only intended to indicate the structure of the calculations performed and not necessarily the structure of the hardware realization. The actual physical system could take many forms, ranging from a totally analog implementation to a totally digital implementation realized with a hard-wired special purpose computer or with software on a general purpose computer. Regardless of exactly how $G(z)$ is realized, additions, multiplications, and storage are required. In the block diagrams the storage will be represented by the blocks labeled z^{-1}. These blocks are one-sample delays.

A causal system that has the pulse transfer function $G(z)$ and a structure in which the parameters $a_0, \ldots, a_M$ and $b_1, \ldots, b_N$ appear explicitly is called a *direct form* realization of $G(z)$. In Section 5.3 we observed that the output $y(nT)$ of a system with pulse transfer function $G(z)$ and input $f(nT)$ can be calculated by the difference equation

$$y(nT) = \sum_{k=0}^{M} a_k f(nT - kT) - \sum_{k=1}^{N} b_k y(nT - kT) \tag{5.5.2}$$

The block diagram of a system for implementing (5.5.2) is shown if Fig. 5.5.1. We will call this structure a *type 0* direct from realization of $G(z)$. This realization requires $M+N$ storage elements.

A direct form realization for $G(z)$ requiring only $\max(M, N)$ storage elements can be derived by considering $G(z)$ to be the cascade of a recursive

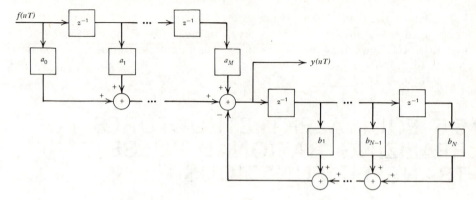

FIGURE 5.5.1. Type 0 direct form realization of $G(z)$.

system with pulse transfer function

$$G_1(z) = \frac{1}{1 + \sum\limits_{k=1}^{N} b_k z^{-k}}$$ (5.5.3)

and a nonrecursive system with pulse transfer function

$$G_2(z) = \sum_{k=0}^{M} a_k z^{-k}$$ (5.5.4)

as shown in Fig. 5.5.2. The relationship between the input and output of $G_1(z)$ is

$$x(nT) = f(nT) - \sum_{k=1}^{N} b_k x(nT - kT)$$ (5.5.5)

The relationship between the input and output of $G_2(z)$ is

$$y(nT) = \sum_{k=0}^{M} a_k x(nT - kT)$$ (5.5.6)

We will call a system that realizes $G(z)$ according to (5.5.5) and (5.5.6) a *type 1 direct form* realization of $G(z)$. The block diagram for a system with this

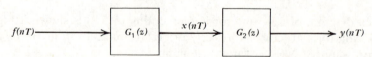

FIGURE 5.5.2. Realizing $G(z)$ as the cascade of a recursive and non-recursive system.

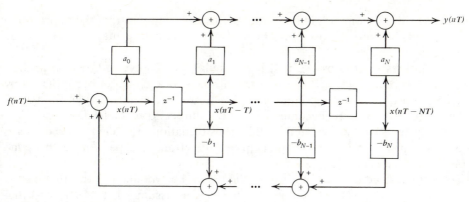

FIGURE 5.5.3. Type 1 direct form realization of $G(z)$ for $M = N$.

structure is shown in Fig. 5.5.3 for the case when $M = N$. The block diagram for the case when $M < N$ is obtained by simply letting $a_{M+1} = \cdots = a_N = 0$. Similarly, the block diagram for the case when $M > N$ is obtained by changing N to M in Fig. 5.5.3 and letting $b_{N+1} = \cdots = b_M = 0$. The type 1 direct form realization requires only max (M, N) storage elements.

Another direct form realization for $G(z)$ is suggested by (5.5.2). For simplicity let us assume that $M = N$. Then we can write (5.2.2) as

$$y(nT) = a_0 f(nT) + \sum_{k=1}^{N} [a_k f(nT - kT) - b_k y(nT - kT)] \tag{5.5.7}$$

The block diagram of a system with the structure suggested by (5.5.7) is shown in Fig. 5.5.4. We will call this a *type 2* direct form realization of $G(z)$. The case

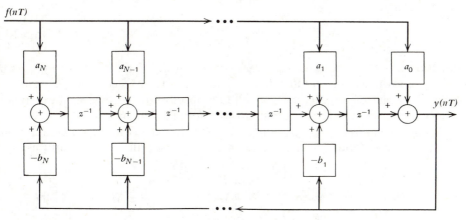

FIGURE 5.5.4. Type 2 direct form realization for $G(z)$ when $M = N$.

when M differs from N can be accounted for by simply setting the appropriate coefficients to zero. This realization also requires only $\max(M, N)$ storage elements.

It is interesting to observe that if the signal flow is reversed in all branches, summation nodes are replaced by branch nodes, branch nodes are replaced by summation nodes, and the input and output are interchanged, then the type 1 direct form realization becomes the type 2 direct form realization and vice-versa. Systems that are related by this transformation are called *transpose pairs* [57]. We prove in Chapter 6 that, in general, transpose pairs have the same pulse transfer function.

Another structure for realizing $G(z)$ can be obtained from the partial fraction expansion of $G(z)$. For simplicity let us assume that $M < N$ and that $G(z)$ has complex first-order poles at $p_1, \ldots, p_r$ and $p_{r+1} = \bar{p}_1, \ldots, p_{2r} = \bar{p}_r$ and real first-order poles at $p_{2r+1}, \ldots, p_N$. Then

$$G(z) = \sum_{k=1}^{N} \frac{A_k}{1 - p_k z^{-1}} \qquad (5.5.8)$$

where

$$A_k = \lim_{z \to p_k} (1 - p_k z^{-1}) G(z) \qquad (5.5.9)$$

If the pairs of terms for the complex conjugate poles of $G(z)$ are combined, (5.5.8) becomes

$$G(z) = \sum_{k=1}^{r} G_k(z) + \sum_{k=2r+1}^{N} G_k(z) \qquad (5.5.10)$$

where for $k = 1, \ldots, r$

$$G_k(z) = \frac{2 \operatorname{Re}\{A_k\} - 2 \operatorname{Re}\{A_k \bar{p}_k\} z^{-1}}{1 - 2 \operatorname{Re}\{p_k\} z^{-1} + |p_k|^2 z^{-2}} \qquad (5.5.11)$$

and for $k = 2r+1, \ldots, N$

$$G_k(z) = \frac{A_k}{1 - p_k z^{-1}} \qquad (5.5.12)$$

Equation 5.5.10 suggests the structure shown in Fig. 5.5.5 for realizing $G(z)$. This is known as the *parallel form* realization. Each section $G_k(z)$ can be realized as one of the direct forms already discussed. In general, the parallel form realization only requires $\max(M, N)$ storage elements.

The last structure discussed in this section is called the *cascade form* realization of $G(z)$. The function $G(z)$ can always be expressed in factored form as

$$G(z) = A \frac{\displaystyle\prod_{k=1}^{M-R} (1 - \alpha_k z^{-1})}{\displaystyle\prod_{k=1}^{N} (1 - \beta_k z^{-1})} z^{-R} \qquad R \leq M \qquad (5.5.13)$$

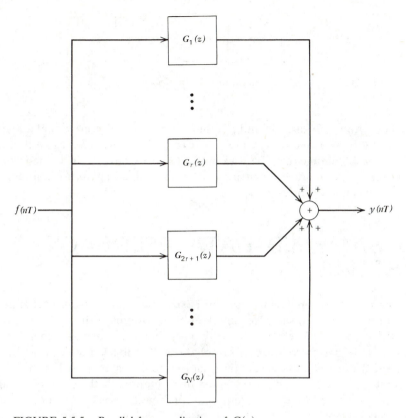

FIGURE 5.5.5.   Parallel form realization of $G(z)$.

Factors for complex conjugate poles or zeros can be combined into second-order factors with real coefficients. The resulting numerator and denominator factors can then be grouped into low-order rational functions $C_1(z), \ldots, C_L(z)$ so that

$$G(z) = \prod_{k=1}^{L} C_k(z) \qquad (5.5.14)$$

A block diagram for this structure is shown in Fig. 5.5.6. Each individual section $C_k(z)$ can be realized as one of the structures already discussed. The cascade form also requires only max (M, N) storage elements.

Each structure presented has various advantages and disadvantages. We see later in this book that the accuracy requirements for system coefficients and arithmetic operations are more stringent for high-order systems with poles near the unit circle than for low-order systems when the direct form realization is

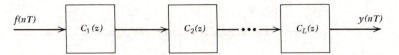

FIGURE 5.5.6. Cascade form realization of $G(z)$.

used. Therefore, the cascade and parallel forms have computational advantages because $G(z)$ is broken up into low-order sections. The cascade form is particularly advantageous when we want $G(z)$ to have well-defined zeros on the unit circle. Some of these structures are examined again in Chapter 6 from the state-variable point of view.

5.6 SAMPLED-DATA FEEDBACK SYSTEMS

The techniques for analyzing and synthesizing continuous-time feedback systems can be applied to sampled-data feedback systems with some modification. Since this book is concerned primarily with signal processing rather than with control system design, we discuss sampled-data feedback systems only briefly.

A typical unity feedback sampled-data system is shown in Fig. 5.6.1. $D(z)$ represents a series compensation filter implemented with a digital computer. The function $\mathcal{G}(s)$ could represent the combined transfer function of a data hold, power amplifier, and plant. From Fig. 5.6.1 we can see that

$$\mathcal{E}(s) = \mathcal{F}(s) - \mathcal{Y}(s) = \mathcal{F}(s) - E(z)D(z)\mathcal{G}(s) \qquad (5.6.1)$$

Applying the star operator to (5.6.1) yields

$$E(z) = F(z) - E(z)D(z)G(z)$$

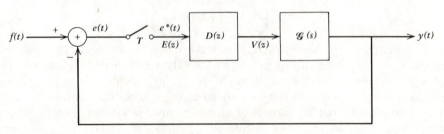

FIGURE 5.6.1. Typical unity feedback sampled-data system.

so that

$$E(z) = \frac{F(z)}{1 + D(z)G(z)} \qquad (5.6.2)$$

Therefore

$$\mathcal{Y}(s) = E(z)D(z)\mathcal{G}(s) = \frac{F(z)D(z)\mathcal{G}(s)}{1 + D(z)G(z)} \qquad (5.6.3)$$

Applying the star operator to (5.6.3) we see that the Z-transform of the output signal is

$$Y(z) = \frac{F(z)D(z)G(z)}{1 + D(z)G(z)} \qquad (5.6.4)$$

Therefore, the input and output samples are related by the pulse transfer function

$$\frac{Y(z)}{F(z)} = \frac{D(z)G(z)}{1 + D(z)G(z)} \qquad (5.6.5)$$

The sequence of output samples can be found from $Y(z)$. However, in many applications it is also necessary to know the behavior of the system between sampling instants to insure that the output does not have excess ripple. In fact, in a pathological situation, the output could contain an unstable oscillation that passes through zero at the sampling instants. This has been called a *hidden oscillation*. Conceptually $y(t)$ can be calculated for all time from (5.6.3). In practice this is difficult numerically. Methods for determining the behavior between sampling instants are discussed later in the chapter.

If only signal samples are of interest, the system in Fig. 5.6.1 can be replaced by the equivalent discrete-time system shown in Fig. 5.6.2. For discrete-time systems the Z-transform plays the same role as the Laplace transform for continuous-time systems. Therefore, discrete- and continuous-time systems whose block diagrams have the same form have pulse transfer and transfer functions of the same form. In general, if all the continuous-time blocks in a

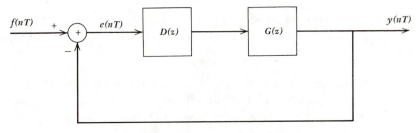

FIGURE 5.6.2. Equivalent discrete-time system for the system in Fig. 5.6.1.

system are separated by samplers, the block diagram of the equivalent discrete-time system can be found by removing all the samplers and replacing all transfer functions by their corresponding pulse transfer functions. Blocks that are already pulse transfer functions are left as is. All the standard techniques for reducing block diagrams can then be applied. In particular, Mason's gain formula [144] can be used to find the overall pulse transfer function.

Example 5.6.1

In Fig. 5.6.1 let $D(z)$ represent an accumulator, so that

$$D(z) = \frac{1}{1 - z^{-1}}$$

and let $\mathcal{G}(s)$ represent a zero-order hold cascaded with a single time constant system, so that

$$\mathcal{G}(s) = \frac{1 - z^{-1}}{s(s + a)}$$

Then

$$G(z) = \frac{1 - z^{-1}}{a}\left[\frac{1}{1 - z^{-1}} - \frac{1}{1 - e^{-aT}z^{-1}}\right]$$

Substituting into (5.6.5) and simplifying shows that

$$Q(z) = \frac{Y(z)}{F(z)} = \frac{(1 - e^{-aT})z^{-1}}{a + (1 - e^{-aT} - ae^{-aT} - a)z^{-1} + ae^{-aT}z^{-2}}$$

If $f(t)$ is a unit step function, then

$$Y(z) = \frac{1}{1 - z^{-1}} Q(z)$$

The final value of the sequence of output samples is

$$\lim_{n \to \infty} y(nT) = \lim_{z \to 1} (1 - z^{-1}) Y(z) = Q(1) = 1$$

Therefore, this system tracks step inputs with zero steady-state error at the sampling instants. It is not difficult to show that the steady-state error between sampling instants is also zero. ◀

Another frequently encountered sampled-data feedback structure is shown in Fig. 5.6.3. The function $\mathcal{H}(s)$ might represent a compensation network, a smoothing filter, or a measuring instrument. In this system

$$\mathcal{E}(s) = \mathcal{F}(s) - \mathcal{Y}(s)\mathcal{H}(s) = \mathcal{F}(s) - E(z)D(z)\mathcal{G}(s)\mathcal{H}(s) \qquad (5.6.6)$$

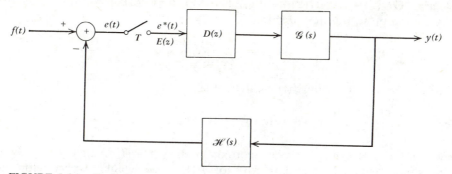

FIGURE 5.6.3.   Typical nonunity feedback sampled-data system.

Applying the star operator to both sides of (5.6.6) gives

$$E(z) = F(z) - E(z)D(z)[\mathcal{G}(s)\mathcal{H}(s)]^* \qquad (5.6.7)$$

In general

$$[\mathcal{G}(s)\mathcal{H}(s)]^* \neq \mathcal{G}^*(s)\mathcal{H}^*(s)$$

For notational convenience, it is customary to denote $[\mathcal{G}(s)\mathcal{H}(s)]^*$ by $\mathcal{G}\mathcal{H}^*(s)$ or $GH(z)$. Using this notation, we can see from (5.6.7) that

$$E(z) = \frac{F(z)}{1 + D(z)GH(z)} \qquad (5.6.8)$$

Therefore

$$\mathcal{Y}(s) = \frac{F(z)D(z)\mathcal{G}(s)}{1 + D(z)GH(z)} \qquad (5.6.9)$$

and

$$Y(z) = F(z)\frac{D(z)G(z)}{1 + D(z)GH(z)} \qquad (5.6.10)$$

Example 5.6.2

In Fig. 5.6.3 let

$$D(z) = 1, \qquad \mathcal{G}(s) = \frac{1}{s + a}, \qquad \text{and} \qquad \mathcal{H}(s) = \frac{1}{s}$$

Then

$$G(z) = \frac{1}{1 - e^{-aT}z^{-1}}, \qquad H(z) = \frac{1}{1 - z^{-1}}$$

and

$$GH(z) = [\mathcal{G}(s)\mathcal{H}(s)]^* = \frac{(1 - e^{-aT})z^{-1}}{a(1 - e^{-aT}z^{-1})(1 - z^{-1})}$$

Clearly, $GH(z) \neq G(z)H(z)$ in this case. From (5.6.10) we find that the closed-loop pulse transfer function is

$$Q(z) = \frac{Y(z)}{F(z)} = \frac{a(1-z^{-1})}{a + (1 - e^{-aT} - ae^{-aT} - a)z^{-1} + ae^{-aT}z^{-2}}$$

If the input is the ramp $f(t) = ctu(t)$, then

$$y_\infty = \lim_{n \to \infty} y(nT) = \lim_{n \to \infty} (1 - z^{-1})F(z)Q(z) = c\frac{aT}{1 - e^{-aT}}$$

When aT is small, e^{-aT} can be approximated by $1 - aT$ so that $y_\infty \cong c$. Thus $y(t)$ is approximately the derivative of $f(t)$ as we would expect with the integrator $1/s$ in the feedback path.　　◄

5.7 STABILITY OF DISCRETE-TIME SYSTEMS

Stability can be defined in a variety of ways. We will use the following definition:

A system is stable if and only if its output is bounded for every bounded input.

This definition is particularly suited to linear systems. For linear systems it is not necessary to test for a bounded output with every bounded input. It is only necessary to examine the pulse response of the system. The condition that the pulse response must satisfy for a time-invariant system is presented in the following theorem.

THEOREM 5.7.1
A linear, time-invariant, discrete-time system with pulse response $g(nT)$ is stable if and only if

$$\sum_{n=-\infty}^{\infty} |g(nT)| < \infty \qquad (5.7.1)$$

Proof Let us assume that $g(nT)$ satisfies (5.7.1) and that $f(nT)$ is any input signal with the property that $|f(nT)| < L < \infty$ for all n. Then the output is

$$y(nT) = \sum_{k=-\infty}^{\infty} g(kT)f(nT - kT)$$

so that

$$|y(nT)| \leq \sum_{k=-\infty}^{\infty} |g(kT)|\,|f(nT - kT)| \leq L \sum_{k=-\infty}^{\infty} |g(kT)|$$

Therefore, if $g(nT)$ is absolutely summable, then any bounded input causes a bounded output.

On the other hand, let us assume that $g(nT)$ is not absolutely summable. We can choose $f(nT)$ as

$$f(nT) = \text{sign } g(rT - nT)$$

where r is an integer and

$$\text{sign } x = \begin{cases} 1 & \text{for} & x > 0 \\ 0 & \text{for} & x = 0 \\ -1 & \text{for} & x < 0 \end{cases}$$

Obviously, $|f(nT)| \leq 1$. With this choice

$$y(rT) = \sum_{k=-\infty}^{\infty} g(kT) \text{ sign } g(kT) = \sum_{k=-\infty}^{\infty} |g(kT)| = \infty$$

Consequently, the system is stable only if $g(nT)$ is absolutely summable.

Q.E.D.

For causal systems with rational pulse transfer functions our definition of stability leads to the following frequently used criterion:

COROLLARY 5.7.1

A causal system with a rational pulse transfer function $G(z)$ is stable if and only if all the poles of $G(z)$ are inside the unit circle.

Proof   If all the poles of $G(z)$ are inside the unit circle, then the region of convergence for

$$\mathscr{Z}\{g(nT)\} = \sum_{n=0}^{\infty} g(nT) z^{-n}$$

includes the unit circle. Therefore, this series converges absolutely for $|z| = 1$ so that

$$\sum_{n=0}^{\infty} |g(nT)| < \infty$$

and the system is stable according to Theorem 5.7.1.

On the other hand, if $G(z)$ has any poles on or outside the unit circle, then the unit circle is not in the region of convergence. In this case, the series diverges for some z_0 with $|z_0| = 1$ so that

$$\infty = \sum_{n=0}^{\infty} g(nT) z_0^{-n} \leq \sum_{n=0}^{\infty} |g(nT)|$$

and the system is not stable according to Theorem 5.7.1.

Q.E.D.

In the next four sections we examine methods for determining the locations of the poles of a rational pulse transfer function relative to the unit circle.

5.8 THE MODIFIED SCHUR-COHN TEST

The pulse transfer function $G(z)$ of a causal system described by a finite-order difference equation can always be arranged in the form

$$G(z) = \frac{N(z)}{D(z)} \tag{5.8.1}$$

where $N(z)$ and $D(z)$ are polynomials in z and the degree of $N(z)$ is no greater than the degree of $D(z)$. The denominator polynomial $D(z)$ is the characteristic polynomial for the system. The poles of $G(z)$ occur at the zeros of $D(z)$, assuming that $N(z)$ and $D(z)$ have no factors in common. Therefore, the system will be stable if all the zeros of $D(z)$ are inside the unit circle.

A simple numerical test similar to the Routh-Hurwitz criterion can be used to check that all the zeros of a polynomial

$$D(z) = d_N z^N + d_{N-1} z^{N-1} + \cdots + d_0 \qquad d_N > 0 \tag{5.8.2}$$

with real coefficients are inside the unit circle. This test was introduced by Tsypkin [134] and Jury [61] and is simpler than an older test discovered by Schur-Cohn [89]. When $D(z)$ has zeros outside the unit circle, the test indicates this fact but does not indicate the number of such zeros. Thus the modified Schur-Cohn criterion is not quite as powerful as the Routh-Hurwitz criterion for continuous time-systems, which indicates the number of zeros in the right half plane.

We will use the division form of the modified Schur-Cohn criterion. The test can also be carried out in tabular form [64]. The first step is to form the polynomial

$$D^{@}(z) = z^N D(z^{-1}) = d_0 z^N + \cdots + d_{N-1} z + d_N \tag{5.8.3}$$

This is called the *reciprocal polynomial* associated with $D(z)$. The roots of $D^{@}(z)$ are the reciprocals of the roots of $D(z)$ and $|D^{@}(z)| = |D(z)|$ on the unit circle. The next step is to divide $D^{@}(z)$ by $D(z)$ starting at the high-power end to obtain a quotient $\alpha_0 = d_0/d_N$ and remainder $D_1^{@}(z)$ of degree $N-1$ or less so that

$$\frac{D^{@}(z)}{D(z)} = \alpha_0 + \frac{D_1^{@}(z)}{D(z)} \tag{5.8.4}$$

The division process is repeated with $D_1^{@}(z)$ and its reciprocal polynomial $D_1(z)$ and the sequence $\alpha_0, \ldots, \alpha_{N-2}$ is generated according to the rule

$$\frac{D_k^{@}(z)}{D_k(z)} = \alpha_k + \frac{D_{k+1}^{@}(z)}{D_k(z)} \qquad \text{for} \qquad k = 0, \ldots, N-2 \tag{5.8.5}$$

where $D_0(z) = D(z)$ and at each step $D_k(z)$ is considered to be a polynomial of degree $N - K$ regardless of its actual degree. Therefore, some α's may be zero.

The zeros of $D(z)$ are all inside the unit circle if and only if the following three conditions are satisfied:

1. $D(1) > 0$

2. $D(-1)\begin{cases} <0 & \text{for} & N & \text{odd} \\ >0 & \text{for} & N & \text{even} \end{cases}$ (5.8.6)

3. $|\alpha_k| < 1$ for $k = 0, \ldots, N-2$

In carrying out the test, it is wise to check conditions 1 and 2 first. If these simple conditions are not satisfied, then $D(z)$ has zeros on or outside the unit circle and condition 3 need not be checked.

Example 5.8.1

Consider a system with the characteristic polynomial

$$D(z) = z^3 + 0.5z^2 + z + 0.5$$

We can see that $D(1) = 3$ and $D(-1) = -1$ so that conditions 1 and 2 are satisfied. Carrying out the division operations we find that $\alpha_0 = 0.5$ and $\alpha_1 = 1$. Therefore, condition 3 is not satisfied since $|\alpha_1| = 1$ and $D(z)$ must have zeros on or outside the unit circle. In fact, this example was fabricated by choosing the zeros of $D(z)$ to be at $\pm j$ and -0.5. ◄

5.9 THE NYQUIST CRITERION FOR DISCRETE-TIME SYSTEMS

The stability of discrete-time feedback systems can be studied by using a simple modification of the Nyquist criterion. We will use the unity feedback system shown if Fig. 5.6.2 to illustrate this technique. The pulse transfer function of the system is

$$H(z) = \frac{D(z)G(z)}{1 + D(z)G(z)} \tag{5.9.1}$$

Assuming that all the system blocks are causal and have rational pulse transfer functions, the open-loop gain $D(z)G(z)$ can be written as

$$D(z)G(z) = \frac{A(z)}{B(z)} \tag{5.9.2}$$

where

$$A(z) = a_L z^L + \cdots + a_0$$
$$B(z) = z^M + b_{M-1} z^{M-1} + \cdots + b_0$$

and

$$L \leq M$$

From (5.9.1) and (5.9.2) we see that

$$H(z) = \frac{A(z)}{B(z) + A(z)} \qquad (5.9.3)$$

Therefore, the characteristic equation for the closed-loop system is

$$B(z) + A(z) = 0$$

or, equivalently

$$1 + D(z)G(z) = 0 \qquad (5.9.4)$$

The system will be stable if $1 + D(z)G(z)$ has no zeros outside the unit circle.

The Nyquist criterion is based on a result in complex variable theory known as the argument principle [72]. This principle can be stated as follows:

THE ARGUMENT PRINCIPLE

Let $V(z)$ be analytic in a domain D and let C be any simple closed path in D along which $V(z)$ is not zero and inside which $V(z)$ has a finite number Z of zeros and is analytic except for a finite number P of poles. A pole or zero of order r is counted as r poles or zeros.

Then, as the path C in the z-plane is traversed once in the counterclockwise direction, the corresponding path in the $V(z)$ plane encircles the origin $N = P - Z$ times in the clockwise direction.

Normally all the components in the system will be individually stable or at most have poles at $z = 1$ as a result of accumulations. In this case, $D(z)G(z)$ and, consequently, $1 + D(z)G(z)$ will have no poles outside the unit circle. If this is not true, we will assume that the number of poles of $D(z)G(z)$ outside the unit circle is known. From (5.9.2) we can see that

$$1 + D(z)G(z) = \frac{B(z) + A(z)}{B(z)} \qquad (5.9.5)$$

Therefore, the numerator and denominator of $1 + D(z)G(z)$ both have degree M except in the rare case when $L = M$ and $a_M = -1$ or, equivalently, when

$$\lim_{z \to \infty} D(z)G(z) = -1$$

We will assume that this case is not present. Then $1 + D(z)G(z)$ has M poles and M zeros in the finite z-plane.

With the assumptions of the previous paragraph, we will now see how to use the argument principle to find the number of zeros of $1 + D(z)G(z)$ outside the unit circle. We will make the additional assumption, for the time being, that $D(z)G(z)$ has no poles on the unit circle. Consider the paths C_1 and C_2 shown in Fig. 5.9.1a. The path C_1 is the unit circle and the path C_2 is a circle with radius large enough to enclose all the poles and zeros of $1 + D(z)G(z)$. Let us assume that $1 + D(z)G(z)$ has P_1 poles and Z_1 zeros inside C_1 and P poles and Z zeros between C_1 and C_2. Since C_2 contains all the poles and zeros of $1 + D(z)G(z)$ by assumption, it must contain $M = P_1 + P$ poles and $M = Z_1 + Z$ zeros. According to the argument principle, the path generated in the $1 + D(z)G(z)$ plane as the path C_1 is traversed once in the counterclockwise direction will encircle the origin in the $1 + D(z)G(z)$ plane $N_1 = P_1 - Z_1$ times in the clockwise direction. The path generated as C_2 is traversed once in the counterclockwise direction will encircle the origin of the $1 + D(z)G(z)$ plane

$$N_2 = (P_1 + P) - (Z_1 + Z) = M - M = 0$$

times in the clockwise direction. Therefore

$$N_2 - N_1 = P - Z$$

or, since $N_2 = 0$

$$Z = P + N_1 \tag{5.9.6}$$

By hypothesis, P is known. N_1 is determined from the map of C_1. Since N_2 is known to be zero, the path corresponding to C_2 need not be generated. If $Z = 0$ the system will be stable.

It is customary to plot the open-loop gain $D(z)G(z)$ rather than $1 + D(z)G(z)$. A typical plot of $D(z)G(z)$ as z varies around the unit circle is shown in Fig. 5.9.1b. We will call this the *Nyquist plot* for the system. As shown in Fig. 5.9.1b, $1 + D(z)G(z)$ simply corresponds to a vector drawn from the -1 point to $D(z)G(z)$. Therefore, N_1 is just the number of clockwise encirclements of the -1 point by the Nyquist plot.

The necessary and sufficient condition for stability of the closed-loop system is that $Z = 0$ or, equivalently

$$N_1 = -P \tag{5.9.7}$$

If the open-loop gain has no poles outside the unit circle, $P = 0$ and the necessary and sufficient condition for stability of the closed-loop system becomes $N_1 = 0$, or, in other words, the Nyquist plot does not encircle the -1 point.

If C_1 is the path $z = e^{j\omega T}$ for $-\pi/T \le \omega \le \pi/T$, then the Nyquist plot is just the plot of the open-loop frequency response $D^*(\omega)G^*(\omega)$. Actually, only the

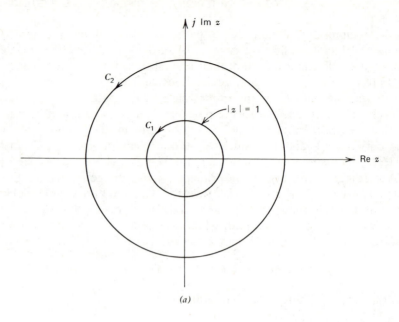

(a)

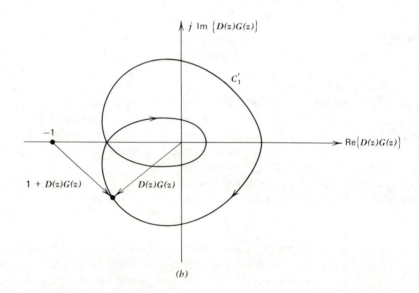

(b)

FIGURE 5.9.1.   The Nyquist criterion for discrete-time systems. (*a*) Paths for the Nyquist plot; (*b*) a typical Nyquist plot.

part of the plot for $0 \le \omega \le \pi/T$ need be calculated. Since

$$D^*(-\omega)G^*(-\omega) = \overline{D^*(\omega)G^*(\omega)}$$

the plot for negative frequencies is the mirror image about the real axis of the plot for positive frequencies. The Nyquist criterion is remarkable in that it relates the closed-loop stability to the open-loop frequency response. In practice, an algebraic expression for the open-loop frequency response may not be known. However, it can be determined experimentally by simple measurements.

Example 5.9.1

The Nyquist plot for

$$D(z)G(z) = \frac{z}{z - 0.5}$$

is shown in Fig. 5.9.2. As z moves counterclockwise around the unit circle from 1 to -1, we can see using the graphical method discussed in Section 5.4 that the Nyquist plot moves clockwise from 2 to $\frac{2}{3}$ through the lower half plane. A transformation of the form

$$w = \frac{az + b}{cz + d}$$

(with $ad \ne bc$) is known as a bilinear or linear fractional transformation. It can be shown [71] that this transformation maps circles into circles. Therefore, the Nyquist plot happens to be a circle for this example. We see that $D(z)G(z)$ has no poles outside the unit circle and the Nyquist plot does not encircle the -1

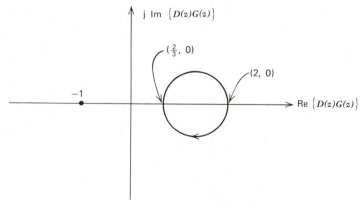

FIGURE 5.9.2. Nyquist plot for Example 5.9.1.

point, so that, $P = N_1 = 0$. Since $Z = P + N_1 = 0$, the closed-loop system is stable. Verify the conclusion for this elementary example by calculating the zeros of $1 + D(z)G(z)$ directly. ◄

If $D(z)G(z)$ has poles on the unit circle, the Nyquist criterion can still be applied by modifying the path C_1 to loop around these poles at an infinitesimal distance. This technique is illustrated in the following example.

Example 5.9.2

Let the open-loop gain of a system be

$$D(z)G(z) = \frac{K}{(z - 0.5)(z - 1)}$$

where K is a positive adjustable gain. $D(z)G(z)$ has a pole at $z = 1$. The z-plane path for the Nyquist plot can be modified as shown in Fig. 5.9.3a to loop around this pole with a semicircle of infinitesimal radius r. The corresponding Nyquist plot is shown in Fig. 5.9.3b. For infinitesimal r the point z_1 shown in Fig. 5.9.3a is essentially

$$z_1 \cong 1 + re^{j\pi/2}$$

This gets mapped to the point

$$z_1' = D(z_1)G(z_1) \cong \frac{Ke^{-j\pi/2}}{0.5r}$$

shown in Fig. 5.9.3b. Thus z_1' has essentially infinite magnitude at angle $-\pi/2$. As z moves around the unit circle counterclockwise from z_1 to z_3, $z - 0.5$ and $z - 1$ both increase in magnitude and both increase in angle to π radians, so that the Nyquist plot decreases in magnitude and rotates clockwise to -2π radians. The plot from z_3 to z_4 is the mirror image of the plot from z_3 to z_1. On Ψ_2

$$z - 0.5 \cong 0.5$$

and

$$z - 1 = re^{j\theta} \qquad \text{for} \qquad -\pi/2 \le \theta \le \pi/2$$

Therefore, on Ψ_2

$$D(z)G(z) \cong \frac{Ke^{-j\theta}}{0.5r}$$

and as z moves from z_4 to z_1, the Nyquist plot swings through the right half plane in a semicircle with essentially infinite radius from z_4' to z_1'.

To determine the values of K for which the system is stable, we must find the

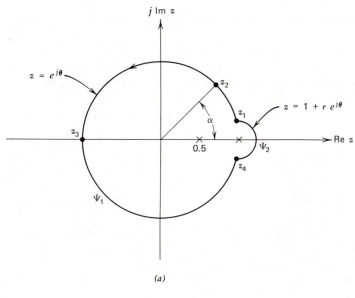

(a)

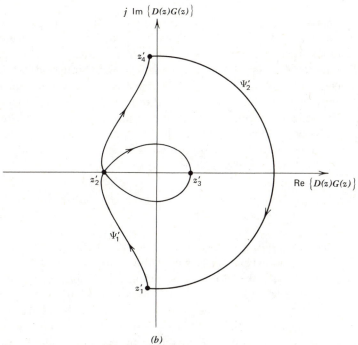

(b)

FIGURE 5.9.3. Nyquist plot for Example 5.9.2. (a) Modified path to enclose pole at $z = 1$; (b) Nyquist plot.

point z_2' at which the Nyquist plot crosses the negative real axis. Letting

$$z_2 = e^{j\alpha} = \cos \alpha + j \sin \alpha$$

the denominator of the open-loop gain can be reduced to the form

$$(z_2 - 0.5)(z_2 - 1) = (2 \cos \alpha - 1.5) \cos \alpha - 0.5$$
$$+ j(2 \cos \alpha - 1.5) \sin \alpha \qquad (5.9.8)$$

If

$$2 \cos \alpha - 1.5 = 0 \qquad \text{or} \qquad \alpha = \text{arc cos } 0.75$$

(5.9.8) reduces to -0.5 and $D(z_2)G(z_2) = -2K$. The z-plane path encloses all the poles of $D(z)G(z)$ so that $P = 0$. If $0 < K < \frac{1}{2}$, the Nyquist plot does not encircle the -1 point so that $Z = N_1 = 0$ and the closed-loop system is stable. If $K > \frac{1}{2}$, the Nyquist plot encircles the -1 point twice in the clockwise direction so that $Z = N_1 = 2$ and the closed-loop system has two poles outside the unit circle. If $K = \frac{1}{2}$, the closed-loop system has poles on the unit circle. ◀

Most feedback control systems are low-pass in nature. The step response of a system can be calculated exactly from its frequency response. However, qualitative statements can be made about the step response of a low-pass system from its frequency response without calculation. Rise-time and bandwidth tend to be inversely related. If the amplitude response is relatively flat or falls off with frequency over the pass-band, then the overshoot and ringing will tend to be small. On the other hand, when the amplitude response has a large resonant peak, the overshoot and ringing will be large. A resonant peak occurs when the Nyquist plot passes near the -1 point. The closed-loop frequency response of the unity feedback system we have been discussing is

$$\frac{D(z)G(z)}{1 + D(z)G(z)} \qquad \text{for} \qquad z = e^{j\omega T}$$

The vectors corresponding to $D(z)G(z)$ and $1 + D(z)G(z)$ are shown in Fig. 5.9.1b. When the Nyquist plot passes near the -1 point, $1 + D(z)G(z)$ becomes small in magnitude and the closed-loop amplitude response becomes large.

All the techniques developed for continuous-time systems using the Nyquist plot can be used for discrete-time systems. However, the Nyquist plot is not as widely used for discrete-time systems as for continuous-time systems. This is largely because discrete-time systems can be easily simulated on a digital computer and more modern time-domain design techniques can be effectively used.

5.10 THE ROOT LOCUS METHOD FOR DETERMINING POLE LOCATIONS

The open-loop gain of a feedback system can usually be easily changed by a scale factor. The location of the roots of the characteristic equation for the closed-loop system and, hence, its time domain behavior depends on this scale factor. The root locus technique [33] is commonly used to set this scale factor for continuous-time systems. The root locus for a discrete-time system can be generated in exactly the same way as for a continuous-time system. Of course, the root locations must be interpreted relative to the unit circle for discrete-time systems. Since the root locus method is presented in detail in almost every classical automatic control system text, only a summary will be given here.

Let us assume that the characteristic equation for a unity feedback system has the form

$$1 + KG(z) = 0 \tag{5.10.1}$$

where $K > 0$, and $G(z) = A(z)/B(z)$ with $A(z)$ a polynomial in z of degree L and $B(z)$ a polynomial in z of degree $M \geq L$. With these definitions, the characteristic equation can be put in the form

$$B(z) + KA(z) = 0 \tag{5.10.2}$$

The locus of solutions of the characteristic equation as K varies from 0 to ∞ is called the *root locus* for the system. The root locus has M branches since the characteristic polynomial on the left-hand side of (5.10.2) has degree M.

From (5.10.1) we can see that any point on the root locus must satisfy the equation

$$G(z) = \frac{-1}{K} \tag{5.10.3}$$

Thus any point for which

$$\arg G(z) = -\pi - 2\pi n \tag{5.10.4}$$

is on the root locus. For these points,

$$K = 1/|G(z)| \tag{5.10.5}$$

Most of the techniques for constructing the root locus are based on searching for points that satisfy (5.10.4).

It is particularly easy to find segments of the root locus that lie on the real axis. Let us define points at which $G(z)$ has poles or zeros to be *critical points*. Then it is easy to see from (5.10.4) that any point on the real axis is part of the root locus if it lies to the left of an odd number of critical points on the real axis.

In the limit as K approaches zero, (5.10.2) converges to $B(z) = 0$, so that the M branches of the root locus converge to the M poles of $G(z)$. From (5.10.3) we see that for points on the root locus

$$\lim_{K \to \infty} G(z) = 0$$

Therefore, the branches of the root locus converge to the zeros of $G(z)$ as K becomes infinite. L branches converge to the zeros in the finite z-plane and the remaining $M - L$ branches go out to infinity because of the zeros there. For large z,

$$G(z) \cong b/z^{M-L}$$

Assuming that b is positive and $z = e^{j\theta}$, (5.10.4) becomes

$$-(M - L)\theta = -\pi + 2\pi n$$

or

$$\theta = \frac{\pi}{M - L} + \frac{2\pi}{M - L} n \qquad (5.10.6)$$

Therefore, the $M - L$ branches approach infinity at angles given by (5.10.6) as K becomes infinite.

Example 5.10.1

Consider a system with the open-loop gain

$$KG(z) = \frac{K}{(z - 0.5)(z - 1)}$$

The root locus for this system is shown in Fig. 5.10.1. The line segment between the poles at 0.5 and 1 is part of the root locus because points on it lie to the left of an odd number of critical points. It is easy to see that points on the perpendicular bisector of this line segment satisfy (5.10.4). For small K the roots are near 0.5 and 1. As K increases the roots move together along the real axis. Eventually, they meet at 0.75 and then become complex and move vertically outward to $0.75 \pm j\infty$. From Fig. 5.10.1 we can see that the root locus meets the unit circle at the point

$$z_2 = 0.75 + j\sqrt{1 - (0.75)^2}$$

The corresponding value for K according to (5.10.5) is

$$K = |z_2 - 0.5| |z_2 - 1| = 0.5$$

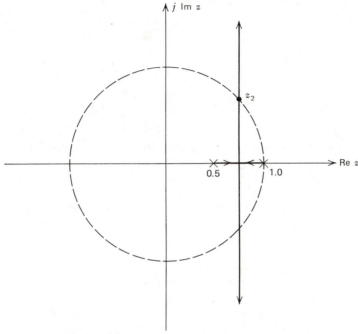

FIGURE 5.10.1.   Root locus for Example 5.10.1.

Therefore, the closed-loop system will be unstable with two poles outside the unit circle for $K > 0.5$. Compare this example with Example 5.9.2.   ◄

5.11 STABILITY ANALYSIS BY CONVENTIONAL CONTINUOUS-TIME TECHNIQUES VIA THE BILINEAR TRANSFORMATION

The bilinear transformation

$$z = \frac{1+s}{1-s} \qquad (5.11.1)$$

is a one-to-one transformation of the extended s-plane onto the extended z-plane. For each z the corresponding unique value of s is

$$s = \frac{z-1}{z+1} \qquad (5.11.2)$$

Letting $s = \sigma + j\omega$, (5.11.1) becomes

$$z = \frac{1 + \sigma + j\omega}{1 - \sigma - j\omega}$$

so that

$$|z|^2 = \frac{(1+\sigma)^2 + \omega^2}{(1-\sigma)^2 + \omega^2} \qquad (5.11.3)$$

For $\sigma < 0$ the denominator on the right-hand side of (5.11.3) is greater than the numerator so that $|z| < 1$. Therefore, the left half s-plane is mapped onto the interior of the unit circle in the z-plane. Similarly, $|z| > 1$ for $\sigma > 0$, so that the right half s-plane is mapped onto the exterior of the unit circle in the z-plane. For $\sigma = 0$, $|z| = 1$, so that the $j\omega$ axis is mapped onto the unit circle in the z-plane. We have already observed that the transformation $z = e^{sT}$ generates a similar mapping. However, unlike the bilinear transformation, this transformation is not one-to-one.

If a discrete-time system has the rational pulse transfer function $H(z)$, then the function

$$H_1(s) = H\left(\frac{1+s}{1-s}\right)$$

is also a rational function of s. Since the bilinear transformation maps the exterior of the unit circle onto the right half plane, the stability of the discrete-time system can be determined by applying the standard tests for determining the stability of continuous-time systems, such as the Routh-Hurwitz criterion and the Nyquist criterion, to the function $H_1(s)$.

The bilinear transformation is not usually used for studying the stability of discrete-time systems because the modified Schur-Cohn criterion, the modified Nyquist criterion, and the root locus methods are more direct. The bilinear transformation is used to convert the designs for continuous-time filters into designs for discrete-time filters. This application is discussed in a later chapter.

5.12 THE MODIFIED Z-TRANSFORM AND THE BEHAVIOR OF A SAMPLED-DATA SYSTEM BETWEEN SAMPLING INSTANTS

The *modified Z-transform* $F(z, \delta)$ of a signal $f(t)$ is defined as

$$F(z, \delta) = \mathscr{Z}\{f(t+\delta)\} = \sum_{n=-\infty}^{\infty} f(nT+\delta)z^{-n} \qquad \text{for} \qquad 0 \le \delta < T \quad (5.12.1)$$

The modified Z-transform is sometimes called the *advanced Z-transform*. Since

$$\mathscr{L}\{f(t+\delta)\} = \mathscr{F}(s)e^{s\delta}$$

the modified Z-transform is also given by

$$F(z, \delta) = [\mathscr{F}(s)e^{s\delta}]^*|_{z=e^{sT}} \qquad \text{for} \qquad 0 \le \delta < T \qquad (5.12.2)$$

Since $f(nT+\delta)$ for $0 \le \delta < T$ can be calculated from $F(z, \delta)$, the values of a signal between sampling instants as well as at the sampling instants can be calculated from the modified Z-transform.

Using the reasoning in Section 4.4 we can see that if $f(t)$ is a causal signal with a Laplace transform $\mathscr{F}(s)$ that converges absolutely for $\sigma = \text{Re } s > \sigma_f$, then

$$[\mathscr{F}(s)e^{s\delta}]^* = \frac{1}{2\pi j} \int_{c-j\infty}^{c+j\infty} \frac{\mathscr{F}(p)e^{p\delta}}{1 - e^{-(s-p)T}} \, dp \qquad (5.12.3)$$

for $0 \le \delta < T$, $\sigma > \sigma_f$, and $\sigma_f < c < \sigma$. This integral can be evaluated by applying Cauchy's Residue Theorem to the closed contour $\Gamma_1\Gamma_2$ shown in Fig. 4.4.1. Since Re $p < 0$ in the left half plane, the integral along Γ_2 vanishes as the radius of the semicircle becomes infinite if $0 < \delta < T$. Therefore

$$[\mathscr{F}(s)e^{s\delta}]^* = \sum_k \text{Res}\left[\frac{\mathscr{F}(p)e^{p\delta}}{1 - e^{-(s-p)T}}, p_k\right] \qquad (5.12.4)$$

for $\sigma > \sigma_f$ and $0 < \delta < T$ with $\{p_k\}$ the set of points at which $\mathscr{F}(p)$ has poles. Letting $z = e^{sT}$, (5.12.4) becomes

$$F(z, \delta) = \sum_k \text{Res}\left[\frac{\mathscr{F}(p)e^{p\delta}}{1 - e^{pT}z^{-1}}, p_k\right] \qquad (5.12.5)$$

for $|z| > e^{\sigma_f T}$, $0 < \delta < T$, and $\{p_k\}$ the set of points at which $\mathscr{F}(p)$ has poles.

Example 5.12.1

Let $\mathscr{F}(s) = 1/(s+a)$. According to (5.12.5)

$$F(z, \delta) = \text{Res}\left[\frac{e^{p\delta}}{(p+a)(1 - e^{pT}z^{-1})}, p = -a\right] = \frac{e^{-a\delta}}{1 - e^{-aT}z^{-1}}$$

This result is also easy to obtain directly from the definition (5.12.1) of $F(z, \delta)$. ◄

Tables of modified Z-transforms can be found in sampled-data books such as References 34, 79, 114, and 131.

The modified Z-transform can be used to find the behavior of a sampled-data system between sampling instants. For example, consider the simple system shown in Fig. 5.12.1. To find $y(t)$ between sampling instants, a fictitious

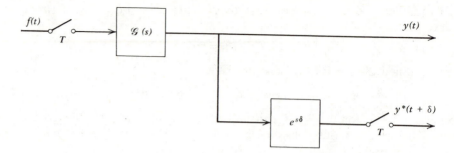

FIGURE 5.12.1.   Finding the output between sampling instants.

advance $e^{s\delta}$ and sampler have been added. The Laplace transform of the output of the fictitious sampler is

$$\mathcal{L}\{y^*(t+\delta)\} = [\mathcal{Y}(s)e^{s\delta}]^* = \mathcal{F}^*(s)[\mathcal{G}(s)e^{s\delta}]^*$$

so that

$$Y(z, \delta) = F(z)G(z, \delta) \qquad (5.12.6)$$

More complicated closed-loop systems can be analyzed in a similar manner. ◀

Example 5.12.2

In the system shown in Fig. 5.12.1, let $f(t) = u(t)$ and $\mathcal{G}(s) = 1/(s+a)$. Then

$$Y(z, \delta) = \frac{1}{1-z^{-1}} \cdot \frac{e^{-a\delta}}{1-e^{-aT}z^{-1}}$$

Taking the inverse Z-transform, we can see that

$$y(nT+\delta) = \frac{e^{-a\delta}}{1-e^{-aT}} (1-e^{-aT}e^{-anT})u(nT)$$

for $0 < \delta < T$.

Notice that the output decays exponentially between sampling instants. This result can also be derived by observing that

$$\dot{y}(t) + ay(t) = f^*(t)$$

If the impulse train $f^*(t)$ is considered to be zero except at the sampling instants, then this differential equation is homogeneous and has the solution

$$y(t) = y(nT+)e^{-a(t-nT)} \qquad \text{for} \qquad nT < t < (n+1)T \qquad ◀$$

5.13 ESTIMATING THE BEHAVIOR BETWEEN SAMPLING INSTANTS BY FAST OUTPUT SAMPLING

So far we have assumed that all the samplers in a system are synchronized and operate at the same rate $\omega_s = 2\pi/T$. The output of such a system between sampling instants nT can be estimated by sampling the output at a faster rate ω_f. Systems containing samplers operating at different rates are known as *multirate systems.*

When the output sampler operates at a rate that is an integral multiple of the internal sampling rate, simple results can be obtained. Consider the system shown in Fig. 5.13.1. We will assume that the output sampler operates with period T/N or rate $\omega_f = N\omega_s$ with N an integer. The continuous-time output can be calculated from the input samples by the equation

$$y(t) = \sum_{n=-\infty}^{\infty} f(nT)g(t-nT) \qquad (5.13.1)$$

Therefore, the output samples are

$$y(kT/N) = \sum_{n=-\infty}^{\infty} f(nT)g(kT/N-nT)$$

$$= \sum_{n=-\infty}^{\infty} f(nT)g[(k-nN)T/N] \qquad (5.13.2)$$

We will represent the sequence of samples of a signal $b(t)$ taken at the fast rate $N\omega_s$ by the generating function

$$B(\lambda:T/N) = \sum_{k=-\infty}^{\infty} b(kT/N)\lambda^{-k} \qquad (5.13.3)$$

and will continue to denote the Z-transform of the slow rate discrete-time signal $b(nT)$ by $B(z)$. If we let $\lambda = e^{sT/N}$, then $B(\lambda:T/N)$ can be considered to be the Laplace transform of the impulse sequence

$$\sum_{k=-\infty}^{\infty} b(kT/N)\,\delta(t-kT/N)$$

FIGURE 5.13.1. System with a slow input and fast output sampler.

From (5.13.2) and (5.13.3) we see that

$$Y(\lambda : T/N) = \sum_{k=-\infty}^{\infty} \sum_{n=-\infty}^{\infty} f(nT)g[(k-nN)T/N]\lambda^{-k}$$

$$= \sum_{n=-\infty}^{\infty} f(nT) \sum_{k=-\infty}^{\infty} g[(k-nN)T/N]\lambda^{-k} \qquad (5.13.4)$$

Making the change of variables $m = k - nN$ in the inner summation, (5.13.4) can be written as

$$Y(\lambda : T/N) = \sum_{n=-\infty}^{\infty} f(nT)(\lambda^N)^{-n} \sum_{m=-\infty}^{\infty} g(mT/N)\lambda^{-m}$$

so that

$$Y(\lambda : T/N) = F(z)|_{z=\lambda^N} G(\lambda : T/N) \qquad (5.13.5)$$

This result is convenient because the input signal and system components contribute separate factors.

Equation 5.13.5 can be derived more intuitively by observing that, as far as its effects on $y(t)$ are concerned, the input sampler can be considered to be operating at the fast rate if $f(t)$ is replaced by a signal $b(t)$ with the properties

$$b(kT/N) = \begin{cases} f(nT) & \text{for} \quad k = nN \\ 0 & \text{elsewhere} \end{cases}$$

It is easy to see that

$$B(\lambda : T/N) = F(z)|_{z=\lambda^N}$$

and that

$$Y(\lambda : T/N) = B(\lambda : T/N)G(\lambda : T/N)$$

Example 5.13.1

In Fig. 5.13.1 let $f(t) = u(t)$, $g(t) = e^{-at}u(t)$, and $N = 2$. Then

$$F(z) = \frac{1}{1 - z^{-1}}$$

and

$$G(\lambda : T/N) = \frac{1}{1 - e^{-aT/2}\lambda^{-1}}$$

Therefore

$$Y(\lambda : T/N) = \frac{1}{(1 - \lambda^{-2})(1 - e^{-aT/2}\lambda^{-1})}$$

The inverse transform of $Y(\lambda : T/2)$ is the sum of the residues of $Y(\lambda : T/2)\lambda^{k-1}$ at $e^{-aT/2}$, 1, and -1 for $k \geq 0$. Letting $k = 2n + r$ where $r = 0$ or 1, this sum can

be put in the form

$$y(kT/2) = \frac{e^{-raT/2}}{1-e^{-aT}}(1-e^{-aT}e^{-anT})u(kT)$$

Compare this result with Example 5.12.2.   ◀

5.14 DATA REDUCTION BY SLOW OUTPUT SAMPLING

In some data processing situations, only the lower frequency components of a discrete-time signal are of interest. To conserve storage, the signal can be passed through a low-pass discrete-time filter and then resampled at a slower rate. Slow output sampling might also be used in a digital computer controlled sampled-data feedback system where the plant has low-pass characteristics to allow the computer to be time-shared among more operations.

As a specific example, consider the system shown in Fig. 5.14.1. The continuous-time output can be expressed as

$$y(t) = \sum_{k=-\infty}^{\infty} f(kT/N)g(t - kT/N)$$

Therefore, the output samples are

$$y(nT) = \sum_{k=-\infty}^{\infty} f(kT/N)g(nT - kT/N) \qquad (5.14.1)$$

Taking the Z-transform of (5.14.1) does not give useful results because, unlike the slow input-fast output case, the transform cannot be factored into the product of an input component and a system component.

If this system is imbedded in a feedback system where the normal sampling period is T, a convenient method of analysis is to use the *switch decomposition* technique. The signal $f_1(t)$ appearing at the output of the fast sampler is

$$f_1(t) = \sum_{k=-\infty}^{\infty} f(kT/N)\,\delta(t - kT/N) \qquad (5.14.2)$$

FIGURE 5.14.1.   System with fast input and slow output sampling.

This summation can be rearranged and written as

$$f_1(t) = \sum_{r=0}^{N-1} \sum_{n=-\infty}^{\infty} f(nT + rT/N)\, \delta(t - nT - rT/N) \tag{5.14.3}$$

Equation 5.14.3 suggests representing the fast sampler by the parallel structure shown in Fig. 5.14.2 which contains only samplers operating with period T. From Fig. 5.14.2 we can see that

$$\mathcal{Y}^*(s) = \sum_{r=0}^{N-1} [\mathcal{F}(s)e^{srT/N}]^*[\mathcal{G}(s)e^{-srT/N}]^* \tag{5.14.4}$$

Example 5.14.1

In Fig. 5.14.1 let $f(t) = u(t)$, $\mathcal{G}(s) = 1/(s+a)$, and $N = 2$. According to (5.14.4)

$$Y(z) = F(z)G(z) + \left[\frac{e^{sT/2}}{s}\right]^* \left[\frac{e^{-sT/2}}{s+a}\right]^*$$

$$= \frac{1}{(1-z^{-1})(1-e^{-aT}z^{-1})} + \frac{1}{1-z^{-1}} \cdot \frac{e^{-aT/2}z^{-1}}{1-e^{-aT}z^{-1}}$$

$$= \frac{1 + e^{-aT/2}z^{-1}}{(1-z^{-1})(1-e^{-aT}z^{-1})}$$

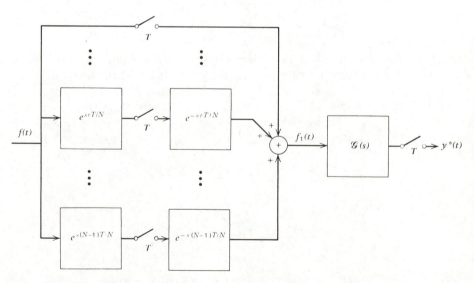

FIGURE 5.14.2. Switch decomposition representation of a fast sampler.

The inverse Z-transform yields

$$y(nT) = \frac{1 - e^{-aT/2}e^{-anT}}{1 - e^{-aT/2}} u(nT) \qquad \blacktriangleleft$$

In many data processing applications the sequence of output samples appearing at the rate ω_s is actually generated by selecting every Nth sample from a sequence appearing at the rate $\omega_f = N\omega_s$. This is sometimes called *skip sampling*. In order to provide a guide for selecting a reasonable sampling rate reduction factor N, we now investigate the frequency domain effects of skip sampling. If the sequence of samples appearing at the rate $N\omega_s$ is represented by the impulse train

$$y^\$(t) = \sum_{k=-\infty}^{\infty} y(kT/N)\, \delta(t - kT/N) \tag{5.14.5}$$

then, according to the aliasing formula

$$Y^\$(\omega) = \frac{N}{T} \sum_{m=-\infty}^{\infty} Y(\omega - mN\omega_s) \tag{5.14.6}$$

Similarly, if the sequence of samples appearing at the rate ω_s is represented by

$$y^*(t) = \sum_{n=-\infty}^{\infty} y(nT)\, \delta(t - nT) \tag{5.14.7}$$

then

$$Y^*(\omega) = \frac{1}{T} \sum_{m=-\infty}^{\infty} Y(\omega - m\omega_s) \tag{5.14.8}$$

The summation in (5.14.8) can be rearranged as

$$Y^*(\omega) = \frac{1}{N} \sum_{r=0}^{N-1} \frac{N}{T} \sum_{m=-\infty}^{\infty} Y(\omega - mN\omega_s - r\omega_s) \tag{5.14.9}$$

Comparing (5.14.6) and (5.14.9), we see that

$$Y^*(\omega) = \frac{1}{N} \sum_{r=0}^{N-1} Y^\$(\omega - r\omega_s) \tag{5.14.10}$$

Thus the spectrum of the slow sampled signal is an aliased version of the spectrum of the fast sampled signal. If, in the interval $|\omega| < N\omega_s/2$, $Y^\$(\omega)$ is identically zero except for $|\omega| < \omega_s/2$, then the aliasing described by (5.14.10) causes no distortion. An example for $N = 2$ is shown in Fig. 5.14.3.

For causal signals skip sampling can be approached in another interesting way. If the fast rate discrete-time signals $p(kT/N)$ and $v(kT/N)$ are defined as

$$p(kT/N) = \begin{cases} 1 & \text{for} \quad k = nN,\ n = 0, 1, \ldots \\ 0 & \text{elsewhere} \end{cases} \tag{5.14.11}$$

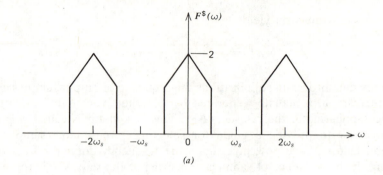

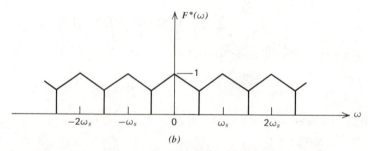

FIGURE 5.14.3. Frequency domain effects of skip sampling. (*a*) Spectrum of original sampled signal; (*b*) spectrum of skip sampled signal for $N = 2$.

and

$$v(kT/N) = y(kT/N)p(kT/N) \tag{5.14.12}$$

then

$$P(\lambda : T/N) = 1/(1 - \lambda^{-N}) \tag{5.14.13}$$

and

$$V(\lambda : T/N) = \sum_{k=0}^{\infty} y(kT/N)p(kT/N)\lambda^{-k} = \sum_{n=0}^{\infty} y(nT)(\lambda^N)^{-n}$$

$$= Y(z)|_{z = \lambda^N} \tag{5.14.14}$$

Also, according to Theorem 4.7.1

$$V(\lambda : T/N) = \frac{1}{2\pi j} \oint_C Y(x : T/N)P(\lambda/x : T/N) \frac{dx}{x} \tag{5.14.15}$$

where C and λ are chosen so that all the poles of $Y(x : T/N)$ lie inside C and all the poles of $P(\lambda/x : T/N)$ lie outside C. From (5.14.13), (5.14.14), and (5.14.15) we see that

$$Y(z) = \frac{1}{2\pi j} \oint_C \frac{Y(x : T/N)}{1 - x^N z^{-1}} \frac{dx}{x} \tag{5.14.16}$$

Using Cauchy's Residue Theorem, this integral can be evaluated as

$$Y(z) = \sum_k \text{Res} \left[\frac{Y(x:T/N)x^{-1}}{1 - x^N z^{-1}}, p_k \right] \tag{5.14.17}$$

where $\{p_k\}$ is the set of points at which $Y(x:T/N)x^{-1}$ has poles.

Example 5.14.2

Consider the system described in Example 5.14.1. From Fig. 5.14.1 we see that

$$Y(\lambda : T/2) = F(\lambda : T/2)G(\lambda : T/2) = \frac{1}{(1 - \lambda^{-1})(1 - e^{-aT/2}\lambda^{-1})}$$

$$= \frac{\lambda^2}{(\lambda - 1)(\lambda - e^{-aT/2})}$$

According to (5.14.17)

$$Y(z) = \sum_{k=1}^{2} \text{Res} \left[\frac{x}{(x-1)(x - e^{-aT/2})(1 - x^2 z^{-1})}, p_k \right] \tag{5.14.18}$$

where $p_1 = 1$ and $p_2 = e^{-aT/2}$. Evaluating (5.14.18) gives the same answer for $Y(z)$ as found by using (5.14.4) in Example 5.14.1.   ◀

Equation 5.14.16 can also be evaluated by taking the negative sum of the residues outside C. The poles of $1/(1 - x^N z^{-1})$ are at

$$x_r = z^{1/N} e^{-jr2\pi/N} \qquad \text{for} \qquad r = 0, 1, \ldots, N-1 \tag{5.14.19}$$

Since $y(kT/N)$ has been assumed to be causal, $Y(x:T/N)$ must be analytic outside a circle of finite radius. Therefore, the integrand in (5.14.16) has a zero of order greater than one at infinity and the residue at infinity must be zero. Consequently

$$Y(z) = -\sum_{r=0}^{N-1} \text{Res} \left[\frac{Y(x:T/N)x^{-1}}{1 - x^N z^{-1}}, x_r \right] \tag{5.14.20}$$

Using Rule 2 in Appendix A to evaluate the residues, we find that

$$Y(z) = \frac{1}{N} \sum_{r=0}^{N-1} Y(z^{1/N} e^{-jr2\pi/N} : T/N) \tag{5.14.21}$$

Letting $z = e^{j\omega T}$, (5.14.21) becomes

$$Y^*(\omega) = \frac{1}{N} \sum_{r=0}^{N-1} Y[e^{j(\omega - r\omega_s)T/N} : T/N] = \frac{1}{N} \sum_{r=0}^{N-1} Y^{\$}(\omega - r\omega_s) \tag{5.14.22}$$

which is identical to (5.14.10).

6

STATE SPACE REPRESENTATION OF SAMPLED-DATA SYSTEMS

6.1 INTRODUCTION

In this chapter the concept of a dynamical system and its state is introduced. Systems that can be described by difference or differential equations are types of dynamical systems. For a dynamical system a set of variables called the state of the system can be found that contains all the information about the past behavior of the system necessary to calculate its future state and output given its present and future input. In particular, we see how to represent an nth order, linear, difference or differential equation by a first-order, linear, matrix difference or differential equation describing the evolution of an n-dimensional state vector and an equation relating the present output to the present state and input. We call these equations the state equation and output equation or, sometimes, simply a state space representation for the difference or differential equation. Methods for solving the state equations are developed. As expected, the solutions to time-invariant, matrix, linear, difference or differential equations can be readily determined by time or frequency domain methods. Different structures for realizing nth order, linear, difference or differential equations are examined from the state space point of view.

The state space approach became popular in the early 1960s with the advent of the aerospace era and as high-speed digital computers became more readily available. Matrix formulation of system equations was convenient for complex systems with multiple inputs and outputs. In addition, this formulation put the equations in a convenient form for generating time-domain solutions directly using high-speed digital computers. Furthermore, it was conceptually helpful to represent a complex system by a simple first-order matrix difference or differential equation when performing theoretical derivations. This chapter is only intended to be a short introduction to state space methods. More detailed treatments can be found in books on system theory and modern control theory such as References 3, 14, 29, and 144.

6.2 THE CONCEPT OF A DYNAMICAL SYSTEM AND ITS STATE

The mathematical time-domain models used to describe sampled-data systems are almost always finite-order difference equations and differential equations whose solutions exist and are unique. We know that the behavior of these systems for $t \geq t_0$ can be uniquely determined if an appropriate set of initial conditions at time t_0 is specified. These initial conditions contain all the information about the past behavior of the system necessary to calculate its future behavior. It is natural to call the set of initial conditions the state of the system at time t_0.

Systems that can be described by difference and differential equations of this type belong to a large class of systems known as dynamical systems. Before trying to define a dynamical system abstractly, let us consider the following two examples.

Example 6.2.1

Suppose that the input $v(n)$ and output $y(n)$ of a discrete-time system are related by the difference equation

$$y(n) + 2y(n-1) + y(n-2) = v(n) \qquad (6.2.1)$$

If $y(n_0-1)$ and $y(n_0-2)$ are specified, then $y(n)$ for $n \geq n_0$ can be calculated recursively from (6.2.1) when $v(n)$ for $n \geq n_0$ is specified. Thus the pair $y(n-1)$ and $y(n-2)$ can be chosen as the state of the system at time n. To show the behavior of the system in terms of its state more explicitly, let us call the vector

$$\mathbf{x}(n) = \begin{bmatrix} x_1(n) \\ x_2(n) \end{bmatrix} = \begin{bmatrix} y(n-2) \\ y(n-1) \end{bmatrix} \qquad (6.2.2)$$

the state vector for the system. Then

$$x_1(n+1) = y(n-1) = x_2(n)$$

and

$$x_2(n+1) = y(n) = v(n) - y(n-2) - 2y(n-1) \tag{6.2.3}$$

or

$$x_2(n+1) = v(n) - x_1(n) - 2x_2(n) \tag{6.2.4}$$

These equations can be written in the matrix form

$$\begin{bmatrix} x_1(n+1) \\ x_2(n+1) \end{bmatrix} = \begin{bmatrix} 0 & 1 \\ -1 & -2 \end{bmatrix} \begin{bmatrix} x_1(n) \\ x_2(n) \end{bmatrix} + \begin{bmatrix} 0 \\ 1 \end{bmatrix} v(n)$$

or

$$\mathbf{x}(n+1) = \mathbf{A}\mathbf{x}(n) + \mathbf{B}v(n) \tag{6.2.5}$$

From (6.2.3) and (6.2.4) we can see that the system output can be expressed in terms of its input and state vector as

$$y(n) = \begin{bmatrix} -1 & -2 \end{bmatrix} \begin{bmatrix} x_1(n) \\ x_2(n) \end{bmatrix} + v(n)$$

or

$$y(n) = \mathbf{C}\mathbf{x}(n) + v(n) \tag{6.2.6}$$

Thus we can describe the system behavior by the vector-matrix difference equation (6.2.5), which describes the evolution of the state and an output equation (6.2.6), rather than by the second-order difference equation (6.2.1).   ◀

Example 6.2.2

Suppose the input $v(t)$ and output $y(t)$ of a continuous-time system are related by the differential equation

$$\dot{y}(t) + \frac{1}{t} y(t) = v(t) \tag{6.2.7}$$

for $0 \le t_0 \le t$. This is equivalent to the equation

$$t\dot{y}(t) + y(t) = tv(t)$$

or

$$\frac{d}{dt}[ty(t)] = tv(t) \tag{6.2.8}$$

Integrating both sides of (6.2.8) we see that

$$ty(t) - t_0 y(t_0) = \int_{t_0}^{t} \tau v(\tau)\, d\tau$$

or

$$y(t) = \frac{t_0}{t} y(t_0) + \int_{t_0}^{t} \frac{\tau}{t} v(\tau)\, d\tau \tag{6.2.9}$$

In this example it is natural to choose $y(t_0)$ as the state of the system at time t_0. The state and output coincide in this case. ◄

More generally, let us suppose that a system has m inputs $v_1(t), \ldots, v_m(t)$ and r outputs $y_1(t), \ldots, y_r(t)$. For convenience we will represent the inputs and outputs by the vectors

$$
\mathbf{v}(t) = \begin{bmatrix} v_1(t) \\ \cdot \\ \cdot \\ \cdot \\ v_m(t) \end{bmatrix} \quad \text{and} \quad \mathbf{y}(t) = \begin{bmatrix} y_1(t) \\ \cdot \\ \cdot \\ \cdot \\ y_r(t) \end{bmatrix}
$$

and the input segment over the interval $[t_0, t)$ by

$$
\mathbf{v}[t_0, t) = \{\mathbf{v}(\tau) : t_0 \leq \tau < t\}
$$

The system is called a dynamical system if its behavior can be described in terms of a state vector $\mathbf{x}(t)$ of minimum dimension and a state transition function $\mathbf{\Psi}\{t, \mathbf{v}[t_0, t), \mathbf{x}_0, t_0\}$ such that

$$
\mathbf{x}(t) = \mathbf{\Psi}\{t, \mathbf{v}[t_0, t), \mathbf{x}_0, t_0\} \qquad \text{for} \qquad t \geq t_0 \tag{6.2.10}
$$

and an output equation of the form

$$
\mathbf{y}(t) = \mathbf{g}\{\mathbf{x}(t), \mathbf{v}(t), t\} \tag{6.2.11}
$$

where $\mathbf{\Psi}\{t, \mathbf{v}[t_0, t), \mathbf{x}_0, t_0\}$ and $\mathbf{g}\{\mathbf{x}, \mathbf{v}, t\}$ are continuous functions of their arguments, and $\mathbf{\Psi}$ has the properties that

$$
\mathbf{x}(t_0) = \lim_{t \to t_0} \mathbf{\Psi}\{t, \mathbf{v}[t_0, t), \mathbf{x}_0, t_0\} = \mathbf{x}_0 \tag{6.2.12}
$$

and

$$
\mathbf{\Psi}\{t, \mathbf{v}[t_0, t), \mathbf{x}_0, t_0\} = \mathbf{\Psi}\{t, \mathbf{v}[t_1, t), \mathbf{\Psi}\{t_1, \mathbf{v}[t_0, t_1), \mathbf{x}_0, t_0\}, t_1\} \tag{6.2.13}
$$

for $t_0 < t_1 \leq t$. The set of values that $\mathbf{x}(t)$ can take on for an allowable class of input signals is called the state space for the system. Notice that (6.2.11) restricts the present output to be an instantaneous function of the present state, input, and time. Also, (6.2.10) implies that dynamical systems are causal.

The state transition function is the rule for calculating the evolution of the state. According to (6.2.12), the argument $\mathbf{x}_0$ is the state of the system at time t_0. The property described by (6.2.13) is known as the transition or semigroup property of $\mathbf{\Psi}$. The transition property says that the trajectory followed by the state for the input segment $\mathbf{v}[t_0, t)$ can be calculated in steps by determining the trajectory from the initial state to the state at some intermediate time t_1 and then calculating the trajectory from this intermediate state to the final state at time t.

It is easy to see that the systems discussed in Examples 6.2.1 and 6.2.2 satisfy our abstract definition of a dynamical system. The state transition function for Example 6.2.1 is given implicitly by (6.2.5). It can be found explicitly by recursively evaluating this equation. The state transition function for Example 6.2.2 is given by (6.2.9).

6.3 LINEAR, DISCRETE-TIME, DYNAMICAL SYSTEMS

By a linear, finite-dimensional, discrete-time, dynamical system we will mean a system with input $\mathbf{v}(t_n)$, output $\mathbf{y}(t_n)$, and state $\mathbf{x}(t_n)$ having a state equation of the form

$$\mathbf{x}(t_{n+1}) = \mathbf{A}(t_n)\mathbf{x}(t_n) + \mathbf{B}(t_n)\mathbf{v}(t_n) \qquad (6.3.1)$$

and an output equation of the form

$$\mathbf{y}(t_n) = \mathbf{C}(t_n)\mathbf{x}(t_n) + \mathbf{D}(t_n)\mathbf{v}(t_n) \qquad (6.3.2)$$

where

$\mathbf{x}(t_n)$ is an N-dimensional column vector
$\mathbf{v}(t_n)$ is an m-dimensional column vector
$\mathbf{y}(t_n)$ is an r-dimensional column vector
$\mathbf{A}(t_n)$ is an $N \times N$ nonsingular matrix
$\mathbf{B}(t_n)$ is an $N \times m$ matrix
$\mathbf{C}(t_n)$ is an $r \times N$ matrix
$\mathbf{D}(t_n)$ is an $r \times m$ matrix

and

$$t_{n_2} > t_{n_1} \qquad \text{for} \qquad n_2 > n_1$$

In most applications we will use uniform sampling and let $t_n = nT$. In this case, the state and output equations can be represented pictorially by the block diagram in Fig. 6.3.1. The double lines indicate the transmission of matrices. All the operations correspond to the appropriate matrix multiplications, additions, and delays in (6.3.1) and (6.3.2).

When the input is identically zero (6.3.1) reduces to

$$\mathbf{x}(t_{n+1}) = \mathbf{A}(t_n)\mathbf{x}(t_n) \qquad (6.3.3)$$

so that

$$\mathbf{x}(t_{n+2}) = \mathbf{A}(t_{n+1})\mathbf{x}(t_{n+1}) = \mathbf{A}(t_{n+1})\mathbf{A}(t_n)\mathbf{x}(t_n)$$

and

$$\mathbf{x}(t_{n+3}) = \mathbf{A}(t_{n+2})\mathbf{A}(t_{n+1})\mathbf{A}(t_n)\mathbf{x}(t_n)$$

and in general

$$\mathbf{x}(t_{n+k}) = \mathbf{A}(t_{n+k-1})\mathbf{A}(t_{n+k-2}) \cdots \mathbf{A}(t_n)\mathbf{x}(t_n) \qquad (6.3.4)$$

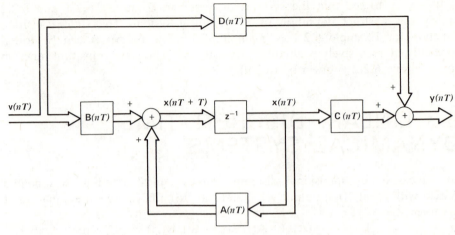

FIGURE 6.3.1.   Pictorial representation of the state and output equations for a uniformly sampled, linear, discrete-time system.

for $k > 0$. Also, if $\mathbf{x}(t_{n+k})$ is given, the past state $\mathbf{x}(t_n)$ must be

$$\mathbf{x}(t_n) = [\mathbf{A}(t_{n+k-1})\mathbf{A}(t_{n+k-2}) \cdots \mathbf{A}(t_n)]^{-1} \mathbf{x}(t_{n+k})$$
$$= \mathbf{A}^{-1}(t_n) \cdots \mathbf{A}^{-1}(t_{n+k-2})\mathbf{A}^{-1}(t_{n+k-1})\mathbf{x}(t_{n+k}) \qquad (6.3.5)$$

If we define the *state transition matrix* for the system to be

$$\boldsymbol{\varphi}(t_{n_2}, t_{n_1}) = \begin{cases} \mathbf{A}(t_{n_2-1})\mathbf{A}(t_{n_2-2}) \cdots \mathbf{A}(t_{n_1}) & \text{for} \quad n_2 > n_1 \\ \mathbf{I} & \text{for} \quad n_2 = n_1 \quad (6.3.6) \\ \mathbf{A}^{-1}(t_{n_1}) \cdots \mathbf{A}^{-1}(t_{n_2-2})\mathbf{A}^{-1}(t_{n_2-1}) & \text{for} \quad n_2 < n_1 \end{cases}$$

then from (6.3.4) and (6.3.5) it follows that the system states at any two times t_{n_2} and t_{n_1} are related by the equation

$$\mathbf{x}(t_{n_2}) = \boldsymbol{\varphi}(t_{n_2}, t_{n_1})\mathbf{x}(t_{n_1}) \qquad (6.3.7)$$

when the input is identically zero. From (6.3.6) we can see that the state transition matrix has the following properties:

1. $$\boldsymbol{\varphi}(t_n, t_n) = \mathbf{I} \qquad (6.3.8)$$

2. $$\boldsymbol{\varphi}(t_{n_2}, t_{n_1}) = \boldsymbol{\varphi}^{-1}(t_{n_1}, t_{n_2}) \qquad (6.3.9)$$

3. $$\boldsymbol{\varphi}(t_{n_3}, t_{n_2})\boldsymbol{\varphi}(t_{n_2}, t_{n_1}) = \boldsymbol{\varphi}(t_{n_3}, t_{n_1}) \qquad (6.3.10)$$

If the input is not identically zero and $\mathbf{x}(t_n)$ is known, then the future states can be calculated recursively from (6.3.1). Performing the recursion we find that

$$\mathbf{x}(t_{n+2}) = \mathbf{A}(t_{n+1})\mathbf{x}(t_{n+1}) + \mathbf{B}(t_{n+1})\mathbf{v}(t_{n+1})$$
$$= \mathbf{A}(t_{n+1})\mathbf{A}(t_n)\mathbf{x}(t_n) + \mathbf{A}(t_{n+1})\mathbf{B}(t_n)\mathbf{v}(t_n) + \mathbf{B}(t_{n+1})\mathbf{v}(t_{n+1})$$
$$= \boldsymbol{\varphi}(t_{n+2}, t_n)\mathbf{x}(t_n) + \boldsymbol{\varphi}(t_{n+2}, t_{n+1})\mathbf{B}(t_n)\mathbf{v}(t_n) + \mathbf{B}(t_{n+1})\mathbf{v}(t_{n+1})$$

and that, in general

$$\mathbf{x}(t_{n+k}) = \boldsymbol{\varphi}(t_{n+k}, t_n)\mathbf{x}(t_n) + \sum_{i=n}^{n+k-1} \boldsymbol{\varphi}(t_{n+k}, t_{i+1})\mathbf{B}(t_i)\mathbf{v}(t_i) \qquad (6.3.11)$$

for $k > 0$.

6.4 TIME-INVARIANT, LINEAR, DISCRETE-TIME, DYNAMICAL SYSTEMS

When $t_n = nT$ and $\mathbf{A}$, $\mathbf{B}$, $\mathbf{C}$, and $\mathbf{D}$ are constant matrices in (6.3.1) and (6.3.2) we say that the system is a time-invariant, linear, discrete-time system. In this case the state equation becomes

$$\mathbf{x}(nT + T) = \mathbf{A}\mathbf{x}(nT) + \mathbf{B}\mathbf{v}(nT) \qquad (6.4.1)$$

and the output equation becomes

$$\mathbf{y}(nT) = \mathbf{C}\mathbf{x}(nT) + \mathbf{D}\mathbf{v}(nT) \qquad (6.4.2)$$

From (6.3.6) we can see that the state transition matrix for the time-invariant system is

$$\boldsymbol{\varphi}(n_2 T, n_1 T) = \mathbf{A}^{n_2 - n_1} \qquad (6.4.3)$$

This is a function only of the time difference $n_2 T - n_1 T$. For this reason it is customary to call the matrix

$$\boldsymbol{\varphi}(nT) = \mathbf{A}^n \qquad (6.4.4)$$

the state transition matrix for the time-invariant system with the understanding that $n = n_2 - n_1$. Using this notation, it follows from (6.3.7) that when the input is identically zero

$$\mathbf{x}(n_2 T) = \boldsymbol{\varphi}(n_2 T - n_1 T)\mathbf{x}(n_1 T) = \mathbf{A}^{n_2 - n_1}\mathbf{x}(n_1 T) \qquad (6.4.5)$$

According to (6.3.11), the solution to the state equation for a nonzero input becomes

$$\mathbf{x}(nT + kT) = \boldsymbol{\varphi}(kT)\mathbf{x}(nT) + \sum_{i=n}^{n+k-1} \boldsymbol{\varphi}[(n + k - i - 1)T]\mathbf{B}\mathbf{v}(iT) \qquad (6.4.6)$$

for $k > 0$.

A closed form for the state transition matrix and solution of the state and output equations can be obtained by Z-transform methods. We will define the one-sided Z-transform of an $r \times s$ matrix function $\mathbf{f}(nT)$ as the $r \times s$ matrix

$$\mathbf{F}(z) = \sum_{n=0}^{\infty} \mathbf{f}(nT) z^{-n} \tag{6.4.7}$$

The elements of $\mathbf{F}(z)$ are just the transforms of the corresponding elements of $\mathbf{f}(nT)$. Taking the transform of both sides of the state equation (6.4.1) gives

$$z\mathbf{X}(z) - z\mathbf{x}(0) = \mathbf{A}\mathbf{X}(z) + \mathbf{B}\mathbf{V}(z)$$

so that

$$\mathbf{X}(z) = (z\mathbf{I} - \mathbf{A})^{-1} z\mathbf{x}(0) + (z\mathbf{I} - \mathbf{A})^{-1} \mathbf{B}\mathbf{V}(z) \tag{6.4.8}$$

From (6.4.2) we see that

$$\mathbf{Y}(z) = \mathbf{C}\mathbf{X}(z) + \mathbf{D}\mathbf{V}(z) \tag{6.4.9}$$

The state $\mathbf{x}(nT)$ and output $\mathbf{y}(nT)$ can be found for $n \geq 0$ by taking the inverse transforms of $\mathbf{X}(z)$ and $\mathbf{Y}(z)$ element by element.

If the input is identically zero (6.4.8) reduces to

$$\mathbf{X}(z) = (z\mathbf{I} - \mathbf{A})^{-1} z\mathbf{x}(0) \tag{6.4.10}$$

so that

$$\mathbf{x}(nT) = \mathscr{L}^{-1}\{z\mathbf{I} - \mathbf{A})^{-1} z\}\mathbf{x}(0) \tag{6.4.11}$$

Letting $n_1 = 0$ and $n_2 = n$, (6.4.5) becomes

$$\mathbf{x}(nT) = \boldsymbol{\varphi}(nT)\mathbf{x}(0) = \mathbf{A}^n \mathbf{x}(0) \tag{6.4.12}$$

Comparing (6.4.11) and (6.4.12), we find that

$$\boldsymbol{\varphi}(nT) = \mathbf{A}^n = \mathscr{L}^{-1}\{(z\mathbf{I} - \mathbf{A})^{-1} z\} \qquad \text{for} \qquad n \geq 0 \tag{6.4.13}$$

or equivalently

$$\boldsymbol{\Phi}(z) = \mathscr{L}_+\{\mathbf{A}^n\} = (z\mathbf{I} - \mathbf{A})^{-1} z \tag{6.4.14}$$

This provides a straightforward method for calculating the state transition matrix by familiar Z-transform techniques. It can also be calculated by using the Cayley-Hamilton Theorem or Sylvester's Theorem [29].

Combining (6.4.8) and (6.4.14) we see that

$$\mathbf{X}(z) = \boldsymbol{\Phi}(z)\mathbf{x}(0) + \boldsymbol{\Phi}(z) z^{-1} \mathbf{B}\mathbf{V}(z) \tag{6.4.15}$$

Using the convolution theorem and the fact that

$$\mathscr{L}^{-1}\{\boldsymbol{\Phi}(z) z^{-1}\} = \boldsymbol{\varphi}(nT - T) u(nT - T)$$

we find that the inverse transform of (6.4.15) can be written as

$$\mathbf{x}(kT) = \boldsymbol{\varphi}(kT)\mathbf{x}(0) + \sum_{i=0}^{k-1} \boldsymbol{\varphi}[(k-i-1)T]\mathbf{Bv}(iT) \qquad (6.4.16)$$

This is identical to (6.4.6) with $n = 0$.

The unforced behavior of the system depends on the location of the poles of the elements of

$$\boldsymbol{\Phi}(z) = (z\mathbf{I} - \mathbf{A})^{-1}z$$

Since

$$(z\mathbf{I} - \mathbf{A})^{-1} = \frac{\text{adj}\,(z\mathbf{I} - \mathbf{A})}{\det\,(z\mathbf{I} - \mathbf{A})}$$

where adj $(\cdot)$ denotes the classical adjoint, these poles can only occur at the roots of the polynomial

$$Q(z) = \det\,(z\mathbf{I} - \mathbf{A}) \qquad (6.4.17)$$

In linear algebra $Q(z)$ is known as the *characteristic polynomial* for $\mathbf{A}$ and its roots are called the *characteristic values* or *eigenvalues* of $\mathbf{A}$ [47]. We will call $Q(z)$ the characteristic polynomial for the system. As we have seen in Chapter 4, the behavior of the elements of $\boldsymbol{\varphi}(nT)$ depends on the location of the roots of the characteristic polynomial relative to the unit circle. If all the roots are inside the unit circle, the system is stable. If any root is outside the unit circle, the system is unstable. The tests developed in Chapter 5 can be applied to $Q(z)$ to check for stability.

Example 6.4.1

Let us consider a system with a single input and output that can be described by the state equation

$$\begin{bmatrix} x_1(nT+T) \\ x_2(nT+T) \end{bmatrix} = \begin{bmatrix} 0 & 1 \\ 0.11 & 1 \end{bmatrix} \begin{bmatrix} x_1(nT) \\ x_2(nT) \end{bmatrix} + \begin{bmatrix} 0 \\ 1 \end{bmatrix} v(nT) \qquad (6.4.18)$$

and the output equation

$$y(nT) = [0.11 \quad 1]\begin{bmatrix} x_1(nT) \\ x_2(nT) \end{bmatrix} + v(nT) \qquad (6.4.19)$$

For this system

$$\mathbf{A} = \begin{bmatrix} 0 & 1 \\ 0.11 & 1 \end{bmatrix}, \quad \mathbf{B} = \begin{bmatrix} 0 \\ 1 \end{bmatrix}, \quad \mathbf{C} = [0.11 \quad 1], \quad \text{and} \quad \mathbf{D} = [1]$$

$$(6.4.20)$$

The characteristic polynomial is

$$Q(z) = \det\,(z\mathbf{I} - \mathbf{A}) = \det\begin{bmatrix} z & -1 \\ -0.11 & z-1 \end{bmatrix}$$

$$= z^2 - z - 0.11 = (z+0.1)(z-1.1)$$

so that

$$\Phi(z) = \frac{z}{(z+0.1)(z-1.1)} \begin{bmatrix} z-1 & 1 \\ 0.11 & z \end{bmatrix} \tag{6.4.21}$$

Since $Q(z)$ has one root outside the unit circle at 1.1, the system is unstable. Taking the inverse transform, we find that

$$\varphi(nT) = \begin{bmatrix} \dfrac{11}{12}(-0.1)^n + \dfrac{1}{12}(1.1)^n & \dfrac{-5}{6}(-0.1)^n + \dfrac{5}{6}(1.1)^n \\ \dfrac{-0.55}{6}(-0.1)^n + \dfrac{0.55}{6}(1.1)^n & \dfrac{1}{12}(-0.1)^n + \dfrac{11}{12}(1.1)^n \end{bmatrix}$$

for $n \geq 0$. As a check it is easy to see that $\varphi(0) = \mathbf{I}$ and $\varphi(T) = \mathbf{A}$.

Suppose that $\mathbf{x}(0) = \mathbf{0}$ and that the input is the unit pulse $v(nT) = \delta_{n0}$ so that $V(z) = 1$. Then, according to (6.4.15)

$$\mathbf{X}(z) = \Phi(z) z^{-1} \mathbf{B} V(z)$$

$$= \frac{1}{(z+0.1)(z-1.1)} \begin{bmatrix} z-1 & 1 \\ 0.11 & z \end{bmatrix} \begin{bmatrix} 0 \\ 1 \end{bmatrix}$$

$$= \frac{1}{(z+0.1)(z-1.1)} \begin{bmatrix} 1 \\ z \end{bmatrix}$$

Taking the inverse transform we find that

$$\mathbf{x}(nT) = \tfrac{5}{6} \begin{bmatrix} (1.1)^{n-1} - (-0.1)^{n-1} \\ (1.1)^n - (-0.1)^n \end{bmatrix} \quad \text{for} \quad n > 0$$

and

$$y(nT) = \mathbf{C}\mathbf{x}(nT) + \mathbf{D}v(nT)$$

$$= \begin{cases} 1 & \text{for} \quad n = 0 \\ \tfrac{5}{6}(1.1)^{n+1} - \tfrac{5}{6}(-0.1)^{n+1} & \text{for} \quad n > 0 \end{cases}$$

Comparing the second row of (6.4.18) with (6.4.19) we see that

$$y(nT) = x_2(nT + T)$$

or

$$x_2(nT) = y(nT - T)$$

From the first row of (6.4.18) we see that

$$x_1(nT + T) = x_2(nT)$$

or

$$x_1(nT) = x_2(nT - T) = y(nT - 2T)$$

Substituting these results into (6.4.19) we find that the input and output are related by the difference equation

$$y(nT) - y(nT - T) - 0.11 y(nT - 2T) = v(nT)$$

Therefore, the pulse transfer function for the system is

$$\frac{Y(z)}{V(z)} = \frac{1}{1 - z^{-1} - 0.11 z^{-2}} \qquad \blacktriangleleft$$

In many applications, one is primarily interested in the pulse transfer functions between the inputs and outputs of a system. Letting $\mathbf{x}(0) = \mathbf{0}$ and substituting (6.4.8) into (6.4.9), we find that

$$\mathbf{Y}(z) = [\mathbf{C}(z\mathbf{I} - \mathbf{A})^{-1}\mathbf{B} + \mathbf{D}]\mathbf{V}(z) \qquad (6.4.22)$$

The matrix

$$\mathbf{G}(z) = \mathbf{C}(z\mathbf{I} - \mathbf{A})^{-1}\mathbf{B} + \mathbf{D} \qquad (6.4.23)$$

is known as the pulse transfer function matrix for the system since its ijth element is the transfer function between the ith output and jth input. For systems with a single input and single output, $\mathbf{G}(z)$ reduces to the ordinary scalar transfer function $Y(z)/V(z)$.

Example 6.4.2

Consider the system described in Example 6.4.1. From (6.4.20), (6.4.21), and (6.4.23) we see that

$$G(z) = \frac{1}{z^2 - z - 0.11} [0.11 \quad 1] \begin{bmatrix} z - 1 & 1 \\ 0.11 & z \end{bmatrix} \begin{bmatrix} 0 \\ 1 \end{bmatrix} + 1$$

$$= \frac{1}{1 - z^{-1} - 0.11 z^{-2}}$$

This same result was found in Example 6.4.1 by a different method. $\qquad \blacktriangleleft$

6.5 STATE SPACE REPRESENTATIONS FOR CONSTANT-COEFFICIENT, LINEAR, DIFFERENCE EQUATIONS

In this section we again examine structures for realizing a system that has the pulse transfer function

$$G(z) = \frac{\sum\limits_{k=0}^{N} a_k z^{-k}}{1 + \sum\limits_{k=1}^{N} b_k z^{-k}} \qquad (6.5.1)$$

or difference equation

$$y(nT) = \sum_{k=0}^{N} a_k v(nT - kT) - \sum_{k=1}^{N} b_k y(nT - kT) \tag{6.5.2}$$

relating its input and output. Some structures of this kind were discussed in Chapter 5. By assigning state variables to the outputs of the delay elements in the block diagrams for the structures, we derive different state space representations for the difference equation (6.5.2). A reason for studying a variety of realizations is to find those that are insensitive to coefficient truncation, finite word length arithmetic, and other deviations from the ideal in actual hardware. These problems are discussed in a later chapter.

In Section 5.5 we saw that the structure shown in Fig. 6.5.1 called the *type 1 direct form realization* has its input and output related by (6.5.2). As shown in Fig. 6.5.1, we will choose the state variables $x_1(nT), \ldots, x_N(nT)$ as the outputs of the delay elements. From the block diagram we see that

$$x_1(nT + T) = x_2(nT)$$
$$x_2(nT + T) = x_3(nT)$$

$$\cdot$$
$$\cdot \tag{6.5.3}$$
$$\cdot$$

$$x_{N-1}(nT + T) = x_N(nT)$$
$$x_N(nT + T) = -b_N x_1(nT) - b_{N-1} x_2(nT) - \cdots - b_1 x_N(nT) + v(nT)$$

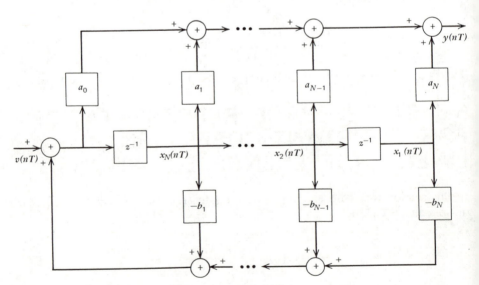

FIGURE 6.5.1. Type 1 direct form realization of $G(z)$.

and

$$y(nT) = a_N x_1(nT) + a_{N-1} x_2(nT) + \cdots + a_1 x_N(nT)$$
$$+ a_0[v(nT) - b_N x_1(nT) - b_{N-1} x_2(nT) - \cdots - b_1 x_N(nT)]$$

or

$$y(nT) = (a_N - a_0 b_N) x_1(nT) + (a_{N-1} - a_0 b_{N-1}) x_2(nT) + \cdots$$
$$+ (a_1 - a_0 b_1) x_N(nT) + a_0 v(nT) \tag{6.5.4}$$

Putting (6.5.3) and (6.5.4) into matrix form, we see that this structure can be described by the state equation

$$
\begin{bmatrix} x_1(nT+T) \\ x_2(nT+T) \\ \vdots \\ \vdots \\ x_{N-1}(nT+T) \\ x_N(nT+T) \end{bmatrix} =
\begin{bmatrix} 0 & 1 & 0 & \cdots & & 0 \\ 0 & 0 & 1 & 0 & \cdots & 0 \\ \vdots & & & & & \\ \vdots & & & & & \\ 0 & 0 & \cdots & & 0 & 1 \\ -b_N & -b_{N-1} & \cdots & & & -b_1 \end{bmatrix}
\begin{bmatrix} x_1(nT) \\ x_2(nT) \\ \vdots \\ \vdots \\ x_{N-1}(nT) \\ x_N(nT) \end{bmatrix} +
\begin{bmatrix} 0 \\ 0 \\ \vdots \\ \vdots \\ 0 \\ 1 \end{bmatrix} v(nT)
\tag{6.5.5}
$$

and output equation

$$y(nT) = [a_N - a_0 b_N \quad a_{N-1} - a_0 b_{N-1} \cdots a_1 - a_0 b_1]
\begin{bmatrix} x_1(nT) \\ x_2(nT) \\ \vdots \\ \vdots \\ x_N(nT) \end{bmatrix} + a_0 v(nT)
\tag{6.5.6}$$

These equations have the desired forms of (6.4.1) and (6.4.2).

In Section 5.5 we also saw that the structure called the *type 2 direct form* realization shown in Fig. 6.5.2 has its input and output related by the difference equation (6.5.2). If we choose the state variables as shown in Fig. 6.5.2, then it follows that

$$y(nT) = x_N(nT) + a_0 v(nT) \tag{6.5.7}$$

and

$$x_1(nT+T) = -b_N y(nT) + a_N v(nT) = -b_N x_N(nT) + (a_N - a_0 b_N) v(nT)$$
$$x_2(nT+T) = x_1(nT) - b_{N-1} x_N(nT) + (a_{N-1} - a_0 b_{N-1}) v(nT)$$

$$\vdots$$

$$\tag{6.5.8}$$

$$x_{N-1}(nT+T) = x_{N-2}(nT) - b_2 x_N(nT) + (a_2 - a_0 b_2) v(nT)$$
$$x_N(nT+T) = x_{N-1}(nT) - b_1 x_N(nT) + (a_1 - a_0 b_1) v(nT)$$

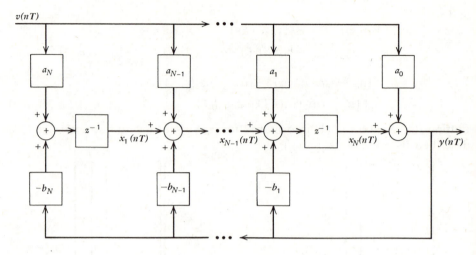

FIGURE 6.5.2. Type 2 direct form realization for $G(z)$.

Putting (6.5.7) and (6.5.8) into matrix form, we see that the type 2 direct form realization is described by the state equation

$$
\begin{bmatrix}
x_1(nT+T) \\
x_2(nT+T) \\
\cdot \\
\cdot \\
\cdot \\
x_{N-1}(nT+T) \\
x_N(nT+T)
\end{bmatrix}
=
\begin{bmatrix}
0 & 0 & \cdots & & 0 & -b_N \\
1 & 0 & \cdots & & 0 & -b_{N-1} \\
\cdot & & & & & \\
\cdot & & & & & \\
\cdot & & & & & \\
0 & 0 & \cdots & 0 & 1 & 0 & -b_2 \\
0 & 0 & \cdots & & 0 & 1 & -b_1
\end{bmatrix}
\begin{bmatrix}
x_1(nT) \\
x_2(nT) \\
\cdot \\
\cdot \\
\cdot \\
x_{N-1}(nT) \\
x_N(nT)
\end{bmatrix}
+
\begin{bmatrix}
a_N - a_0 b_N \\
a_{N-1} - a_0 b_{N-1} \\
\cdot \\
\cdot \\
\cdot \\
a_2 - a_0 b_2 \\
a_1 - a_0 b_1
\end{bmatrix}
v(nT)
$$

$$(6.5.9)$$

and output equation

$$
y(nT) = [0 \cdots 0 \quad 1]
\begin{bmatrix}
x_1(nT) \\
\cdot \\
\cdot \\
\cdot \\
x_N(nT)
\end{bmatrix}
+ a_0 v(nT)
\qquad (6.5.10)
$$

Another structure that can be used to realize the difference equation (6.5.2) is shown in Fig. 6.5.3. This is sometimes called the *standard form* realization. We will now see how to choose the parameters $\alpha_0, \ldots, \alpha_N$ and $\beta_1, \ldots, \beta_N$ to obtain the proper input–output relationship. From the block diagram it is clear

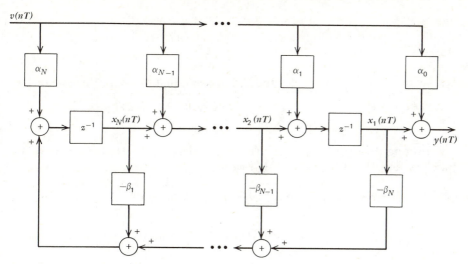

FIGURE 6.5.3.   Standard form realization.

that

$$x_1(nT+T) = x_2(nT) + \alpha_1 v(nT)$$

$$\cdot$$
$$\cdot$$
$$\cdot$$

$$x_k(nT+T) = x_{k+1}(nT) + \alpha_k v(nT) \qquad \text{for} \qquad 1 \le k \le N-1 \qquad (6.5.11)$$

$$\cdot$$
$$\cdot$$
$$\cdot$$

$$x_N(nT+T) = -\beta_N x_1(nT) - \beta_{N-1} x_2(nT) - \cdots - \beta_1 x_N(nT) + \alpha_N v(nT)$$

and

$$y(nT) = x_1(nT) + \alpha_0 v(nT) \qquad (6.5.12)$$

From (6.5.12) we see that

$$y(nT+T) = x_1(nT+T) + \alpha_0 v(nT+T)$$

and using the expression for $x_1(nT+T)$ from (6.5.11) that

$$y(nT+T) = x_2(nT) + \alpha_1 v(nT) + \alpha_0 v(nT+T) \qquad (6.5.13)$$

Similarly

$$y(nT+2T) = x_3(nT) + \alpha_2 v(nT) + \alpha_1 v(nT+T) + \alpha_0 v(nT+2T)$$

$$\cdot$$
$$\cdot$$

$$(6.5.14)$$

$$y[nT+(N-1)T] = x_N(nT) + \alpha_{N-1} v(nT) + \alpha_{N-2} v(nT+T) + \cdots$$
$$+ \alpha_0 v[nT+(N-1)T]$$

and

$$y(nT+NT) = -\beta_N x_1(nT) - \beta_{N-1} x_2(nT) - \cdots - \beta_1 x_N(nT)$$
$$+ \alpha_N v(nT) + \cdots + \alpha_0 v(nT+NT) \qquad (6.5.15)$$

Replacing n by $n+N$, the desired difference equation (6.5.2) becomes

$$y(nT+NT) = -b_1 y[nT+(N-1)T] - b_2 y[nT+(N-2)T] - \cdots - b_N y(nT)$$
$$+ a_0 v(nT+NT) + \cdots + a_N v(nT) \qquad (6.5.16)$$

Substituting the expressions for $y(nT), \ldots, y[nT+(N-1)T]$ given by (6.5.12), (6.5.13), and (6.5.14) into (6.5.16) we get

$$y(nT+NT) = -b_1\{x_N(nT) + \alpha_{N-1} v(nT) + \alpha_{N-2} v(nT+T) + \cdots$$
$$+ \alpha_0 v[nT+(N-1)T]\}$$
$$- b_2\{x_{N-1}(nT) + \alpha_{N-2} v(nT) + \cdots + \alpha_0 v[nT+(N-2)T]\}$$

$$\cdot$$
$$\cdot$$
$$\cdot$$

$$- b_N\{x_1(nT) + \alpha_0 v(nT)\}$$
$$+ a_0 v(nT+NT) + \cdots + a_N v(nT) \qquad (6.5.17)$$

Equating the coefficients of $x_1(nT), \cdots, x_N(nT)$ and $v(nT), \ldots, v(nT+NT)$ in (6.5.15) and (6.5.17), we see that the standard form realization parameters must be

$$\beta_k = b_k \qquad \text{for} \qquad k = 1, \ldots, N \qquad (6.5.18)$$

and

$$\alpha_0 = a_0$$
$$\alpha_1 = a_1 - b_1 \alpha_0$$
$$\alpha_2 = a_2 - b_2 \alpha_0 - b_1 \alpha_1 \qquad (6.5.19)$$
$$\cdot$$
$$\cdot$$
$$\cdot$$
$$\alpha_N = a_N - b_N \alpha_0 - b_{N-1} \alpha_1 - \cdots - b_1 \alpha_{N-1}$$

The set of equations in (6.5.19) is equivalent to

$$
\begin{bmatrix} a_0 \\ a_1 \\ a_2 \\ \cdot \\ \cdot \\ \cdot \\ a_N \end{bmatrix}
=
\begin{bmatrix}
1 & 0 & 0 & 0 & \cdots & 0 \\
b_1 & 1 & 0 & 0 & \cdots & 0 \\
b_2 & b_1 & 1 & 0 & \cdots & 0 \\
\cdot & & & & & \cdot \\
\cdot & & & & & \cdot \\
\cdot & & & & & \cdot \\
b_N & b_{N-1} & b_{N-2} & b_{N-3} & \cdots & b_1 & 1
\end{bmatrix}
\begin{bmatrix} \alpha_0 \\ \alpha_1 \\ \alpha_2 \\ \cdot \\ \cdot \\ \cdot \\ \alpha_N \end{bmatrix}
$$

so that

$$
\begin{bmatrix} \alpha_0 \\ \alpha_1 \\ \alpha_2 \\ \cdot \\ \cdot \\ \cdot \\ \alpha_N \end{bmatrix} = \begin{bmatrix} 1 & 0 & 0 & 0 & \cdots & 0 \\ b_1 & 1 & 0 & 0 & \cdots & 0 \\ b_2 & b_1 & 1 & 0 & \cdots & 0 \\ \cdot & & & & & \\ \cdot & & & & & \\ \cdot & & & & & \\ b_N & b_{N-1} & b_{N-2} & b_{N-3} & \cdots & b_1 \ 1 \end{bmatrix}^{-1} \begin{bmatrix} a_0 \\ a_1 \\ a_2 \\ \cdot \\ \cdot \\ \cdot \\ a_N \end{bmatrix}
\tag{6.5.20}
$$

Equations 6.5.19 provide a convenient iterative solution to (6.5.20). Putting (6.5.11) and (6.5.12) into matrix form, we find that the standard form realization has the state equation

$$
\begin{bmatrix} x_1(nT+T) \\ x_2(nT+T) \\ \cdot \\ \cdot \\ x_{N-1}(nT+T) \\ x_N(nT+T) \end{bmatrix} = \begin{bmatrix} 0 & 1 & 0 & \cdots & & 0 \\ 0 & 0 & 1 & 0 & \cdots & 0 \\ \cdot & \cdot & & & & \cdot \\ \cdot & \cdot & & & & \cdot \\ 0 & 0 & \cdots & & 0 & 1 \\ -b_N & -b_{N-1} & \cdots & & -b_2 & -b_1 \end{bmatrix} \begin{bmatrix} x_1(nT) \\ x_2(nT) \\ \cdot \\ \cdot \\ x_{N-1}(nT) \\ x_N(nT) \end{bmatrix} + \begin{bmatrix} \alpha_1 \\ \alpha_2 \\ \cdot \\ \cdot \\ \alpha_{N-1} \\ \alpha_N \end{bmatrix} v(nT)
\tag{6.5.21}
$$

and output equation

$$
y(nT) = \begin{bmatrix} 1 & 0 \cdots 0 \end{bmatrix} \begin{bmatrix} x_1(nT) \\ \cdot \\ \cdot \\ \cdot \\ x_N(nT) \end{bmatrix} + \alpha_0 v(nT)
\tag{6.5.22}
$$

Another state space representation can be obtained by the partial fraction technique. This method results in a *parallel form* structure. First, let us assume that $G(z)$ has N simple poles located at $p_1, \ldots, p_N$. Then $G(z)$ can be expressed as

$$
G(z) = d_0 + \sum_{k=1}^{N} \frac{d_k}{z - p_k}
\tag{6.5.23}
$$

where

$$
d_0 = \lim_{z \to \infty} G(z) = a_0
$$

and

$$
d_k = \lim_{z \to p_k} (z - p_k) G(z) \qquad \text{for} \qquad k = 1, \ldots, N
$$

Therefore

$$
Y(z) = G(z) V(z) = a_0 V(z) + \sum_{k=1}^{N} d_k \frac{V(z)}{z - p_k}
\tag{6.5.24}
$$

Letting

$$X_k(z) = \frac{V(z)}{z - p_k} \qquad \text{for} \qquad k = 1, \ldots, N \tag{6.5.25}$$

$Y(z)$ becomes

$$Y(z) = a_0 V(z) + \sum_{k=1}^{N} d_k X_k(z) \tag{6.5.26}$$

The time-domain equivalents of (6.5.25) and (6.5.26) are

$$x_k(nT + T) = p_k x_k(nT) + v(nT) \qquad \text{for} \qquad k = 1, \ldots, N \tag{6.5.27}$$

and

$$y(nT) = \sum_{k=1}^{N} d_k x_k(nT) + a_0 v(nT) \tag{6.5.28}$$

Putting (6.5.27) and (6.5.28) into matrix form, we find that the difference equation (6.5.2) can be represented by the state equation

$$\begin{bmatrix} x_1(nT+T) \\ x_2(nT+T) \\ \cdot \\ \cdot \\ \cdot \\ x_N(nT+T) \end{bmatrix} = \begin{bmatrix} p_1 & 0 & 0 & \cdots & 0 \\ 0 & p_2 & 0 & \cdots & 0 \\ \cdot & \cdot & \cdot & & \\ \cdot & \cdot & \cdot & & \\ \cdot & \cdot & \cdot & & \\ 0 & 0 & 0 & \cdots & p_N \end{bmatrix} \begin{bmatrix} x_1(nT) \\ x_2(nT) \\ \cdot \\ \cdot \\ \cdot \\ x_N(nT) \end{bmatrix} + \begin{bmatrix} 1 \\ 1 \\ \cdot \\ \cdot \\ \cdot \\ 1 \end{bmatrix} v(nT) \tag{6.5.29}$$

and output equation

$$y(nT) = [d_1 \quad d_2 \cdots d_N] \begin{bmatrix} x_1(nT) \\ x_2(nT) \\ \cdot \\ \cdot \\ \cdot \\ x_N(nT) \end{bmatrix} + a_0 v(nT) \tag{6.5.30}$$

This is known as the *normal form* representation of (6.5.2). In this representation the "**A**" matrix is diagonal so that the state variables are uncoupled.

The partial fraction technique can still be used if $G(z)$ has some poles that are not simple. To illustrate the method, let us assume that $G(z)$ has a pole of order r at p_1 and simple poles at $p_{r+1}, \ldots, p_N$. Then $G(z)$ can be expressed as

$$G(z) = d_0 + \sum_{k=1}^{r} \frac{d_k}{(z - p_1)^{r-k+1}} + \sum_{k=r+1}^{N} \frac{d_k}{(z - p_k)} \tag{6.5.31}$$

where

$$d_0 = \lim_{z \to \infty} G(z) = a_0$$

and

$$d_k = \begin{cases} \lim\limits_{z \to p_1} \dfrac{1}{(k-1)!} \dfrac{d^{k-1}}{dz^{k-1}}[(z-p_1)^r G(z)] & \text{for} \quad 1 \le k \le r \\[2ex] \lim\limits_{z \to p_k} (z-p_k)G(z) & \text{for} \quad r+1 \le k \le N \end{cases}$$

Therefore

$$Y(z) = G(z)V(z) = a_0 V(z) + \sum_{k=1}^{r} d_k \frac{V(z)}{(z-p_1)^{r-k+1}} + \sum_{k=r+1}^{N} d_k \frac{V(z)}{z-p_k}$$

(6.5.32)

Letting

$$X_k(z) = \frac{V(z)}{(z-p_1)^{r-k+1}} \qquad \text{for} \quad 1 \le k \le r \tag{6.5.33}$$

and

$$X_k(z) = \frac{V(z)}{z-p_k} \qquad \text{for} \quad r+1 \le k \le N \tag{6.5.34}$$

$Y(z)$ becomes

$$Y(z) = a_0 V(z) + \sum_{k=1}^{N} d_k X_k(z) \tag{6.5.35}$$

Notice that

$$X_r(z) = \frac{V(z)}{z-p_1} \tag{6.5.36}$$

and

$$X_k(z) = \frac{X_{k+1}(z)}{z-p_1} \qquad \text{for} \quad 1 \le k \le r-1 \tag{6.5.37}$$

The time domain equivalents of (6.5.34), (6.5.35), (6.5.36), and (6.5.37) are

$$x_k(nT+T) = \begin{cases} p_1 x_k(nT) + x_{k+1}(nT) & \text{for} \quad 1 \le k \le r-1 \\ p_1 x_r(nT) + v(nT) & \text{for} \quad k = r \\ p_k x_k(nT) + v(nT) & \text{for} \quad r+1 \le k \le N \end{cases} \tag{6.5.38}$$

and

$$y(nT) = \sum_{k=1}^{N} d_k x_k(nT) + a_0 v(nT) \tag{6.5.39}$$

or

$$\begin{bmatrix} x_1(nT+T) \\ x_2(nT+T) \\ \vdots \\ x_{r-1}(nT+T) \\ x_r(nT+T) \\ x_{r+1}(nT+T) \\ x_{r+2}(nT+T) \\ \vdots \\ x_{N-1}(nT+T) \\ x_N(nT+T) \end{bmatrix} = \left[\begin{array}{ccccc:ccccc} p_1 & 1 & 0 & \cdots & & & & & & 0 \\ 0 & p_1 & 1 & 0 & \cdots & & & & & 0 \\ \vdots & & & & & & & \mathbf{0} & & \\ 0 & \cdots & & 0 & p_1 & 1 & & & & \\ 0 & \cdots & & & 0 & p_1 & & & & \\ \hdashline & & & & & p_{r+1} & 0 & \cdots & & 0 \\ & & & & & 0 & p_{r+2} & 0 & \cdots & 0 \\ & & \mathbf{0} & & & & & & & \\ & & & & & 0 & \cdots & 0 & p_{N-1} & 0 \\ & & & & & 0 & \cdots & & 0 & p_N \end{array} \right] \begin{bmatrix} x_1(nT) \\ x_2(nT) \\ \vdots \\ x_{r-1}(nT) \\ x_r(nT) \\ x_{r+1}(nT) \\ x_{r+2}(nT) \\ \vdots \\ x_{N-1}(nT) \\ x_N(nT) \end{bmatrix} + \begin{bmatrix} 0 \\ 0 \\ \vdots \\ 0 \\ 1 \\ 1 \\ 1 \\ \vdots \\ 1 \\ 1 \end{bmatrix} v(nT)$$

(6.5.40)

and

$$y(nT) = [d_1 \cdots d_N] \begin{bmatrix} x_1(nT) \\ \cdot \\ \cdot \\ \cdot \\ x_N(nT) \end{bmatrix} + a_0 v(nT) \qquad (6.5.41)$$

The $r \times r$ block in the upper left-hand corner of the "**A**" matrix is called a *Jordan* block. The block diagram corresponding to this system realization is shown in Fig. 6.5.4. This structure is called a parallel form realization for obvious reasons.

The *cascade form* realization is frequently used in practice. As discussed in Chapter 5, this structure results when $G(z)$ is expressed as the product of low-order rational factors and is realized as a cascade of sections corresponding to these factors. This structure is particularly appropriate when $G(z)$ has zeros on or near the unit circle. To illustrate one form of cascade realization and the corresponding state space representation, let us assume that a_0 is not zero in

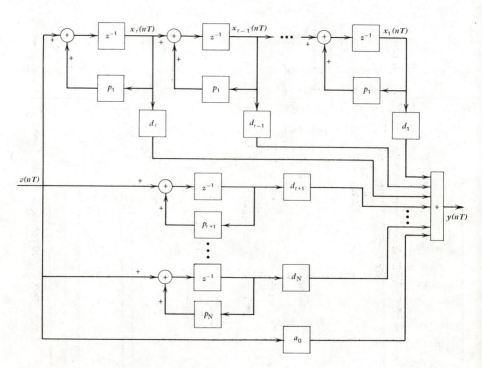

FIGURE 6.5.4. Parallel form realization.

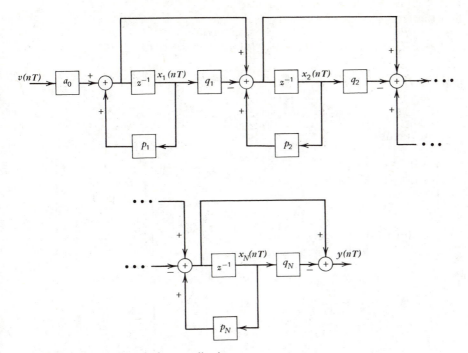

FIGURE 6.5.5.   A cascade form realization.

(6.5.1). Then $G(z)$ can be factored and written as

$$G(z) = a_0 \prod_{k=1}^{N} \frac{z - q_k}{z - p_k} \qquad (6.5.42)$$

When each first-order rational factor of (6.5.42) is realized by a type 1 direct form section and the output and input adders of adjacent sections are combined, we obtain the structure shown in Fig. 6.5.5. Choosing the state variables as shown in this figure we find that

$$x_1(nT + T) = p_1 x_1(nT) + a_0 v(nT) \qquad (6.5.43)$$

$$x_k(nT + T) = p_k x_k(nT) - q_{k-1} x_{k-1}(nT) + x_{k-1}(nT + T)$$

$$\text{for} \qquad 2 \le k \le N \quad (6.5.44)$$

and

$$y(nT) = x_N(nT + T) - q_N x_N(nT) \qquad (6.5.45)$$

Starting with $k = 2$, using (6.5.43), and recursively evaluating (6.5.44) we also find that

$$x_k(nT+T) = p_k x_k(nT) + \sum_{r=1}^{k-1} (p_r - q_r)x_r(nT) + a_0 v(nT) \qquad \text{for} \qquad 2 \le k \le N$$

(6.5.46)

and

$$y(nT) = \sum_{k=1}^{N} (p_k - q_k)x_k(nT) + a_0 v(nT)$$

(6.5.47)

or, equivalently

$$
\begin{bmatrix}
x_1(nT+T) \\
x_2(nT+T) \\
\vdots \\
\vdots \\
x_{N-1}(nT+T) \\
x_N(nT+T)
\end{bmatrix}
=
\begin{bmatrix}
p_1 & 0 & 0 & \cdots & & 0 \\
p_1-q_1 & p_2 & 0 & & & 0 \\
\vdots & & & & & \vdots \\
\vdots & & & & & \vdots \\
p_1-q_1 & p_2-q_2 & \cdots & & p_{N-1} & 0 \\
p_1-q_1 & p_2-q_2 & \cdots & & p_{N-1}-q_{N-1} & p_N
\end{bmatrix}
\begin{bmatrix}
x_1(nT) \\
x_2(nT) \\
\vdots \\
\vdots \\
x_{N-1}(nT) \\
x_N(nT)
\end{bmatrix}
+
\begin{bmatrix}
a_0 \\
a_0 \\
\vdots \\
\vdots \\
a_0 \\
a_0
\end{bmatrix}
v(nT)
$$

(6.5.48)

and

$$
y(nT) = [p_1-q_1 \cdots p_N-q_N]
\begin{bmatrix}
x_1(nT) \\
\vdots \\
\vdots \\
\vdots \\
x_N(nT)
\end{bmatrix}
+ a_0 v(nT)
$$

(6.5.49)

In practice, the state variables would most likely be calculated recursively by (6.5.44) rather than directly from (6.5.48). This corresponds to calculating the outputs of the adders in Fig. 6.5.5 sequentially from left to right.

If $G(z)$ has any complex poles or zeros, the cascade form realization shown in Fig. 6.5.5 requires complex arithmetic. The parallel form realization shown in Fig. 6.5.4 also requires complex arithmetic when $G(z)$ has complex poles. The need for complex arithmetic is frequently eliminated by combining complex conjugate terms into low-order sections with real coefficients. These sections are then implemented as direct or standard form realizations.

Various other structures have been suggested for realizing rational pulse transfer functions. In particular, there has been recent interest in realizations using various types of ladder structures [19,27,40,50,94,95]. These will not be discussed further here.

It should be clear by now that there are an infinite number of realizations for $G(z)$. Some have basically different structures while others differ simply by scale factors. In general, an input–output relationship does not uniquely describe the internal structure of a system. If a realization is described by the equations

$$\mathbf{x}(nT+T) = \mathbf{A}\mathbf{x}(nT) + \mathbf{B}\mathbf{v}(nT)$$

(6.5.50)

and

$$\mathbf{y}(nT) = \mathbf{C}\mathbf{x}(nT) + \mathbf{D}\mathbf{v}(nT) \tag{6.5.51}$$

then for any $N \times N$ nonsingular matrix $\mathbf{F}$ the transformation

$$\mathbf{x}(nT) = \mathbf{F}\mathbf{x}'(nT)$$

results in a new realization described by the equations

$$\mathbf{x}'(nT + T) = \mathbf{A}'\mathbf{x}'(nT) + \mathbf{B}'\mathbf{v}(nT) \tag{6.5.52}$$

and

$$\mathbf{y}(nT) = \mathbf{C}'\mathbf{x}'(nT) + \mathbf{D}'\mathbf{v}(nT) \tag{6.5.53}$$

where

$$\mathbf{A}' = \mathbf{F}^{-1}\mathbf{A}\mathbf{F}$$

$$\mathbf{B}' = \mathbf{F}^{-1}\mathbf{B}$$

$$\mathbf{C}' = \mathbf{C}\mathbf{F}$$

and

$$\mathbf{D}' = \mathbf{D}$$

6.6 TRANSPOSE SYSTEMS

Suppose that a system with m inputs and r outputs is described by the state equation

$$\mathbf{x}(nT + T) = \mathbf{A}\mathbf{x}(nT) + \mathbf{B}\mathbf{v}(nT) \tag{6.6.1}$$

and output equation

$$\mathbf{y}(nT) = \mathbf{C}\mathbf{x}(nT) + \mathbf{D}\mathbf{v}(nT) \tag{6.6.2}$$

Then its transpose configuration is defined to be the system with r inputs and m outputs described by the state equation

$$\mathbf{x}'(nT + T) = \mathbf{A}^t\mathbf{x}'(nT) + \mathbf{C}^t\mathbf{v}'(nT) \tag{6.6.3}$$

and output equation

$$\mathbf{y}'(nT) = \mathbf{B}^t\mathbf{x}'(nT) + \mathbf{D}^t\mathbf{v}'(nT) \tag{6.6.4}$$

where the superscript t denotes transpose [57]. From (6.5.5), (6.5.6), (6.5.9), and (6.5.10) we see that the type 1 and type 2 direct form realizations are transpose pairs.

According to (6.4.23), the pulse transfer function matrix for the original system is

$$\mathbf{H}(z) = \mathbf{C}(z\mathbf{I} - \mathbf{A})^{-1}\mathbf{B} + \mathbf{D} \tag{6.6.5}$$

and for the transpose configuration is

$$\mathbf{H}_t(z) = \mathbf{B}^t(z\mathbf{I} - \mathbf{A}^t)^{-1}\mathbf{C}^t + \mathbf{D}^t = \mathbf{H}^t(z) \tag{6.6.6}$$

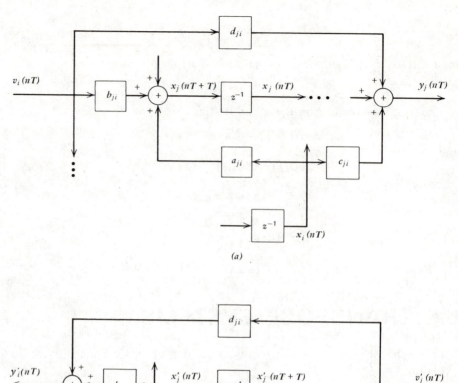

FIGURE 6.6.1. Partial block diagrams for a system and its transpose configuration. (a) Original system; (b) transpose configuration.

Thus the pulse transfer function matrices for a system and its transpose configuration are transposes. Consequently, a system with a single input and a single output and its transpose configuration have identical pulse transfer functions.

It is interesting to observe that a block diagram for the transpose configuration can be generated directly from the block diagram for the original system. First, the block diagram for the original system must be arranged so that the inputs to the adders all have positive polarity. A block diagram for the transpose configuration is then obtained by reversing the signal flow direction in all branches and by replacing adders by branch nodes and branch nodes by adders. The new block diagram must be relabeled by changing the original outputs $y_k(nT)$ to the new inputs $v'_k(nT)$ for $1 \le k \le r$, the original inputs $v_k(nT)$ to the new outputs $y'_k(nT)$ for $1 \le k \le m$, the original state variables $x_k(nT)$ to the new advanced state variables $x'_k(nT+T)$ for $1 \le k \le N$, and the original advanced state variables $x_k(nT+T)$ to the new state variables $x'_k(nT)$ for $1 \le k \le N$. This transformation is illustrated by the partial system block diagrams in Fig. 6.6.1. Also consider Figs. 6.5.1 and 6.5.2 showing the type 1 and type 2 direct form realizations for another example. From Fig. 6.6.1b we can see that for the system corresponding to this new block diagram

$$x'_i(nT+T) = \sum_{j=1}^{N} a_{ji}x'_j(nT) + \sum_{j=1}^{r} c_{ji}v'_j(nT) \qquad \text{for} \qquad i=1,\ldots,N \quad (6.6.7)$$

and

$$y'_i(nT) = \sum_{j=1}^{N} b_{ji}x'_j(nT) + \sum_{j=1}^{r} d_{ji}v'_j(nT) \qquad \text{for} \qquad i=1,\ldots,m \quad (6.6.8)$$

The set of equations (6.6.7) and (6.6.8) can also be written as the pair of matrix equations (6.6.3) and (6.6.4) that were used to define the transpose configuration.

We have studied the concept of the transpose configuration because it sometimes provides an easy method for generating a new structure for realizing a pulse transfer function. However, the transpose configuration may differ trivially or may be identical to the original structure.

6.7 LINEAR, CONTINUOUS-TIME, DYNAMICAL SYSTEMS

The set of continuous-time systems that can be described by linear differential equations is an important class of dynamical systems. These systems are frequently called linear, continuous-time, dynamical systems. If, as usual, we denote the input vector by $\mathbf{v}(t)$, the output vector by $\mathbf{y}(t)$, and the state vector by $\mathbf{x}(t)$, then a linear, continuous-time, dynamical system is described by a state equation that is a linear differential equation of the form

$$\dot{\mathbf{x}}(t) = \mathbf{A}(t)\mathbf{x}(t) + \mathbf{B}(t)\mathbf{v}(t) \tag{6.7.1}$$

and an output equation of the form

$$\mathbf{y}(t) = \mathbf{C}(t)\mathbf{x}(t) + \mathbf{D}(t)\mathbf{v}(t) \tag{6.7.2}$$

where

$\mathbf{x}(t)$ is an N-dimensional column vector
$\mathbf{v}(t)$ is an m-dimensional column vector
$\mathbf{y}(t)$ is an r-dimensional column vector
$\mathbf{A}(t)$ is a $N \times N$ matrix
$\mathbf{B}(t)$ is an $N \times m$ matrix
$\mathbf{C}(t)$ is an $r \times N$ matrix
$\mathbf{D}(t)$ is an $r \times m$ matrix

The derivative of a matrix is defined to be that matrix whose elements are the derivatives of the corresponding elements of the original matrix. In particular

$$\dot{\mathbf{x}}(t) = \frac{d}{dt} \begin{bmatrix} x_1(t) \\ \cdot \\ \cdot \\ \cdot \\ x_N(t) \end{bmatrix} = \begin{bmatrix} \dot{x}_1(t) \\ \cdot \\ \cdot \\ \cdot \\ \dot{x}_N(t) \end{bmatrix}$$

The integral of a matrix is defined in a similar way. The state and output equations can be represented by the block diagram shown in Fig. 6.7.1. Notice the strong similarity of this block diagram to the block diagram for a linear discrete-time system given in Fig. 6.3.1.

It can be shown [22] that when $\mathbf{A}(t)$ is continuous over the interval (t_1, t_2) and $\mathbf{B}(t)$ and $\mathbf{v}(t)$ are piecewise continuous over (t_1, t_2), the differential equation (6.7.1) has a unique solution over (t_1, t_2) passing through a specified state $\mathbf{x}_0$ at time $t_0 \in (t_1, t_2)$ and, in addition, the solution is continuous over (t_1, t_2).

When the input is identically zero, the state equation reduces to the homogeneous differential equation

$$\dot{\mathbf{x}}(t) = \mathbf{A}(t)\mathbf{x}(t) \tag{6.7.3}$$

It can be shown [22] that if $\mathbf{A}(t)$ is continuous over the interval (t_1, t_2), then N nontrivial linearly independent solutions of (6.7.3) over (t_1, t_2) can be found. Let us denote a set of these solutions by $\mathbf{x}_1(t), \ldots, \mathbf{x}_N(t)$. This set is not unique. By the solutions being linearly independent we mean that the only set of

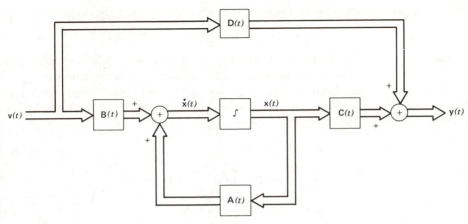

FIGURE 6.7.1.   Block diagram for a system described by a linear differential equation.

scalars $c_1, \ldots, c_N$ such that

$$c_1\mathbf{x}_1(t) + \cdots + c_N\mathbf{x}_N(t) = \mathbf{0}$$

for all $t \in (t_1, t_2)$ is $c_1 = \cdots = c_N = 0$. Furthermore, it can be shown that for each $t \in (t_1, t_2)$, the set of N solutions spans an N-dimensional vector space. This means that for each $t \in (t_1, t_2)$ and each N-dimensional column vector $\mathbf{f}$ a set of scalars $c_1(t), \ldots, c_N(t)$ can be found such that

$$\mathbf{f} = \sum_{k=1}^{N} c_k(t)\mathbf{x}_k(t)$$

Any solution of the homogeneous differential equation can be expressed as a linear combination of $\mathbf{x}_1(t), \ldots, \mathbf{x}_N(t)$.

The $N \times N$ matrix $\mathbf{\Psi}(t)$ whose columns are the N linearly independent solutions, that is

$$\mathbf{\Psi}(t) = [\mathbf{x}_1(t) \quad \mathbf{x}_2(t) \cdots \mathbf{x}_N(t)] \qquad (6.7.4)$$

is called a *fundamental* matrix for the system. The statement that the solutions span an N dimensional vector space for each $t \in (t_1, t_2)$ is equivalent to the statement

$$\det \mathbf{\Psi}(t) \neq 0 \qquad \text{for} \qquad t \in (t_1, t_2) \qquad (6.7.5)$$

From (6.7.3) we can see that a fundamental matrix for the system must satisfy the differential equation

$$\dot{\mathbf{\Psi}}(t) = \mathbf{A}(t)\mathbf{\Psi}(t) \qquad \text{for} \qquad t \in (t_1, t_2) \qquad (6.7.6)$$

In general, the solution to the state equation (6.7.1) has the form

$$\mathbf{x}(t) = \mathbf{h}(t) + \mathbf{p}(t) \tag{6.7.7}$$

where $\mathbf{h}(t)$ is a solution to the homogeneous equation (6.7.3) and $\mathbf{p}(t)$ is a solution to (6.7.1). The function $\mathbf{h}(t)$ is called a *homogeneous* solution and $\mathbf{p}(t)$ is called a *particular* solution. If, somehow, a fundamental matrix $\mathbf{\Psi}(t)$ has been found, then a particular solution can be found by a method known as *variation of parameters*. Let us assume that

$$\mathbf{p}(t) = \mathbf{\Psi}(t)\,\mathbf{d}(t) \tag{6.7.8}$$

where

$$\mathbf{d}(t) = [d_1(t) \quad d_2(t) \cdots d_N(t)]^t$$

Then

$$\dot{\mathbf{p}}(t) = \dot{\mathbf{\Psi}}(t)\,\mathbf{d}(t) + \mathbf{\Psi}(t)\,\dot{\mathbf{d}}(t) \tag{6.7.9}$$

and

$$\mathbf{A}(t)\mathbf{p}(t) = \mathbf{A}(t)\mathbf{\Psi}(t)\,\mathbf{d}(t) \tag{6.7.10}$$

Subtracting (6.7.10) from (6.7.9) gives

$$\dot{\mathbf{p}}(t) - \mathbf{A}(t)\mathbf{p}(t) = [\dot{\mathbf{\Psi}}(t) - \mathbf{A}(t)\mathbf{\Psi}(t)]\,\mathbf{d}(t) + \mathbf{\Psi}(t)\,\dot{\mathbf{d}}(t) \tag{6.7.11}$$

Since $\mathbf{p}(t)$ is a solution of (6.7.1), the left-hand side of (6.7.11) must be $\mathbf{B}(t)\mathbf{v}(t)$. From (6.7.6) we can see that the first term on the right-hand side of (6.7.11) is zero. Therefore

$$\mathbf{B}(t)\mathbf{v}(t) = \mathbf{\Psi}(t)\dot{\mathbf{d}}(t)$$

or

$$\dot{\mathbf{d}}(t) = \mathbf{\Psi}^{-1}(t)\mathbf{B}(t)\mathbf{v}(t) \tag{6.7.12}$$

If we require that $\mathbf{d}(t_0) = \mathbf{0}$, then by integrating (6.7.12) we find that

$$\mathbf{d}(t) = \int_{t_0}^{t} \mathbf{\Psi}^{-1}(\tau)\mathbf{B}(\tau)\mathbf{v}(\tau)\, d\tau \tag{6.7.13}$$

Thus

$$\mathbf{p}(t) = \mathbf{\Psi}(t)\,\mathbf{d}(t) = \int_{t_0}^{t} \mathbf{\Psi}(t)\mathbf{\Psi}^{-1}(\tau)\mathbf{B}(\tau)\mathbf{v}(\tau)\, d\tau \tag{6.7.14}$$

is a particular solution with $\mathbf{p}(t_0) = \mathbf{0}$. The homogeneous solution must be a linear combination of the columns of $\mathbf{\Psi}(t)$ and consequently must have the form

$$\mathbf{h}(t) = \mathbf{\Psi}(t)\mathbf{f} \tag{6.7.15}$$

where $\mathbf{f}$ is a constant N-dimensional column vector. Therefore, the total solution takes the form

$$\mathbf{x}(t) = \mathbf{\Psi}(t)\mathbf{f} + \mathbf{p}(t)$$

so that

$$\mathbf{x}(t_0) = \mathbf{\Psi}(t_0)\mathbf{f}$$

and
$$\mathbf{f} = \mathbf{\Psi}^{-1}(t_0)\mathbf{x}(t_0) \tag{6.7.16}$$

since we have chosen $\mathbf{p}(t_0)$ to be zero. Combining these results, we find that the total solution is

$$\mathbf{x}(t) = \mathbf{\Psi}(t)\mathbf{\Psi}^{-1}(t_0)\mathbf{x}(t_0) + \int_{t_0}^{t} \mathbf{\Psi}(t)\mathbf{\Psi}^{-1}(\tau)\mathbf{B}(\tau)\mathbf{v}(\tau)\,d\tau \tag{6.7.17}$$

The $N \times N$ matrix
$$\boldsymbol{\varphi}(t, \tau) = \mathbf{\Psi}(t)\mathbf{\Psi}^{-1}(\tau) \tag{6.7.18}$$

is called the state transition matrix for the system. From this definition we can see that the state transition matrix has the following properties:

1. $\quad\quad\quad\quad \boldsymbol{\varphi}(t, t) = \mathbf{I}$ $\hspace{4cm}$ (6.7.19)
2. $\quad\quad\quad\quad \boldsymbol{\varphi}(t, \tau) = \boldsymbol{\varphi}^{-1}(\tau, t)$ $\hspace{3cm}$ (6.7.20)
3. $\quad\quad\quad\quad \boldsymbol{\varphi}(t_3, t_2)\boldsymbol{\varphi}(t_2, t_1) = \boldsymbol{\varphi}(t_3, t_1)$ $\hspace{2cm}$ (6.7.21)

These properties are the same as those of the state transition matrix for a linear, discrete-time, dynamical system. Since

$$\dot{\boldsymbol{\varphi}}(t, \tau) = \dot{\mathbf{\Psi}}(t)\mathbf{\Psi}^{-1}(\tau) = \mathbf{A}(t)\mathbf{\Psi}(t)\mathbf{\Psi}^{-1}(\tau) = \mathbf{A}(t)\boldsymbol{\varphi}(t, \tau) \tag{6.7.22}$$

the state transition matrix for each τ is the unique fundamental matrix satisfying the condition $\boldsymbol{\varphi}(\tau, \tau) = \mathbf{I}$. The solution (6.7.17) of the state equation can be written in terms of the state transition matrix as

$$\mathbf{x}(t) = \boldsymbol{\varphi}(t, t_0)\mathbf{x}(t_0) + \int_{t_0}^{t} \boldsymbol{\varphi}(t, \tau)\mathbf{B}(\tau)\mathbf{v}(\tau)\,d\tau \tag{6.7.23}$$

This is the continuous time analog of the solution (6.3.11) of the state equation for a linear, discrete-time, dynamical system.

Example 6.7.1

Consider a system described by the scalar second-order differential equation

$$\ddot{y}(t) + \frac{2}{t}\,\dot{y}(t) - y(t) = \frac{1}{t}\,v(t) \tag{6.7.24}$$

for $t > 0$. If we let $x_1(t) = y(t)$ and $x_2(t) = \dot{y}(t)$, then

$$\begin{bmatrix} \dot{x}_1(t) \\ \dot{x}_2(t) \end{bmatrix} = \begin{bmatrix} 0 & 1 \\ 1 & -2/t \end{bmatrix}\begin{bmatrix} x_1(t) \\ x_2(t) \end{bmatrix} + \begin{bmatrix} 0 \\ 1/t \end{bmatrix}v(t) \tag{6.7.25}$$

and

$$y(t) = \begin{bmatrix} 1 & 0 \end{bmatrix}\begin{bmatrix} x_1(t) \\ x_2(t) \end{bmatrix} \tag{6.7.26}$$

By direct substitution, it can be shown that a fundamental matrix for (6.7.25) is

$$\mathbf{\Psi}(t) = \begin{bmatrix} \dfrac{e^{-t}}{t} & \dfrac{e^{t}}{t} \\[3mm] \dfrac{-e^{-t}}{t}\left(1+\dfrac{1}{t}\right) & \dfrac{e^{t}}{t}\left(1-\dfrac{1}{t}\right) \end{bmatrix} \qquad (6.7.27)$$

Therefore, the state transition matrix is

$$\boldsymbol{\varphi}(t, \tau) = \mathbf{\Psi}(t)\mathbf{\Psi}^{-1}(\tau)$$

$$= \frac{1}{2}\begin{bmatrix} \dfrac{\tau-1}{t}e^{\tau-t}+\dfrac{\tau+1}{t}e^{t-\tau} & \dfrac{-\tau}{t}e^{\tau-t}+\dfrac{\tau}{t}e^{t-\tau} \\[3mm] \dfrac{(1-\tau)(t+1)}{t^2}e^{\tau-t}+\dfrac{(\tau+1)(t-1)}{t^2}e^{t-\tau} & \dfrac{\tau(t+1)}{t^2}e^{\tau-t}+\dfrac{\tau(t-1)}{t^2}e^{t-\tau} \end{bmatrix}$$

$$(6.7.28)$$

As an exercise show that if $\mathbf{x}(t_0) = \mathbf{0}$ and $\mathbf{v}(t) = 1$ for $t > t_0$, then

$$y(t) = (e^{t-t_0} - e^{t_0-t})/t \qquad \text{for} \qquad t \geq t_0 > 0 \qquad \blacktriangleleft$$

6.8 TIME-INVARIANT, LINEAR, CONTINUOUS-TIME, DYNAMICAL SYSTEMS

When $\mathbf{A}$, $\mathbf{B}$, $\mathbf{C}$, and $\mathbf{D}$ are constant matrices in (6.7.1) and (6.7.2), we will call the system a time-invariant, linear, continuous-time, dynamical system. In this case the state equation becomes

$$\dot{\mathbf{x}}(t) = \mathbf{A}\mathbf{x}(t) + \mathbf{B}\mathbf{v}(t) \qquad (6.8.1)$$

and the output equation becomes

$$\mathbf{y}(t) = \mathbf{C}\mathbf{x}(t) + \mathbf{D}\mathbf{v}(t) \qquad (6.8.2)$$

Let us define the matrix exponential function $e^{\mathbf{A}t}$ by the series

$$e^{\mathbf{A}t} = \sum_{n=0}^{\infty} \frac{(\mathbf{A}t)^n}{n!} \qquad (6.8.3)$$

where by convention $(\mathbf{A}t)^0 = \mathbf{I}$. It can be shown that this series converges for all $\mathbf{A}t$. Using this series definition, it is not difficult to show that

$$e^{\mathbf{A}t_1}e^{\mathbf{A}t_2} = e^{\mathbf{A}(t_1+t_2)} \qquad (6.8.4)$$

and

$$(e^{\mathbf{A}t})^{-1} = e^{-\mathbf{A}t} \qquad (6.8.5)$$

We will see shortly how to find a closed form for $e^{\mathbf{A}t}$ by using the Laplace transform. Differentiating (6.8.3) we find that

$$\frac{d}{dt} e^{\mathbf{A}t} = \mathbf{A} \sum_{n=1}^{\infty} \frac{(\mathbf{A}t)^{n-1}}{(n-1)!} = \mathbf{A} e^{\mathbf{A}t} \tag{6.8.6}$$

Therefore, $e^{\mathbf{A}t}$ is a fundamental matrix for the system since it is the unique $N \times N$ matrix satisfying the homogeneous differential equation

$$\dot{\boldsymbol{\Psi}}(t) = \mathbf{A}\boldsymbol{\Psi}(t) \tag{6.8.7}$$

with $\boldsymbol{\Psi}(0) = \mathbf{I}$.

According to (6.7.18), the state transition matrix for the system is

$$\boldsymbol{\varphi}(t, \tau) = e^{\mathbf{A}t} e^{-\mathbf{A}\tau} = e^{\mathbf{A}(t-\tau)} \tag{6.8.8}$$

This is a function only of the time difference $t - \tau$. Consequently, it is customary to call the matrix

$$\boldsymbol{\varphi}(t) = e^{\mathbf{A}t} \tag{6.8.9}$$

the state transition matrix for the time-invariant system. From (6.7.23) we see that the total solution to the state equation is

$$\mathbf{x}(t) = e^{\mathbf{A}(t-t_0)} \mathbf{x}(t_0) + \int_{t_0}^{t} e^{\mathbf{A}(t-\tau)} \mathbf{B}\mathbf{v}(\tau) \, d\tau \tag{6.8.10}$$

The solution to the state equation can also be found by Laplace transform methods. Taking the one-sided Laplace transform of (6.8.1) yields

$$s\mathbf{X}(s) - \mathbf{x}(0) = \mathbf{A}\mathbf{X}(s) + \mathbf{B}\mathbf{V}(s)$$

so that

$$\mathbf{X}(s) = (s\mathbf{I} - \mathbf{A})^{-1}\mathbf{x}(0) + (s\mathbf{I} - \mathbf{A})^{-1}\mathbf{B}\mathbf{V}(s) \tag{6.8.11}$$

If we let

$$\boldsymbol{\Phi}(s) = (s\mathbf{I} - \mathbf{A})^{-1} \tag{6.8.12}$$

then the inverse transform of (6.8.11) can be written as

$$\mathbf{x}(t) = \boldsymbol{\varphi}(t)\mathbf{x}(0) + \int_{0}^{t} \boldsymbol{\varphi}(t - \tau)\mathbf{B}\mathbf{v}(\tau) \, d\tau \tag{6.8.13}$$

Comparing (6.8.13) to (6.8.10) with $t_0 = 0$, we see that

$$\boldsymbol{\varphi}(t) = e^{\mathbf{A}t}$$

Therefore, (6.8.12) is a closed form for the Laplace transform of the state transition matrix.

The natural behavior of the system depends on the poles of $\boldsymbol{\Phi}(s)$. Since

$$(s\mathbf{I} - \mathbf{A})^{-1} = \frac{\text{adj}\,(s\mathbf{I} - \mathbf{A})}{\det\,(s\mathbf{I} - \mathbf{A})} \tag{6.8.14}$$

the poles can only occur at the roots of the polynomial

$$Q(s) = \det(s\mathbf{I} - \mathbf{A}) \tag{6.8.15}$$

As in the discrete-time case, $Q(s)$ is called the characteristic polynomial for the system and its roots are called the eigenvalues of $\mathbf{A}$. The system will be stable if all the eigenvalues are in the left half plane.

Letting $\mathbf{x}(0) = \mathbf{0}$ in (6.8.11) and taking the transform of the output equation (6.8.2), we find that

$$\mathbf{Y}(s) = [\mathbf{C}(s\mathbf{I} - \mathbf{A})^{-1}\mathbf{B} + \mathbf{D}]\mathbf{V}(s) \tag{6.8.16}$$

Therefore, the transfer function matrix for the system is

$$\mathbf{G}(s) = \mathbf{C}(s\mathbf{I} - \mathbf{A})^{-1}\mathbf{B} + \mathbf{D} \tag{6.8.17}$$

Notice that $\mathbf{G}(s)$ is identical to the pulse transfer function matrix $\mathbf{G}(z)$ given by (6.4.23) if s is replaced by z.

Example 6.8.1

Suppose that a system with a single input and a single output is described by the state equation

$$\begin{bmatrix} \dot{x}_1(t) \\ \dot{x}_2(t) \end{bmatrix} = \begin{bmatrix} 0 & -2 \\ 1 & -3 \end{bmatrix} \begin{bmatrix} x_1(t) \\ x_2(t) \end{bmatrix} + \begin{bmatrix} 0 \\ -6 \end{bmatrix} v(t) \tag{6.8.18}$$

and the output equation

$$y(t) = \begin{bmatrix} 0 & 1 \end{bmatrix} \begin{bmatrix} x_1(t) \\ x_2(t) \end{bmatrix} + v(t) \tag{6.8.19}$$

The characteristic polynomial for the system is

$$Q(s) = \det(s\mathbf{I} - \mathbf{A}) = \det \begin{bmatrix} s & 2 \\ -1 & s+3 \end{bmatrix} = s^2 + 3s + 2$$
$$= (s+1)(s+2)$$

Therefore, the system eigenvalues are -1 and -2 so that the system is stable. According to (6.8.12)

$$\mathbf{\Phi}(s) = (s\mathbf{I} - \mathbf{A})^{-1} = \frac{1}{s^2 + 3s + 2} \begin{bmatrix} s+3 & -2 \\ 1 & s \end{bmatrix}$$

Taking the inverse Laplace transform, we find that the state transition matrix is

$$\boldsymbol{\varphi}(t) = e^{\mathbf{A}t} = \begin{bmatrix} 2e^{-t} - e^{-2t} & -2e^{-t} + 2e^{-2t} \\ e^{-t} - e^{-2t} & -e^{-t} + 2e^{-2t} \end{bmatrix}$$

Using (6.8.17), we find that the transfer function for the system is

$$G(s) = \frac{Y(s)}{V(s)} = \frac{s^2 - 3s + 2}{s^2 + 3s + 2}$$

This corresponds to the differential equation

$$\ddot{y}(t) + 3\dot{y}(t) + 2y(t) = \ddot{v}(t) - 3\dot{v}(t) + 2v(t)$$

Notice that $|G(j\omega)| = 1$ so that the system is an all-pass filter. ◀

6.9 STATE SPACE REPRESENTATIONS FOR CONSTANT-COEFFICIENT, LINEAR, DIFFERENTIAL EQUATIONS

A system whose input $v(t)$ and output $y(t)$ are related by the constant-coefficient, linear, differential equation

$$\frac{d^N}{dt^N} y(t) + b_1 \frac{d^{N-1}}{dt^{N-1}} y(t) + \cdots + b_N y(t)$$

$$= a_0 \frac{d^N}{dt^N} v(t) + a_1 \frac{d^{N-1}}{dt^{N-1}} v(t) + \cdots + a_N v(t) \quad (6.9.1)$$

has the transfer function

$$G(s) = \frac{Y(s)}{V(s)} = \frac{a_0 s^N + a_1 s^{N-1} + \cdots + a_N}{s^N + b_1 s^{N-1} + \cdots + b_N} \quad (6.9.2)$$

The pulse transfer function $G(z)$ given by (6.5.1) can be put in the form

$$G(z) = \frac{a_0 z^N + a_1 z^{N-1} + \cdots + a_N}{z^N + b_1 z^{N-1} + \cdots + b_N} \quad (6.9.3)$$

Notice that (6.9.2) and (6.9.3) are identical if z is replaced by s. Therefore, each of the structures discussed in Section 6.5 realizes the transfer function $G(s)$ when z is replaced by s. Each delay element labeled z^{-1} becomes an element labeled s^{-1} which is an integrator. The outputs of the integrators can be chosen as the state variables. To maintain the correspondence between the continuous and discrete-time systems, we will choose the state variables so that $x_k(t)$ corresponds to $x_k(nT)$. The input to the kth delay element $x_k(nT + T)$ becomes the input to the kth integrator and so must be relabeled $\dot{x}_k(t)$. This transformation is shown for the type 1 direct form realization in Fig. 6.9.1.

Clearly, the state and output equations for the continuous-time structure obtained by the simple transformation described in the previous paragraph can

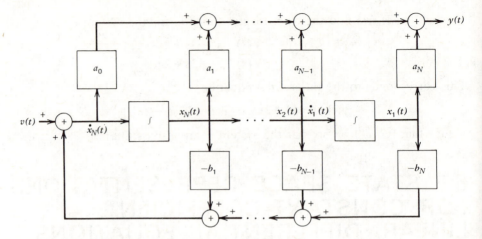

FIGURE 6.9.1.   Type 1 direct form realization of $G(s)$.

be determined from the equations for the original discrete-time structure simply by replacing $v(nT)$ by $v(t)$, $y(nT)$ by $y(t)$, $\mathbf{x}(nT)$ by $\mathbf{x}(t)$, and $\mathbf{x}(nT+T)$ by $\dot{\mathbf{x}}(t)$. For example, from (6.5.5) and (6.5.6) we see that the state and output equations for the type 1 direct form realization of $G(s)$ are

$$
\begin{bmatrix} \dot{x}_1(t) \\ \dot{x}_2(t) \\ \cdot \\ \cdot \\ \cdot \\ \dot{x}_{N-1}(t) \\ \dot{x}_N(t) \end{bmatrix} = \begin{bmatrix} 0 & 1 & 0 & \cdots & & 0 \\ 0 & 0 & 1 & 0 & \cdots & 0 \\ \cdot & \cdot & & & & \cdot \\ \cdot & \cdot & & & & \cdot \\ \cdot & \cdot & & & & \cdot \\ 0 & 0 & \cdots & & 0 & 1 \\ -b_N & -b_{N-1} & \cdots & & & -b_1 \end{bmatrix} \begin{bmatrix} x_1(t) \\ x_2(t) \\ \cdot \\ \cdot \\ \cdot \\ x_{N-1}(t) \\ x_N(t) \end{bmatrix} + \begin{bmatrix} 0 \\ 0 \\ \cdot \\ \cdot \\ \cdot \\ 0 \\ 1 \end{bmatrix} v(t) \quad (6.9.4)
$$

and

$$
y(t) = [a_N - a_0 b_N \cdots a_1 - a_0 b_1] \begin{bmatrix} x_1(t) \\ \cdot \\ \cdot \\ \cdot \\ x_2(t) \end{bmatrix} + a_0 v(t) \quad (6.9.5)
$$

Because of the direct analogy with the discrete-time structures discussed in Section 6.5 we will not elaborate on continuous-time structures for realizing $G(s)$ further.

6.10 LINEAR, CONTINUOUS-TIME, DYNAMICAL SYSTEMS WITH SAMPLED INPUTS

State space techniques can be used to analyze the behavior of linear, continuous-time, dynamical sytems with sampled inputs. The combined Laplace and Z-transform approach discussed in Chapter 5 is essentially limited to analyzing time-invariant systems with uniform sampling. State space techniques can be used for analyzing time-varying or time-invariant systems with uniform or nonuniform sampling. In addition, the behavior of a system between sampling instants can be easily calculated using the state space approach.

First let us consider the case when the input to the continuous-time system is the impulse train

$$\mathbf{v}(t) = \sum_{n=-\infty}^{\infty} \mathbf{f}(t_n)\, \delta(t-t_n) \qquad (6.10.1)$$

obtained by impulse sampling a signal $\mathbf{f}(t)$ at the set of possibly nonuniformly spaced times $\{t_n\}$. As usual, we will assume that the state equation for the system has the form

$$\dot{\mathbf{x}}(t) = \mathbf{A}(t)\mathbf{x}(t) + \mathbf{B}(t)\mathbf{v}(t) \qquad (6.10.2)$$

where $\mathbf{A}(t)$ is a continuous matrix so that the state transition matrix $\boldsymbol{\varphi}\,(t, \tau)$ is a unique continuous function of t and τ. If the state

$$\mathbf{x}(t_n{-}) = \lim_{\substack{t \to t_n \\ t < t_n}} \mathbf{x}(t)$$

is known, then according to (6.7.23) the state at any time t can be calculated as

$$\mathbf{x}(t) = \boldsymbol{\varphi}(t, t_n{-})\mathbf{x}(t_n{-}) + \int_{t_n{-}}^{t} \boldsymbol{\varphi}(t, \tau)\mathbf{B}(\tau)\mathbf{v}(\tau)\, d\tau \qquad (6.10.3)$$

Because the state transition matrix is continuous, its argument $t_n{-}$ can be replaced by t_n. For $t_n < t < t_{n+1}$ only the impulse in $\mathbf{v}(t)$ at time t_n is included in the integration interval in (6.10.3) so that

$$\mathbf{x}(t) = \boldsymbol{\varphi}(t, t_n)\mathbf{x}(t_n{-}) + \boldsymbol{\varphi}(t, t_n)\mathbf{B}(t_n)\mathbf{f}(t_n) \qquad \text{for} \qquad t_n < t < t_{n+1} \quad (6.10.4)$$

The state immediately after the impulse at time t_n is

$$\mathbf{x}(t_n{+}) = \lim_{\substack{t \to t_n \\ t > t_n}} \mathbf{x}(t) = \mathbf{x}(t_n{-}) + \mathbf{B}(t_n)\mathbf{f}(t_n) \qquad (6.10.5)$$

Therefore (6.10.4) can also be written as

$$\mathbf{x}(t) = \boldsymbol{\varphi}(t, t_n)\mathbf{x}(t_n{+}) \qquad \text{for} \qquad t_n < t < t_{n+1} \qquad (6.10.6)$$

Notice that (6.10.6) is the solution to the homogeneous state equation passing through $\mathbf{x}(t_n+)$. This result becomes obvious when we observe that $\mathbf{v}(t)$ is zero between sampling instants. The evolution of the state between sampling instants is given by either (6.10.4) or (6.10.6). From these two equations we find that the state just prior to the next sampling instant t_{n+1} is

$$\mathbf{x}(t_{n+1}-) = \boldsymbol{\varphi}(t_{n+1}, t_n)\mathbf{x}(t_n-) + \boldsymbol{\varphi}(t_{n+1}, t_n)\mathbf{B}(t_n)\mathbf{f}(t_n) \tag{6.10.7}$$

or

$$\mathbf{x}(t_{n+1}-) = \boldsymbol{\varphi}(t_{n+1}, t_n)\mathbf{x}(t_n+) \tag{6.10.8}$$

Equation 6.10.7 is a linear difference equation with the form of the state equation (6.3.1) for a discrete-time dynamical system.

Example 6.10.1

Suppose that the input and output of a system are related by the differential equation

$$\ddot{y}(t) + \dot{y}(t) = v(t) \tag{6.10.9}$$

The type 2 direct form realization for the system is shown in Fig. 6.10.1. This realization has the state equation

$$\begin{bmatrix} \dot{x}_1(t) \\ \dot{x}_2(t) \end{bmatrix} = \begin{bmatrix} 0 & 0 \\ 1 & -1 \end{bmatrix} \begin{bmatrix} x_1(t) \\ x_2(t) \end{bmatrix} + \begin{bmatrix} 1 \\ 0 \end{bmatrix} v(t) \tag{6.10.10}$$

and output equation

$$y(t) = \begin{bmatrix} 0 & 1 \end{bmatrix} \begin{bmatrix} x_1(t) \\ x_2(t) \end{bmatrix} = x_2(t) \tag{6.10.11}$$

The Laplace transform of the state transition matrix is

$$\boldsymbol{\Phi}(s) = (s\mathbf{I} - \mathbf{A})^{-1} = \begin{bmatrix} s & 0 \\ -1 & s+1 \end{bmatrix}^{-1}$$

$$= \begin{bmatrix} \dfrac{1}{s} & 0 \\ \dfrac{1}{s(s+1)} & \dfrac{1}{s+1} \end{bmatrix}$$

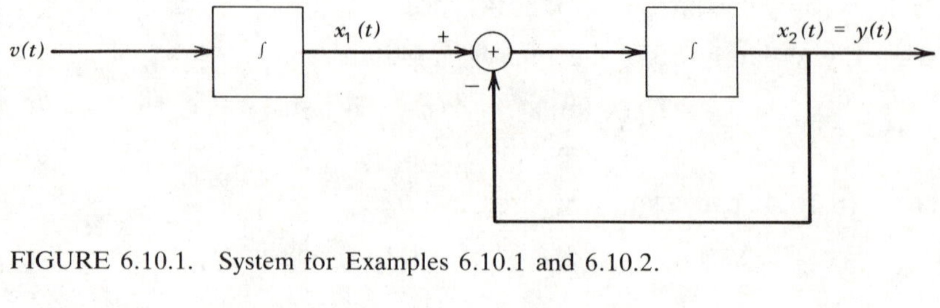

FIGURE 6.10.1.   System for Examples 6.10.1 and 6.10.2.

so that

$$\boldsymbol{\varphi}(t) = \begin{bmatrix} 1 & 0 \\ 1-e^{-t} & e^{-t} \end{bmatrix} \qquad \text{for} \qquad t \geq 0 \qquad (6.10.12)$$

Let us assume that

$$v(t) = \sum_{n=-\infty}^{\infty} f(nT) \, \delta(t-nT) \qquad (6.10.13)$$

If $\mathbf{x}(nT_-)$ is given, then according to (6.10.5)

$$\mathbf{x}(nT_+) = \mathbf{x}(nT_-) + \mathbf{B}f(nT)$$
$$= \begin{bmatrix} x_1(nT_-) + f(nT) \\ x_2(nT_-) \end{bmatrix} \qquad (6.10.14)$$

From Fig. 6.10.1 it is obvious that an impulse $f(nT)\,\delta(t-nT)$ applied to the input causes a jump of $f(nT)$ in x_1 but no change in x_2 at time nT. Since $y(t) = x_2(t)$, the output is continuous at the sampling instants. According to (6.10.4) or (6.10.6), the behavior of the state between sampling instants is

$$\begin{bmatrix} x_1(nT+r) \\ x_2(nT+r) \end{bmatrix} = \begin{bmatrix} x_1(nT-) + f(nT) \\ (1-e^{-r})[x_1(nT-) + f(nT)] + e^{-r}x_2(nT-) \end{bmatrix} \quad (6.10.15)$$

for $0 < r < T$. The entire behavior of the system can be easily calculated from (6.10.15). ◀

Frequently, the input to a continuous-time system is the staircase function appearing at the output of a zero-order hold. In this case

$$\mathbf{v}(t) = \mathbf{f}(t_n) \qquad \text{for} \qquad t_n \leq t \leq t_{n+1} \qquad (6.10.16)$$

Since $\mathbf{v}(t)$ is piecewise continuous, the solution to the state equation must be continuous. According to (6.7.23)

$$\mathbf{x}(t) = \boldsymbol{\varphi}(t, t_n)\mathbf{x}(t_n) + \int_{t_n}^{t} \boldsymbol{\varphi}(t, \tau)\mathbf{B}(\tau) \, d\tau\mathbf{f}(t_n) \qquad (6.10.17)$$

for $t_n \leq t < t_{n+1}$. Letting $t = t_{n+1}$, (6.10.17) becomes

$$\mathbf{x}(t_{n+1}) = \boldsymbol{\varphi}(t_{n+1}, t_n)\mathbf{x}(t_n) + \int_{t_n}^{t_{n+1}} \boldsymbol{\varphi}(t_{n+1}, \tau)\mathbf{B}(\tau) \, d\tau\mathbf{f}(t_n) \qquad (6.10.18)$$

Therefore, when only signal samples are of interest, the cascade of the zero-order hold and continuous-time system can be represented by an equivalent discrete-time system with the state equation

$$\mathbf{x}(t_{n+1}) = \mathbf{A}'(t_n)\mathbf{x}(t_n) + \mathbf{B}'(t_n)\mathbf{f}(t_n) \qquad (6.10.19)$$

where

$$\mathbf{A}'(t_n) = \boldsymbol{\varphi}(t_{n+1}, t_n)$$

and

$$\mathbf{B}'(t_n) = \int_{t_n}^{t_{n+1}} \boldsymbol{\varphi}(t_{n+1}, \tau) \mathbf{B}(\tau) \, d\tau$$

Example 6.10.2

Let us consider a system with uniform sampling consisting of a zero-order hold followed by the continuous-time system of Example 6.10.1. From (6.10.12) and (6.10.17) we find that

$$\mathbf{x}(nT + r) = \begin{bmatrix} 1 & 0 \\ 1 - e^{-r} & e^{-r} \end{bmatrix} \mathbf{x}(nT) + \begin{bmatrix} r \\ r + e^{-r} - 1 \end{bmatrix} f(nT) \qquad (6.10.20)$$

for $0 \le r \le T$. Therefore

$$\mathbf{x}(nT + T) = \begin{bmatrix} 1 & 0 \\ 1 - e^{-T} & e^{-T} \end{bmatrix} \mathbf{x}(nT) + \begin{bmatrix} T \\ T + e^{-T} - 1 \end{bmatrix} f(nT) \qquad (6.10.21)$$

According to (6.10.11), the system output samples are

$$y(nT) = [0 \quad 1] \mathbf{x}(nT) \qquad (6.10.22)$$

From (6.4.22) it follows that the pulse transfer function for the equivalent discrete-time system with the state equation (6.10.21) and output equation (6.10.22) is

$$\frac{Y(z)}{F(z)} = \frac{(T + e^{-T} - 1)z^{-1} + (1 - e^{-T} - Te^{-T})z^{-2}}{(1 - z^{-1})(1 - e^{-T}z^{-1})} \qquad (6.10.23)$$

As an exercise, find this pulse transfer function using the Z-transform methods of Chapter 5. ◄

7

LINEAR SYSTEMS AND UNIFORMLY SAMPLED RANDOM PROCESSES

7.1 INTRODUCTION

In this chapter we find that all the classical analysis techniques and results for continuous-time stationary random processes and linear systems have direct analogs for discrete-time systems with uniform sampling. First the concept of sampled correlation functions and power spectral densities is introduced. Then the effects of discrete-time linear filters on the sampled correlation functions and power spectal densities are examined. Next the discrete-time version of the Wiener filter problem is discussed. The problem of optimum linear prediction one sample into the future using the last N samples is examined in detail. This linear predictor has recently found important applications in speech analysis and compression. In addition, the solution to the problem leads to an interesting synthesis procedure for discrete-time filters in terms of a cascaded lattice network structure with guaranteed stability. Analysis methods for mixed continuous and discrete-time systems are developed and applied to the problem of optimally reconstructing a continuous-time signal from noisy samples. Throughout the chapter the criterion for optimality is the minimum mean square error criterion. Finally, formulas for the mean square reconstruction error of the zero-order hold, first-order hold, and linear point connector are derived. These are commonly used suboptimum reconstruction filters.

It will be assumed that the reader has a knowledge of probability and random processes. In particular, familiarity with correlation functions and power spectral densities will be assumed. The notation $E\{X\}$ will be used to indicate the statistical expectation of a random variable X. The cross-correlation function for two random processes $x(t)$ and $y(t)$ will be defined as

$$R_{xy}(t_1, t_2) = E\{x(t_1)y(t_2)\}$$

If $x(t)$ and $y(t)$ are jointly wide-sense stationary their cross-correlation function will be defined as

$$R_{xy}(\tau) = E\{x(t+\tau)y(t)\}$$

The Fourier transform of $R_{xy}(\tau)$ will be called the cross power spectral density $S_{xy}(\omega)$. As usual, when $y(t)$ is identical to $x(t)$ we will call $R_{xx}(\tau)$ the autocorrelation function and $S_{xx}(\omega)$ the power spectral density for $x(t)$.

7.2 POWER SPECTRAL DENSITIES FOR DISCRETE-TIME RANDOM PROCESSES

Let us consider a discrete-time random process $x(nT)$ that might have been obtained by sampling a continuous-time random process $x(t)$ with the autocorrelation function $R_{xx}(\tau)$ and power spectral density $S_{xx}(\omega)$. We will denote the Z-transform of $R_{xx}(\tau)$ by

$$S_{xx}(z) = \sum_{n=-\infty}^{\infty} R_{xx}(nT)z^{-n} \tag{7.2.1}$$

As is customary, we will make dual use of the symbol $S_{xx}(\cdot)$ to represent the Fourier or Z-transform of $R_{xx}(\tau)$ with the argument ω or z resolving the ambiguity. The *sampled power spectral density* for $x(nT)$ is defined to be

$$S_{xx}^*(\omega) = S_{xx}(z)\big|_{z=e^{j\omega T}} = \sum_{n=-\infty}^{\infty} R_{xx}(nT)e^{-j\omega nT} \tag{7.2.2}$$

We have seen in Chapter 2 that $S_{xx}^*(\omega)$ can also be expressed as

$$S_{xx}^*(\omega) = \frac{1}{T}\sum_{n=-\infty}^{\infty} S_{xx}(\omega - n\omega_s) \tag{7.2.3}$$

Since the power spectral density $S_{xx}(\omega)$ must be real, nonnegative, and even, it follows from (7.2.3) that $S_{xx}^*(\omega)$ must also be real, nonnegative and even. If the envelope of $R_{xx}(\tau)$ decays exponentially with positive and negative τ, then the region of convergence for $S_{xx}(z)$ must include the unit circle. In the case where $R_{xx}(\tau)$ has undamped periodic components it follows from (7.2.3) that the series in (7.2.2) converges in the distribution sense to a function that contains impulse functions since $S_{xx}(\omega)$ contains impulses. In the future we will refer to either $S_{xx}^*(\omega)$ or $S_{xx}(z)$ as the sampled power spectral density for $x(nT)$.

The average power in $x(nT)$ is

$$E\{x^2(nT)\} = R_{xx}(0) = \frac{1}{2\pi j} \oint_C S_{xx}(z) \frac{dz}{z} \tag{7.2.4}$$

where C is a simple closed contour in the region of convergence. If C is chosen to be the unit circle by making the change of variables $z = e^{j\omega T}$, then the average power can be expressed as

$$R_{xx}(0) = \frac{1}{\omega_s} \int_{-\omega_s/2}^{\omega_s/2} S_{xx}^*(\omega)\, d\omega \tag{7.2.5}$$

Thus it appears that the average power in the incremental frequency band $\omega_0 \le \omega \le \omega_0 + d\omega$ is $S_{xx}^*(\omega)\, d\omega/\omega_s$. This is verified later.

Example 7.2.1

Suppose $R_{xx}(nT) = a^{|n|}$ with $0 < a < 1$. Then

$$S_{xx}(z) = \sum_{n=-\infty}^{\infty} a^{|n|} z^{-n} = \sum_{n=-\infty}^{0} a^{-n} z^{-n} + \sum_{n=0}^{\infty} a^n z^{-n} - 1$$

The first summation on the right-hand side converges to $1/(1 - az)$ for $|z| < 1/a$, and the second summation converges to $1/(1 - az^{-1})$ for $|z| > a$. Combining these results we find that

$$S_{xx}(z) = \frac{1 - a^2}{(1 - az^{-1})(1 - az)} \quad \text{for} \quad a < |z| < 1/a \qquad \blacktriangleleft$$

The sampled *cross power spectral density* $S_{xy}(z)$ for two jointly wide-sense stationary random processes $x(t)$ and $y(t)$ is defined to be the Z-transform of their cross-correlation function. Thus

$$S_{xy}(z) = \sum_{n=-\infty}^{\infty} R_{xy}(nT) z^{-n} \tag{7.2.6}$$

Since $R_{xy}(nT) = R_{yx}(-nT)$, it follows that

$$S_{xy}(z) = S_{yx}(z^{-1}) \tag{7.2.7}$$

In particular, if $x(t) = y(t)$ then

$$S_{xx}(z) = S_{xx}(z^{-1}) \tag{7.2.8}$$

or, equivalently

$$S_{xx}^*(\omega) = S_{xx}^*(-\omega) \tag{7.2.9}$$

If $S_{xx}(z)$ is rational it can be factored in many ways. The following theorem presents a canonical factorization that will be used in deriving the discrete-time version of the Wiener filter.

THEOREM 7.2.1. The Spectral Factorization Theorem

Let $x(nT)$ be a real, wide-sense stationary, random process with the rational sampled power spectral density

$$S_{xx}(z) = \sum_{n=-\infty}^{\infty} R_{xx}(nT)z^{-n} = \frac{N(z)}{D(z)}$$

where $N(z)$ and $D(z)$ are polynomials in z. Then $S_{xx}(z)$ can be represented as

$$S_{xx}(z) = \sigma^2 B(z)B(z^{-1}) \tag{7.2.10}$$

where

$$\sigma^2 > 0$$

and

$$B(z) = \frac{\displaystyle\prod_{k=1}^{L}(1 - \alpha_k z^{-1})}{\displaystyle\prod_{k=1}^{M}(1 - \beta_k z^{-1})} = \frac{\displaystyle\sum_{k=0}^{L} a_k z^{-k}}{\displaystyle\sum_{k=0}^{M} b_k z^{-k}}$$

with $|\alpha_k| < 1$ and $|\beta_k| < 1$ for all approximate k, and a_k and b_k real for all appropriate k.

Proof $N(z)$ and $D(z)$ must have real coefficients since $R_{xx}(nT)$ is real. Suppose z_0 is a root of

$$N(z) = \sum_{k} n_k z^{-k}$$

Then

$$0 = N(z_0) = \sum_{k} n_k z^{-k} \tag{7.2.11}$$

Taking the complex conjugate of both sides of (7.2.11) yields

$$0 = \sum_{k} n_k (\bar{z}_0)^{-k} = N(\bar{z}_0)$$

Thus the complex conjugate of each zero of $S_{xx}(z)$ is also a zero. The same conclusion holds for its poles. Since $S_{xx}(z) = S_{xx}(z^{-1})$, the reciprocal of each zero must also be a zero and the reciprocal of each pole must also be a pole. The numerator of $B(z)$ is constructed by selecting the set of zeros $\{\alpha_k\}_{k=1}^{L}$ of $S_{xx}(z)$ inside the unit circle and forming the product of factors shown in the theorem statement. Since the conjugate of any complex zero is also a zero, this product must be a polynomical of degree L with real coefficients. The denominator of $B(z)$ is formed in a similar fashion. $B(z^{-1})$ automatically includes all poles and zeros outside the unit circle. The parameter σ^2 is simply the required scale factor.

Q.E.D.

Example 7.2.2

Suppose that

$$S_{xx}(z) = \frac{1-a^2}{(1-az^{-1})(1-az)}$$

as in Example 7.2.1. Then

$$\sigma^2 = 1 - a^2$$

and

$$B(z) = 1/(1-az^{-1}) \qquad \blacktriangleleft$$

7.3 FINDING $S_{xx}(z)$ FROM $S_{xx}(\omega)$ FOR RATIONAL SPECTRAL DENSITIES

We will now derive a formula for determining $S_{xx}(z)$ directly from $S_{xx}(\omega)$ that is particularly convenient for rational spectral densities. The series for $S_{xx}(z)$ can be grouped and written as

$$S_{xx}(z) = S_+(z) + S_-(z) - R_{xx}(0) \tag{7.3.1}$$

where

$$S_+(z) = \sum_{n=0}^{\infty} R_{xx}(nT)z^{-n}$$

and

$$S_-(z) = \sum_{n=-\infty}^{0} R_{xx}(nT)z^{-n}$$

Notice that

$$S_-(z) = S_+(z^{-1}) \tag{7.3.2}$$

since $R_{xx}(nT) = R_{xx}(-nT)$. In Chapter 4 we saw that

$$S_+(z)|_{z=e^{sT}} = \mathscr{L}\left\{R_{xx}(\tau)\sum_{n=0}^{\infty}\delta(\tau-nT)\right\} \tag{7.3.3}$$

The frequency domain convolution theorem can be used to evaluate (7.3.3). Let us assume that the Laplace transform of $R_{xx}(\tau)$ converges to $S_{xx}(s/j)$ for $|\mathrm{Re}\,s| < A$. The Laplace transform of the semi-infinite impulse train in (7.3.3) converges to

$$G(s) = 1/(1 - e^{-sT}) \quad \text{for} \quad \text{Re } s > 0$$

Therefore

$$S_+(e^{sT}) = \frac{1}{2\pi j} \int_{c-j\infty}^{c+j\infty} S_{xx}(p/j) G(s-p) \, dp$$

$$= \frac{1}{2\pi j} \int_{c-j\infty}^{c+j\infty} \frac{S_{xx}(p/j)}{1 - e^{-(s-p)T}} \, dp \qquad (7.3.4)$$

The variable s must be chosen so that the regions of convergence for $G(s-p)$ and $S_{xx}(p/j)$ overlap. Since the region of convergence for $G(s-p)$ is the half plane $\text{Re } p < \text{Re } s$, s must be chosen so that $\text{Re } s > -A$. The contour of integration must be taken in the overlapping regions of convergence. Therefore c must be chosen so that

$$-A < c < \min(A, \text{Re } s)$$

These relationships are illustrated in Fig. 7.3.1. The convolution integral can be evaluated using Cauchy's Residue Theorem and the closed contour shown in Fig. 7.3.1. Since $x(t)$ has finite power, $S_{xx}(p/j)$ must approach zero as

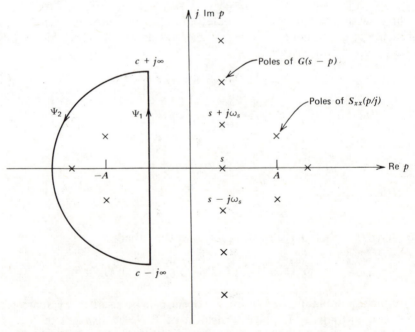

FIGURE 7.3.1.   Contour for finding $S_{xx}(z)$ from $S_{xx}(p/j)$.

$1/p^2$ for large p and the integral along Ψ_2 is zero. Therefore

$$S_+(z) = \sum_k \text{Res} \left[\frac{S_{xx}(p/j)}{1 - e^{pT}z^{-1}}, p_k \right] \qquad (7.3.5)$$

where $\{p_k\}$ is the set of left half plane poles of $S_{xx}(p)$. According to the Laplace transform inversion formula

$$R_{xx}(0) = \frac{1}{2\pi j} \int_{c-j\infty}^{c+j\infty} S_{xx}(p/j) \, dp$$

Therefore

$$R_{xx}(0) = \sum_k \text{Res} \left[S_{xx}(p/j), p_k \right] \qquad (7.3.6)$$

Using (7.3.2), (7.3.5), and (7.3.6) in (7.3.1) we find that

$$S_{xx}(z) = \sum_k \text{Res} \left[S_{xx}(p/j) \left(\frac{1}{1 - e^{pT}z^{-1}} + \frac{1}{1 - e^{pT}z} - 1 \right), p_k \right] \qquad (7.3.7)$$

or

$$S_{xx}(z) = \sum_k \text{Res} \left[S_{xx}(p/j) \frac{1 - e^{2pT}}{(1 - e^{pT}z^{-1})(1 - e^{pT}z)}, p_k \right] \qquad (7.3.8)$$

where $\{p_k\}$ is the set of left half plane poles of $S_{xx}(p/j)$.

Example 7.3.1

Suppose that $S_{xx}(p/j) = 2a/(a^2 - p^2)$. This has a left half plane pole at $p_1 = -a$. Therefore,

$$S_{xx}(z) = \text{Res} \left[\frac{2a(1 - e^{2pT})}{(a+p)(a-p)(1 - e^{pT}z^{-1})(1 - e^{pT}z)}, -a \right]$$

$$= \frac{1 - e^{-2aT}}{(1 - e^{-aT}z^{-1})(1 - e^{-aT}z)} \qquad \blacktriangleleft$$

7.4 DISCRETE-TIME RANDOM PROCESSES PASSED THROUGH LINEAR DISCRETE-TIME FILTERS

Frequently it is necessary to find the effects of filtering on the correlation functions or power spectral densities of discrete-time random processes. Consider the situation shown in Fig. 7.4.1. We will assume that $x(nT)$ and $y(nT)$ are real, jointly wide-sense stationary, random processes and that $R_{xx}(nT)$, $R_{yy}(nT)$, and $R_{xy}(nT)$ are known. The blocks labeled $A(z)$ and $B(z)$ represent

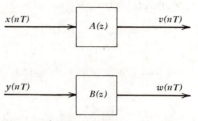

FIGURE 7.4.1. Filtering discrete-time random processes.

filters with the real pulse responses $a(nT)$ and $b(nT)$ respectively. The filter outputs can be expressed as

$$v(nT) = \sum_{k=-\infty}^{\infty} a(kT)x(nT-kT) \qquad (7.4.1)$$

and

$$w(nT) = \sum_{k=-\infty}^{\infty} b(kT)y(nT-kT) \qquad (7.4.2)$$

First we will find $R_{vy}(mT) = E\{v(mT+nT)y(nT)\}$. Replacing n by $n+m$ in (7.4.1) and multiplying by $y(nT)$ we find that

$$R_{vy}(mT) = E\left\{ \sum_{k=-\infty}^{\infty} a(kT)x(mT+nT-kT)y(nT) \right\}$$

$$= \sum_{k=-\infty}^{\infty} a(kT)E\{x(mT+nT-kT)y(nT)\}$$

$$= \sum_{k=-\infty}^{\infty} a(kT)R_{xy}(mT-kT) \qquad (7.4.3)$$

Notice that $R_{vy}(mT)$ is just the convolution of $a(mT)$ with $R_{xy}(mT)$. Therefore,

$$S_{vy}(z) = A(z)S_{xy}(z) \qquad (7.4.4)$$

Now let us find $R_{vw}(mT) = E\{v(mT+nT)w(nT)\}$. Multiplying (7.4.2) by $v(mT+nT)$ and taking expected values gives

$$R_{vw}(mT) = E\left\{ \sum_{k=-\infty}^{\infty} b(kT)y(nT-kT)v(mT+nT) \right\}$$

$$= \sum_{k=-\infty}^{\infty} b(kT)R_{vy}(mT+kT) \qquad (7.4.5)$$

Taking Z-transforms, we find that

$$S_{vw}(z) = B(z^{-1})S_{vy}(z) \qquad (7.4.6)$$

Combining (7.4.6) and (7.4.4) we obtain the desired result that

$$S_{vw}(z) = A(z)B(z^{-1})S_{xy}(z) \qquad (7.4.7)$$

This is analogous to the result for continuous time systems. Usually it is easiest to find the power spectral density, $S_{vw}(z)$, first and then calculate $R_{vw}(mT)$ from $S_{vw}(z)$.

An important special case is when $x(nT) = y(nT)$ and $A(z) = B(z)$. Then (7.4.7) becomes

$$S_{vv}(z) = A(z)A(z^{-1})S_{xx}(z) \qquad (7.4.8)$$

or, on the unit circle

$$S_{vv}^*(\omega) = A^*(\omega)A^*(-\omega)S_{xx}^*(\omega) = |A^*(\omega)|^2 S_{xx}^*(\omega) \qquad (7.4.9)$$

Now let us assume that in the Nyquist band $|\omega| \le \omega_s/2$, $A^*(\omega)$ has the ideal band-pass characteristic

$$A^*(\omega) = \begin{cases} 1 & \text{for} \quad 0 < \omega_0 < |\omega| < \omega_0 + d \\ 0 & \text{elsewhere} \end{cases}$$

Then, according to (7.2.5) and the fact that $S_{vv}^*(\omega)$ is even, the power in the pass-band is

$$E\{v^2(nT)\} = R_{vv}(0) = \frac{2}{\omega_s} \int_{\omega_0}^{\omega_0 + d} S_{xx}^*(\omega)\, d\omega \qquad (7.4.10)$$

This confirms the conclusion in Section 7.2 that $S_{xx}^*(\omega)$ represents the frequency domain resolution of the average power in $x(nT)$.

We can also argue from (7.4.10) that $S_{xx}^*(\omega)$ must be nonnegative. Clearly, $R_{vv}(0)$ is nonnegative. If we assume that $S_{xx}^*(\omega)$ can be negative, then ω_0 and d can be chosen so that the integral is negative resulting in a contradiction.

Example 7.4.1

Suppose that $R_{xx}(mT) = \delta_{m0}$ so that $S_{xx}(z) = 1$. The process $x(nT)$ is frequently called white noise because its sampled power spectral density is constant. If $x(nT)$ is passed through a filter with the pulse transfer function $A(z) = 1/(1 - az^{-1})$ resulting in the output $v(nT)$, then

$$S_{vv}(z) = A(z)A(z^{-1})S_{xx}(z) = \frac{1}{(1 - az^{-1})(1 - az)}$$

Taking the inverse Z-transform, we find that

$$R_{vv}(mT) = \frac{1}{1 - a^2} a^{|m|}$$

and so

$$R_{vv}(0) = E\{v^2(nT)\} = 1/(1 - a^2)$$

Also

$$S_{vx}(z) = A(z)S_{xx}(z) = 1/(1 - az^{-1})$$

so that

$$R_{vx}(mT) = a^m u(m)$$ ◀

7.5 OPTIMUM LINEAR FILTERING OF DISCRETE-TIME RANDOM PROCESSES USING THE INFINITE PAST

We will now investigate the problem of how to linearly filter an observed, wide-sense stationary, discrete-time, random process $y(nT)$ to obtain the best estimate in the minimum mean square error sense of a related discrete-time random process $x(nT)$. It appears that this problem was first investigated by the Russian, A. N. Kolmogorov [75]. At about the same time Norbert Wiener was investigating the continuous-time version of the same problem in the United States. His work initially was published by the Government Printing Office and was later published in book form [141]. Because of the wider publicity of Wiener's work, the resulting filter came to be known as the Wiener filter in the United States. A detailed derivation of the Wiener filter can be found in most communication theory books published in the last 20 years.

The problem is illustrated in Fig. 7.5.1. Here $x(nT)$ is a desired wide-sense stationary, discrete-time, random process and $y(nT)$ is an observed wide-sense stationary random process related to $x(nT)$. For example, $y(nT)$ might be the sum of $x(nT)$ and some undesired noise $v(nT)$. We will assume that $S_{xy}(z)$ and

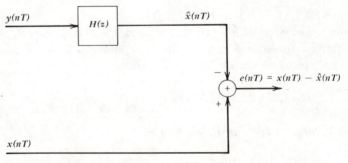

FIGURE 7.5.1. Pictorial representation of the optimum filtering problem.

FIGURE 7.5.2.   Estimation of $x(nT)$ from the innovations process $a(nT)$.

$S_{yy}(z)$ are known rational functions of z. The problem is to find the pulse transfer function $H(z)$ that minimizes the mean square error $E\{e^2(nT)\}$.

We will solve this problem somewhat indirectly. According to Theorem 7.2.1, $S_{yy}(z)$ can be written as $S_{yy}(z) = \sigma^2 B(z)B(z^{-1})$ where all the poles and zeros of $B(z)$ are inside the unit circle. Therefore, $B(z)$ and $1/B(z)$ both correspond to physically realizable, stable filters. Let us first consider the problem of estimating $x(nT)$ from the sequence $a(nT)$ obtained by passing $y(nT)$ through a filter with the pulse transfer function $1/B(z)$. This is illustrated in Fig. 7.5.2. The signals $a(nT)$ and $y(nT)$ contain the same information since $y(nT)$ can be recovered uniquely by passing $a(nT)$ through a filter with the pulse transfer function $B(z)$. We will assume that the filters have been operating on $a(nT)$ or $y(nT)$ since the infinite past so that transients can be ignored. According to (7.4.8),

$$S_{aa}(z) = \frac{1}{B(z)B(z^{-1})} S_{yy}(z) = \sigma^2 \tag{7.5.1}$$

so that

$$R_{aa}(mT) = \sigma^2 \delta_{m0} \tag{7.5.2}$$

Thus $a(nT)$ is an uncorrelated or white noise sequence with variance σ^2. The signal $a(nT)$ is known as the *innovations* process associated with $y(nT)$ [65,66]. If the pulse response of the estimation filter $G(z)$ shown in Fig. 7.5.2 is $g(nT)$, then

$$\hat{x}(nT) = \sum_{k=-\infty}^{\infty} g(kT)a(nT-kT) \tag{7.5.3}$$

The mean square error is

$$E\{e^2(nT)\} = E\left\{\left[x(nT) - \sum_{k=-\infty}^{\infty} g(kT)a(nT-kT)\right]^2\right\}$$

$$= E\{x^2(nT)\} - 2E\left\{\sum_{k=-\infty}^{\infty} g(kT)x(nT)a(nT-kT)\right\}$$

$$+ E\left\{\left[\sum_{k=-\infty}^{\infty} g(kT)a(nT-kT)\right]^2\right\} \tag{7.5.4}$$

The first term on the right-hand side of (7.5.4) is simply $R_{xx}(0)$. When the summation and expectation operations are interchanged the second term reduces to

$$-2 \sum_{k=-\infty}^{\infty} g(kT)R_{xa}(kT)$$

The third term is equivalent to

$$E\left\{ \sum_{k=-\infty}^{\infty} \sum_{r=-\infty}^{\infty} g(kT)g(rT)a(nT-kT)a(nT-rT) \right\}$$

$$= \sum_{k=-\infty}^{\infty} \sum_{r=-\infty}^{\infty} g(kT)g(rT)R_{aa}(rT-kT) = \sigma^2 \sum_{k=-\infty}^{\infty} g^2(kT)$$

Combining these terms we find that

$$E\{e^2(nT)\} = R_{xx}(0) - 2 \sum_{k=-\infty}^{\infty} g(kT)R_{xa}(kT) + \sigma^2 \sum_{k=-\infty}^{\infty} g^2(kT) \qquad (7.5.5)$$

Completing the square (7.5.5) becomes

$$E\{e^2(nT)\} = R_{xx}(0) + \sum_{k=-\infty}^{\infty} \left[\sigma g(kT) - \frac{R_{xa}(kT)}{\sigma} \right]^2 - \frac{1}{\sigma^2} \sum_{k=-\infty}^{\infty} R_{xa}^2(kT)$$

$$(7.5.6)$$

First we will find the optimum filter when no physical realizability constraint is imposed. Notice that the first and last terms on the right-hand side of (7.5.6) are independent of $g(nT)$. The middle term is nonnegative. Thus the mean square error is minimized by choosing $g(nT)$ to make this term identically zero. Therefore, the pulse response of the optimum nonrealizable filter must be

$$g(nT) = \frac{1}{\sigma^2} R_{xa}(nT) \qquad \text{for} \qquad -\infty < n < \infty$$

and its pulse transfer function must be

$$G(z) = \frac{1}{\sigma^2} S_{xa}(z)$$

But according to (7.4.7)

$$S_{xa}(z) = S_{xy}(z)/B(z^{-1}) \qquad (7.5.7)$$

so that

$$G(z) = \frac{1}{\sigma^2} \frac{S_{xy}(z)}{B(z^{-1})} \qquad (7.5.8)$$

From Fig. 7.5.2 it is clear that the optimum filter for estimating $x(nT)$ from $y(nT)$ is

$$H(z) = \frac{1}{B(z)} G(z) = \frac{S_{xy}(z)}{\sigma^2 B(z)B(z^{-1})}$$

or

$$H(z) = \frac{S_{xy}(z)}{S_{yy}(z)} \tag{7.5.9}$$

The mean square error for the optimum nonrealizable filter is

$$E\{e^2(nT)\} = R_{xx}(0) - \frac{1}{\sigma^2} \sum_{k=-\infty}^{\infty} R_{xa}^2(kT) \tag{7.5.10}$$

Using Parseval's Theorem (7.5.10) can be written as

$$E\{e^2(nT)\} = \frac{1}{2\pi j} \oint \left[S_{xx}(z) - \frac{1}{\sigma^2} S_{xa}(z) S_{xa}(z^{-1}) \right] \frac{dz}{z}$$

$$= \frac{1}{2\pi j} \oint \left[S_{xx}(z) - \frac{S_{xy}(z) S_{xy}(z^{-1})}{\sigma^2 B(z) B(z^{-1})} \right] \frac{dz}{z}$$

$$= \frac{1}{2\pi j} \oint \left[S_{xx}(z) - \frac{S_{xy}(z) S_{xy}(z^{-1})}{S_{yy}(z)} \right] \frac{dz}{z} \tag{7.5.11}$$

Finally, using (7.5.9) the mean square error can be put in the form

$$E\{e^2(nT)\} = \frac{1}{2\pi j} \oint [S_{xx}(z) - H(z) S_{xy}(z^{-1})] \frac{dz}{z} \tag{7.5.12}$$

The unit circle can be taken as the contour of integration.

Example 7.5.1

Suppose that $y(nT) = x(nT) + v(nT)$ where $x(nT)$ is a desired transmitted signal and $v(nT)$ is additive noise uncorrelated with $x(nT)$. Then it follows that

$$S_{xy}(z) = S_{xx}(z)$$

and

$$S_{yy}(z) = S_{xx}(z) + S_{vv}(z)$$

Therefore, the optimum nonrealizable filter is

$$H(z) = \frac{S_{xx}(z)}{S_{xx}(z) + S_{vv}(z)}$$

and on the unit circle

$$H^*(\omega) = \frac{S_{xx}^*(\omega)}{S_{xx}^*(\omega) + S_{vv}^*(\omega)}$$

Notice that at frequencies where the noise spectral density is negligible $H^*(\omega)$ is close to 1. When the noise spectral density is large relative to the signal spectral density, $H^*(\omega)$ is small. This is the type of filter one would design intuitively. ◄

In some applications an estimate of a signal $w(nT)$ obtained by passing $x(nT)$ through a linear filter $L(z)$ may be desired. For example, $L(z)$ might represent an approximate integral or derivative. It follows from (7.5.9) that the pulse transfer function of the optimum nonrealizable filter for estimating $w(nT)$ from $y(nT)$ is

$$F(z) = \frac{S_{wy}(z)}{S_{yy}(z)}$$

But

$$S_{wy}(z) = L(z)S_{xy}(z) \tag{7.5.13}$$

so that

$$F(z) = L(z)\frac{S_{xy}(z)}{S_{yy}(z)} = L(z)H(z) \tag{7.5.14}$$

where $H(z)$ is the optimum nonrealizable filter for estimating $x(nT)$ from $y(nT)$. Thus the optimum estimate of $w(nT)$ can be found by first filtering $y(nT)$ with $H(z)$ to find the optimum estimate of $x(nT)$ and then filtering the optimum estimate of $x(nT)$ with $L(z)$. Replacing x by w in (7.5.11) and denoting the filter output by $\hat{w}(nT)$ we see that the mean square error is

$$\begin{aligned}
E\{[w(nT) - \hat{w}(nT)]^2\} &= \frac{1}{2\pi j}\oint \left[S_{ww}(z) - \frac{S_{wy}(z)S_{wy}(z^{-1})}{S_{yy}(z)} \right]\frac{dz}{z}\\
&= \frac{1}{2\pi j}\oint L(z)L(z^{-1})\left[S_{xx}(z) - \frac{S_{xy}(z)S_{xy}(z^{-1})}{S_{yy}(z)} \right]\frac{dz}{z}
\end{aligned} \tag{7.5.15}$$

Now we will find the optimum physically realizable filter for estimating $x(nT)$ from $y(nT)$. The filter $1/B(z)$ shown in Fig. 7.5.2 is physically realizable and has no overall delay. Therefore, $G(z)$ must be realizable to make the cascade realizable. Thus we must require that $g(nT) = 0$ for negative n. With this requirement (7.5.6) reduces to

$$E\{e^2(nT)\} = R_{xx}(0) + \sum_{k=0}^{\infty}\left[\sigma g(kT) - \frac{R_{xa}(kT)}{\sigma} \right]^2 - \frac{1}{\sigma^2}\sum_{k=0}^{\infty} R_{xa}^2(kT) \tag{7.5.16}$$

The mean square error is minimized by choosing $g(kT)$ to make the first summation on the right-hand side of (7.5.16) zero. Therefore, the pulse response of the optimum realizable filter for estimating $x(nT)$ from $a(nT)$ is

$$g(nT) = \begin{cases} R_{xa}(nT)/\sigma^2 & \text{for} \quad n \geq 0 \\ 0 & \text{elsewhere} \end{cases} \tag{7.5.17}$$

We will use the notation $[F(z)]_+$ to represent the one-sided Z-transform of

$f(nT)$, that is,

$$[F(z)]_+ = \sum_{n=0}^{\infty} f(nT)z^{-n} = \mathcal{L}\{f(nT)u(nT)\}$$

Thus

$$G(z) = \frac{1}{\sigma^2}[S_{xa}(z)]_+ = \frac{1}{\sigma^2}\left[\frac{S_{xy}(z)}{B(z^{-1})}\right]_+ \qquad (7.5.18)$$

From Fig. 7.5.2 it is clear that the optimum filter for estimating $x(nT)$ from $y(nT)$ is

$$H(z) = \frac{1}{\sigma^2 B(z)}\left[\frac{S_{xy}(z)}{B(z^{-1})}\right]_+ \qquad (7.5.19)$$

For the optimum realizable filter the mean square error given by (7.5.16) reduces to

$$E\{e^2(nT)\} = R_{xx}(0) - \frac{1}{\sigma^2}\sum_{k=0}^{\infty} R_{xa}^2(kT) \qquad (7.5.20)$$

Comparing (7.5.20) with (7.5.10) we find that the mean square error for the optimum physically realizable filter is greater than that for the nonrealizable filter by

$$\frac{1}{\sigma^2}\sum_{k=-\infty}^{-1} R_{xa}^2(kT)$$

Equation (7.5.20) can also be written as

$$E\{e^2(nT)\} = R_{xx}(0) - \frac{1}{\sigma^2}\sum_{k=-\infty}^{\infty}[R_{xa}(kT)u(kT)]R_{xa}(kT) \qquad (7.5.21)$$

Using Parseval's Theorem (7.5.21) can be expressed as

$$E\{e^2(nT)\} = \frac{1}{2\pi j}\oint\left[S_{xx}(z) - \frac{1}{\sigma^2}[S_{xa}(z)]_+S_{xa}(z^{-1})\right]\frac{dz}{z}$$

$$= \frac{1}{2\pi j}\oint\left[S_{xx}(z) - \frac{1}{\sigma^2}\left[\frac{S_{xy}(z)}{B(z^{-1})}\right]_+\frac{S_{xy}(z^{-1})}{B(z)}\right]\frac{dz}{z}$$

$$= \frac{1}{2\pi j}\oint[S_{xx}(z) - H(z)S_{xy}(z^{-1})]\frac{dz}{z} \qquad (7.5.22)$$

The unit circle can be taken as the contour of integration. Equation 7.5.22 has the same form as the mean square error for the optimum nonrealizable filter except that $H(z)$ is now given by (7.5.19).

Example 7.5.2

Suppose that $y(nT) = x(nT) + v(nT)$
with

$$S_{xx}(z) = \frac{0.36}{(1 - 0.8z^{-1})(1 - 0.8z)}$$

$$S_{vv}(z) = 1$$

and

$$S_{xv}(z) = 0$$

The signal $x(nT)$ represents a desired signal and $v(nT)$ is additive white noise. Since the signal and noise are uncorrelated, it follows that $S_{xy}(z) = S_{xx}(z)$ and $S_{yy}(z) = S_{xx}(z) + S_{vv}(z)$. After some algebra, $S_{yy}(z)$ can be put in the form

$$S_{yy}(z) = 1.6 \frac{(1 - 0.5z^{-1})(1 - 0.5z)}{(1 - 0.8z^{-1})(1 - 0.8z)}$$

Thus

$$\sigma^2 = 1.6 \quad \text{and} \quad B(z) = \frac{1 - 0.5z^{-1}}{1 - 0.8z^{-1}}$$

According to (7.5.19), the pulse transfer function of the optimum realizable filter for estimating $x(nT)$ from $y(nT)$ is

$$H(z) = \frac{1 - 0.8z^{-1}}{1.6(1 - 0.5z^{-1})} \left[\frac{0.36}{(1 - 0.8z^{-1})(1 - 0.5z)} \right]_+$$

Now

$$\mathscr{L}^{-1} \left[\frac{0.36}{(1 - 0.8z^{-1})(1 - 0.5z)} \right] = \tfrac{3}{5}(0.8)^n \quad \text{for} \quad n \geq 0$$

so that

$$\left[\frac{0.36}{(1 - 0.8z^{-1})(1 - 0.5z)} \right]_+ = \frac{\tfrac{3}{5}}{1 - 0.8z^{-1}}$$

and

$$H(z) = \frac{\tfrac{3}{8}}{1 - 0.5z^{-1}}$$

Using (7.5.22), the mean square error can be computed to be $\tfrac{3}{8}$. The mean square error before filtering is $E\{[y(nT) - x(nT)]^2\} = E\{v^2(nT)\} = 1$. Thus the optimum realizable filter reduces the mean square error by a factor of $\tfrac{8}{3}$.

According to (7.5.9), the optimum nonrealizable filter is

$$H(z) = \frac{S_{xy}(z)}{S_{xx}(z)} = \frac{0.225}{(1 - 0.5z^{-1})(1 - 0.5z)}$$

Using (7.5.12), the corresponding mean square error is found to be $\tfrac{3}{10}$. Thus the optimum nonrealizable filter is only slightly better than the optimum realizable filter in this particular case. ◄

If $w(nT)$ is the signal obtained by passing $x(nT)$ through the filter $L(z)$, the pulse transfer function of the optimum realizable filter for estimating $w(nT)$ from $y(nT)$ is

$$F(z) = \frac{1}{\sigma^2 B(z)} \left[\frac{S_{wy}(z)}{B(z^{-1})} \right]_+ = \frac{1}{\sigma^2 B(z)} \left[\frac{L(z)S_{xy}(z)}{B(z^{-1})} \right]_+. \qquad (7.5.23)$$

Denoting the optimum estimate by $\hat{w}(nT)$ and replacing x by w in (7.5.22), we find that the corresponding mean square error is

$$E\{[w(nT) - \hat{w}(nT)]^2\} = \frac{1}{2\pi j} \oint [S_{ww}(z) - F(z)S_{wy}(z^{-1})] \frac{dz}{z}$$

$$= \frac{1}{2\pi j} \oint [L(z)L(z^{-1})S_{xx}(z) - F(z)L(z^{-1})S_{xy}(z^{-1})] \frac{dz}{z}$$

$$(7.5.24)$$

Example 7.5.3 Pure Prediction

Suppose that $y(nT) = x(nT)$. Then $S_{xx}(z) = S_{yy}(z) = S_{xy}(z) = \sigma^2 B(z)B(z^{-1})$. When $x(nT)$ is passed through the physically nonrealizable filter $L(z) = z^N$, the signal $x(nT + NT)$ is obtained. From (7.5.23) it follows that the optimum realizable filter for predicting N samples into the future is specified by

$$F(z) = \frac{1}{B(z)} [z^N B(z)]_+ \qquad (7.5.25)$$

According to (7.5.24), the mean square error is

$$\Delta = E\{[x(nT + NT) - \hat{x}(nT + NT)]^2\}$$

$$= \frac{1}{2\pi j} \oint [\sigma^2 B(z)B(z^{-1}) - \sigma^2 [z^N B(z)]_+ z^{-N} B(z^{-1})] \frac{dz}{z} \qquad (7.5.26)$$

Using Parseval's Theorem (7.5.26) can be written as

$$\Delta = \sigma^2 \sum_{n=-\infty}^{\infty} b^2(nT) - \sigma^2 \sum_{n=-\infty}^{\infty} [b(nT + NT)u(nT)]b(nT + NT) \qquad (7.5.27)$$

Since $b(nT)$ is a causal sequence (7.5.27) reduces to

$$\Delta = \sigma^2 \sum_{n=0}^{N-1} b^2(nT) \qquad (7.5.28)$$

As a specific example, suppose that

$$S_{xx}(z) = \frac{1-a^2}{(1-az^{-1})(1-az)} \qquad \text{with} \qquad -1 < a < 1$$

Then according to (7.5.25)

$$F(z) = (1-az^{-1})[z^N/(1-az^{-1})]_+$$

But

$$\mathscr{L}^{-1}\{z^N/(1-az^{-1})\} = a^N a^n u(n+N)$$

so that

$$[z^N/(1-az^{-1})]_+ = a^N/(1-az^{-1})$$

and

$$F(z) = a^N$$

Thus optimum prediction consists of simply multiplying the present sample by a^N in this case. Since $b(nT) = a^n u(nT)$ and $\sigma^2 = 1 - a^2$, the mean square error according to (7.5.28) is

$$\Delta = (1-a^2) \sum_{n=0}^{N-1} a^{2n} = 1 - a^{2N} \qquad \blacktriangleleft$$

7.6 ONE-STEP LINEAR PREDICTION USING THE LAST *N* SAMPLES

In Example 7.5.3 a formula for optimum prediction using the infinite past was derived. In this section we examine the problem of optimally predicting one sample into the future using a linear combination of only the last N samples. The solution to this problem has recently found an interesting application in speech analysis, compression, and synthesis [2,54,55,70]. In this application a relatively short memory is required since the speech signal is not a stationary random process. However, it can be considered to be stationary over approximately 20 ms time intervals. A number of other problems that at first appear different turn out to have solutions that are identical in form to the solution of the optimum, finite memory, linear prediction problem. These include maximum likelihood estimation of the parameters in an autoregressive model for a random process [138] or, equivalently, maximum likelihood estimation of the parameters in an all-pole model for a sampled power spectral density [125], optimum inverse filtering [116], and orthogonalization of polynomials on the unit circle relative to a given distribution [130].

Let us assume that the linear minimum mean square error estimate of $x(nT)$

given the past N samples is

$$\hat{x}(nT) = - \sum_{k=1}^{N} d_{k,N} x(nT - kT) \qquad (7.6.1)$$

The prediction error or residual, as it is frequently called, is

$$e_N(nT) = x(nT) - \hat{x}(nT) = \sum_{k=0}^{N} d_{k,N} x(nT - kT) \qquad (7.6.2)$$

where by convention $d_{0,N} = 1$. According to the convolution theorem, the residual can be generated by passing $x(nT)$ through the filter

$$D_N(z) = \sum_{k=0}^{N} d_{k,N} z^{-k} \qquad \text{with} \qquad d_{0,N} = 1 \qquad (7.6.3)$$

We will call $D_N(z)$ the *analysis* filter of order N. Assuming that all the zeros of $D_N(z)$ are inside the unit circle, $x(nT)$ can be recovered by passing $e_N(nT)$ through the filter $1/D_N(z)$. We call $1/D_N(z)$ the *synthesis* filter of order N. It can be realized in direct form by the difference

$$x(nT) = e_N(nT) + \hat{x}(nT) = e_N(nT) - \sum_{k=1}^{N} d_{k,N} x(nT - kT) \qquad (7.6.4)$$

The utility of the linear predictor in speech compression comes from the fact that a synthesis filter of order about $N = 10$ to 12 provides an excellent model for the transfer function of the vocal tract. The residual for voiced speech can be approximated by a sequence of pulses spaced at the pitch period and for unvoiced speech by an uncorrelated noise sequence. Thus a short segment of speech on the order of 20 ms can be characterized by about 10 to 12 predictor coefficients, and pitch, amplitude, and voicing information. This information can be digitally encoded to give a bit rate significantly less than that required by ordinary pulse-code or delta modulation for equivalent voice quality.

The mean square prediction error is

$$p_N = E\{e_N^2(nT)\} = E\left\{ \sum_{i=0}^{N} \sum_{j=0}^{N} d_{i,N} d_{j,N} x(iT) x(jT) \right\}$$

$$= \sum_{i=0}^{N} \sum_{j=0}^{N} d_{i,N} d_{j,N} R_{xx}(iT - jT) \qquad (7.6.5)$$

The optimum predictor coefficients must satisfy the set of equations

$$\frac{\partial p_N}{\partial d_{k,N}} = 2 \sum_{i=0}^{N} d_{i,N} R_{xx}(iT - kT) = 0 \qquad \text{for} \qquad k = 1, \ldots, N$$

or

$$\sum_{i=0}^{N} d_{i,N} R_{xx}(iT - kT) = 0 \qquad \text{for} \qquad k = 1, \ldots, N \qquad (7.6.6)$$

Since $d_{0,N} = 1$ (7.6.6) is equivalent to

$$\sum_{i=1}^{N} d_{i,N} R_{xx}(iT - kT) = -R_{xx}(kT) \qquad \text{for} \qquad k = 1, \ldots, N \qquad (7.6.7)$$

or in matrix form

$$\begin{bmatrix} R_{xx}(0) & R_{xx}(T) & \cdots & R_{xx}(NT-T) \\ R_{xx}(T) & R_{xx}(0) & \cdots & R_{xx}(NT-2T) \\ \cdot & & & \cdot \\ \cdot & & & \cdot \\ \cdot & & & \cdot \\ R_{xx}(NT-T) & \cdots & & R_{xx}(0) \end{bmatrix} \begin{bmatrix} d_{1,N} \\ d_{2,N} \\ \cdot \\ \cdot \\ \cdot \\ d_{N,N} \end{bmatrix} = - \begin{bmatrix} R_{xx}(T) \\ R_{xx}(2T) \\ \cdot \\ \cdot \\ \cdot \\ R_{xx}(NT) \end{bmatrix} \qquad (7.6.8)$$

or

$$\mathbf{R}_N \mathbf{d}_N = -\mathbf{r}_N$$

with the obvious matrix correspondences. Therefore, the optimum coefficient vector is

$$\mathbf{d}_N = -\mathbf{R}_N^{-1} \mathbf{r}_N \qquad (7.6.10)$$

From (7.6.6) we see that the inner summation in (7.6.5) is zero for $i = 1, \ldots, N$, so that the mean square error reduces to

$$p_N = \sum_{j=0}^{N} d_{j,N} R_{xx}(jT) \qquad (7.6.11)$$

To completely characterize the linear predictor it will be necessary to find the optimum linear filter for "predicting" one sample backward from the future N samples. Suppose that the optimum estimate of $x(nT - NT - T)$ given the next N samples is

$$\tilde{x}(nT - NT - T) = -\sum_{k=1}^{N} c_{k,N} x(nT - kT) \qquad (7.6.12)$$

The backward prediction error is

$$w_N(nT) = x(nT - NT - T) - \tilde{x}(nT - NT - T) = \sum_{k=1}^{N+1} c_{k,N} x(nT - kT) \qquad (7.6.13)$$

with $c_{N+1,N} = 1$. The backward prediction residual $w_N(nT)$ can be generated by passing $x(nT)$ through the filter

$$C_N(z) = \sum_{k=1}^{N+1} c_{k,N} z^{-k} \qquad \text{with} \qquad c_{N+1,N} = 1 \qquad (7.6.14)$$

The mean square backward prediction error is

$$q_N = E\{w_N^2(nT)\} = \sum_{i=1}^{N+1} \sum_{j=1}^{N+1} c_{i,N} c_{j,N} R_{xx}(iT - jT) \qquad (7.6.15)$$

The optimum coefficients must satisfy the set of equations

$$\frac{\partial q_N}{\partial c_{k,N}} = 2 \sum_{i=1}^{N+1} c_{i,N} R_{xx}(iT - kT) = 0 \qquad \text{for} \qquad k = 1, \ldots, N$$

or

$$\sum_{i=1}^{N+1} c_{i,N} R_{xx}(iT - kT) = 0 \qquad \text{for} \qquad k = 1, \ldots, N \qquad (7.6.16)$$

Since $c_{N+1,N} = 1$ (7.6.16) can be put in the matrix form

$$\begin{bmatrix} R_{xx}(0) & R_{xx}(T) & \cdots & R_{xx}(NT-T) \\ R_{xx}(T) & R_{xx}(0) & & R_{xx}(NT-2T) \\ \cdot & & & \cdot \\ \cdot & & & \cdot \\ \cdot & & & \cdot \\ R_{xx}(NT-T) & \cdots & & R_{xx}(0) \end{bmatrix} \begin{bmatrix} c_{1,N} \\ c_{2,N} \\ \cdot \\ \cdot \\ \cdot \\ c_{N,N} \end{bmatrix} = - \begin{bmatrix} R_{xx}(NT) \\ \cdot \\ \cdot \\ \cdot \\ R_{xx}(2T) \\ R_{xx}(T) \end{bmatrix}$$

$$(7.6.17)$$

or

$$\mathbf{R}_N \mathbf{c}_N = -\boldsymbol{\rho}_N \qquad (7.6.18)$$

Thus the optimum coefficient vector is

$$\mathbf{c}_N = -\mathbf{R}_N^{-1} \boldsymbol{\rho}_N \qquad (7.6.19)$$

Notice that (7.6.17) differs from (7.6.8) only in that the elements of the column vector on the right-hand side appear in the reverse order. A little thought will show that because of the symmetry of the correlation matrix $\mathbf{R}_N$

$$\begin{bmatrix} c_{1,N} \\ \vdots \\ c_{N,N} \end{bmatrix} = \begin{bmatrix} d_{N,N} \\ \vdots \\ d_{1,N} \end{bmatrix} \qquad (7.6.20)$$

Therefore

$$C_N(z) = d_{N,N} z^{-1} + \cdots + d_{2,N} z^{-(N-1)} + d_{1,N} z^{-N} + z^{-(N+1)}$$
$$= z^{-1}(d_{N,N} + \cdots + d_{1,N} z^{-N+1} + z^{-N}) = z^{-(N+1)} D_N(z^{-1})$$

$$(7.6.21)$$

where $D_N(z)$ is defined by (7.6.3). Notice that $z^{-N} D_N(z^{-1})$ is simply $D_N(z)$ with its coefficients reversed and is similar to the reciprocal polynomial defined in Section 5.8. The sampled power spectral density for the backward prediction residual is

$$S_{w_N w_N}(z) = C_N(z) C_N(z^{-1}) S_{xx}(z) = D_N(z) D_N(z^{-1}) S_{xx}(z) = S_{e_N e_N}(z) \quad (7.6.22)$$

Thus the backward and forward prediction residuals have the same power spectral densities, and the two prediction mean square errors are identical, that is,

$$p_N = q_N \quad (7.6.23)$$

Another useful quantity is the correlation between the forward and backward prediction residuals. Using (7.6.2) and (7.6.13) we find that

$$v_N = E\{e_N(nT) w_N(nT)\} = \sum_{k=1}^{N+1} c_{k,N} \sum_{i=0}^{N} d_{i,N} R_{xx}(iT - kT) \quad (7.6.24)$$

According to (7.6.6), the inner summation in (7.6.24) is zero for $k = 1, \ldots, N$, so that

$$v_N = \sum_{i=0}^{N} d_{i,N} R_{xx}(iT - NT - T) \quad (7.6.25)$$

The parameter estimation formula (7.6.10) can be evaluated using standard matrix inversion techniques that apply to arbitrary, nonsingular matrices. However, we will now develop a more efficient recursive solution algorithm that exploits the symmetry of the correlation matrix $\mathbf{R}_N$. Equations 7.6.11, 7.6.6, and 7.6.25 can be combined into the single matrix equation

$$\begin{bmatrix} R_{xx}(0) & R_{xx}(T) & \cdots & R_{xx}(NT) & R_{xx}(NT+T) \\ R_{xx}(T) & R_{xx}(0) & \cdots & R_{xx}(NT-T) & R_{xx}(NT) \\ \vdots & & & \vdots & \vdots \\ R_{xx}(NT) & & & R_{xx}(0) & R_{xx}(T) \\ R_{xx}(NT+T) & \cdots & & R_{xx}(T) & R_{xx}(0) \end{bmatrix} \begin{bmatrix} 1 \\ d_{1,N} \\ \vdots \\ d_{N,N} \\ 0 \end{bmatrix} = \begin{bmatrix} p_N \\ 0 \\ \vdots \\ 0 \\ v_N \end{bmatrix}$$

$$(7.6.26)$$

This set of equations can also be written as

$$\begin{bmatrix} R_{xx}(0) & \cdots & R_{xx}(NT) & R_{xx}(NT+T) \\ R_{xx}(T) & & R_{xx}(NT-T) & R_{xx}(NT) \\ \vdots & & \vdots & \vdots \\ R_{xx}(NT) & & R_{xx}(0) & R_{xx}(T) \\ R_{xx}(NT+T) & \cdots & R_{xx}(T) & R_{xx}(0) \end{bmatrix} \begin{bmatrix} 0 \\ d_{N,N} \\ \vdots \\ d_{1,N} \\ 1 \end{bmatrix} = \begin{bmatrix} v_N \\ 0 \\ \vdots \\ 0 \\ p_N \end{bmatrix} \quad (7.6.27)$$

Multiplying (7.6.27) by an arbitrary constant k_{N+1} and subtracting from (7.6.26) gives

$$
\begin{bmatrix}
R_{xx}(0) & \cdots & R_{xx}(NT+T) \\
R_{xx}(T) & \cdots & R_{xx}(NT) \\
\vdots & & \vdots \\
R_{xx}(NT) & \cdots & R_{xx}(T) \\
R_{xx}(NT+T) & \cdots & R_{xx}(0)
\end{bmatrix}
\begin{bmatrix}
1 \\
d_{1,N}-k_{N+1}d_{N,N} \\
\vdots \\
d_{N,N}-k_{N+1}d_{1,n} \\
-k_{N+1}
\end{bmatrix}
=
\begin{bmatrix}
p_N-k_{N+1}v_N \\
0 \\
\vdots \\
0 \\
v_N-k_{N+1}p_N
\end{bmatrix}
$$

$$(7.6.28)$$

For the predictor of order $N+1$ (7.6.26) becomes

$$
\begin{bmatrix}
 & & \vdots & R_{xx}(NT+2T) \\
 & & \vdots & R_{xx}(NT+T) \\
 & \mathbf{R}_{N+2} & \vdots & \vdots \\
 & & \vdots & R_{xx}(T) \\
\hline
R_{xx}(NT+2T) & \cdots & \vdots & R_{xx}(0)
\end{bmatrix}
\begin{bmatrix}
1 \\
d_{1,N+1} \\
\vdots \\
d_{N+1,N+1} \\
0
\end{bmatrix}
=
\begin{bmatrix}
p_{N+1} \\
0 \\
\vdots \\
0 \\
v_{N+1}
\end{bmatrix}
$$

$$(7.6.29)$$

where $\mathbf{R}_{N+2}$ is the correlation matrix in (7.6.26). Now suppose that k_{N+1} is chosen so that the bottom element of the column vector on the right-hand side of (7.6.28) is zero, that is

$$k_{N+1} = v_N/p_N \qquad (7.6.30)$$

Then the $N+1$ equations consisting of all but the first row of (7.6.28) have the same form as the $N+1$ equations consisting of all but the first and last rows of (7.6.29). Since the solution to these sets of equations must be unique, we see that

$$d_{k,N+1} = d_{k,N} - k_{N+1}d_{N+1-k,N} \qquad \text{for} \qquad k=1,\ldots,N \qquad (7.6.31)$$

$$d_{N+1,N+1} = -k_{N+1} \qquad (7.6.32)$$

$$p_{N+1} = p_N - k_{N+1}v_N = p_N(1-k_{N+1}^2) \qquad (7.6.33)$$

and

$$v_{N+1} = \sum_{k=0}^{N+1} R_{xx}(NT+2T-kT)d_{k,N+1} \qquad (7.6.34)$$

From (7.6.31), (7.6.32), and (7.6.21) it follows that

$$D_{N+1}(z) = D_N(z) - k_{N+1}z^{-(N+1)}D_N(z^{-1}) = D_N(z) - k_{N+1}C_N(z)$$

$$(7.6.35)$$

and

$$C_{N+1}(z) = z^{-1}[C_N(z) - k_{N+1}D_N(z)] \tag{7.6.36}$$

Equations 7.6.30 to 7.6.36 provide a means for recursively calculating the $N+1$st order predictor coefficients from p_N, v_N, and the Nth order predictor coefficients. First k_{N+1} can be calculated by (7.6.30). Then the $N+1$st order predictor coefficients can be calculated using (7.6.31) and (7.6.32) or (7.6.35). Finally, p_{N+1} and v_{N+1} can be calculated from (7.6.33) and (7.6.34). The recursion can be started with $N=0$ by observing that a linear combination of $N=0$ samples must be zero so that the forward prediction residual is $x(nT)$ and the backward prediction residual is $x(nT-T)$. Therefore, the required initial conditions are

$$p_0 = E\{x^2(nT)\} = R_{xx}(0) \tag{7.6.37}$$
$$v_0 = E\{x(nT)x(nT-T)\} = R_{xx}(T) \tag{7.6.38}$$

and

$$d_{0,0} = 1 \quad \text{or} \quad D_0(z) = 1 \tag{7.6.39}$$

The lower order predictors can also be calculated recursively from the higher order predictors. Replacing z by z^{-1} in (7.6.35) and multiplying both sides by $k_{N+1}z^{-(N+1)}$ we obtain

$$k_{N+1}z^{-(N+1)}D_{N+1}(z^{-1}) = k_{N+1}z^{-(N+1)}D_N(z^{-1}) - k_{N+1}^2 D_N(z) \tag{7.6.40}$$

Adding (7.6.35) and (7.6.40) gives

$$D_N(z) = [D_{N+1}(z) + k_{N+1}z^{-(N+1)}D_{N+1}(z^{-1})]/(1 - k_{N+1}^2) \tag{7.6.41}$$

Also, from (7.6.32) we see that

$$k_N = -d_{N,N} \tag{7.6.42}$$

Since $p_N = q_N$, the parameter k_{N+1} can also be written as

$$k_{N+1} = \frac{v_N}{(p_N q_N)^{\frac{1}{2}}} \tag{7.6.43}$$

This is simply the correlation coefficient between the forward and backward prediction residuals. Thus $|k_{N+1}| \le 1$. The k_N's are known as the *partial correlation coefficients*. They are related to the reflection coefficients in a cascaded concentric pipe model for an acoustic tube [73,92]. From (7.6.33) we see that the mean square error p_{N+1} is zero if $|k_{N+1}| = 1$. This is a rare special case. From (7.6.22) it follows that

$$p_N = \frac{1}{\omega_s} \int_{-\omega_s/2}^{\omega_s/2} |D_N(e^{j\omega T})|^2 S_{xx}^*(\omega)\, d\omega \tag{7.6.44}$$

Thus p_N must be strictly positive if $S_{xx}^*(\omega)$ is a bounded, piecewise continuous function that is nonzero on a positive length interval. In this case $|k_N|<1$ for all finite N.

The synthesis filter $1/D_N(z)$ will be useful only if it is stable. We will now show that this filter is stable if and only if $|k_n|<1$ for $n \le N$. First let us assume that $D_n(z)$ has no zeros on or outside the unit circle so that $1/D_n(z)$ is stable. Consider the quotient

$$\frac{D_{n+1}(z)}{D_n(z)} = 1 + \frac{D_{n+1}(z) - D_n(z)}{D_n(z)} \qquad (7.6.45)$$

The number of zeros of $D_{n+1}(z)$ on or outside the unit circle can be determined by applying the Nyquist criterion discussed in Section 5.9 to the function

$$G(z) = \frac{D_{n+1}(z) - D_n(z)}{D_n(z)} \qquad (7.6.46)$$

Using (7.6.35) we find that

$$G(z) = -k_{n+1} z^{-(n+1)} D_n(z^{-1})/D_n(z) \qquad (7.6.47)$$

For $z = e^{j\omega T}$

$$|G(z)| = |k_{n+1}| \qquad (7.6.48)$$

since $D_n(z^{-1})$ and $D_n(z)$ are complex conjugates on the unit circle. Therefore, the Nyquist plot will not encircle the -1 point if $|k_{n+1}|<1$ and $D_{n+1}(z)$ will have no zeros on or outside the unit circle. If $|k_{n+1}|=1$ the Nyquist plot will pass through the -1 point indicating that $D_{n+1}(z)$ has a zero on the unit circle. If we assume that k_{n+1} is simply an arbitrary constant in the recursive equations rather than a correlation coefficient and let $|k_{n+1}|>1$, then the Nyquist plot will encircle the -1 point indicating that $D_{n+1}(z)$ has zeros outside the unit circle. To complete the proof, observe from the recursion formulas (7.6.30 to 7.6.35) and the initial conditions (7.6.37 to 7.6.39) that $k_1 = R_{xx}(T)/R_{xx}(0)$ and $D_1(z) = 1 - k_1 z^{-1}$. Thus the zero of $D_1(z)$ is inside the unit circle if we rule out the rare case of partial correlation coefficients with unity magnitude. Notice that we have essentially just derived the modified Schur-Cohn test for stability outlined in Section 5.7.

It is interesting to observe that both the analysis and synthesis filters can be realized as cascaded lattice networks in which the filter coefficients are the partial correlation coefficients. A realization of the analysis filter is shown in Fig. 7.6.1. This structure is simply a block diagram showing the recursive calculation of $C_N(z)$ and $D_N(z)$ using (7.6.35) and (7.6.36). The input to the kth adder from the left along the top path is the forward prediction residual $e_{k-1}(nT)$ and along the bottom path is the backward prediction residual $w_{k-1}(nT)$.

A realization of the synthesis filter is shown in Fig. 7.6.2. The fact that its

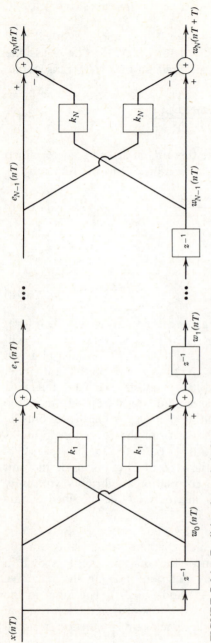

FIGURE 7.6.1. Realizing the analysis filter by cascaded lattice networks.

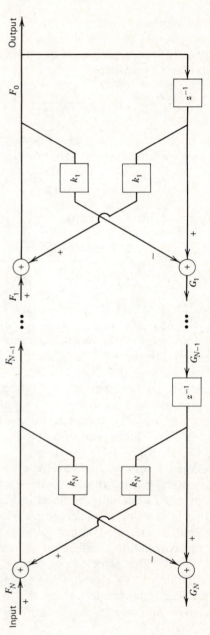

FIGURE 7.6.2. Realizing the synthesis filter by cascaded lattice networks.

transfer function is $1/D_N(z)$ can be proven by induction. If we assume that

$$\frac{G_{N-1}(z)}{F_{N-1}(z)} = \frac{zC_{N-1}(z)}{D_{N-1}(z)} \tag{7.6.49}$$

then it follows from the filter structure and (7.6.35) and (7.6.36) that

$$\frac{F_{N-1}(z)}{F_N(z)} = \frac{D_{N-1}(z)}{D_N(z)} \tag{7.6.50}$$

and

$$\frac{G_N(z)}{F_N(z)} = \frac{zC_N(z)}{D_N(z)} \tag{7.6.51}$$

Also it is easy to show that

$$\frac{F_0(z)}{F_1(z)} = \frac{1}{1 - k_1 z^{-1}} = \frac{D_0(z)}{D_1(z)} \tag{7.6.52}$$

and

$$\frac{G_1(z)}{F_1(z)} = \frac{-k_1 + z^{-1}}{1 - k_1 z^{-1}} = \frac{zC_1(z)}{D_1(z)} \tag{7.6.53}$$

The cascaded lattice network realization of the synthesis filter has several advantages over a direct form realization. For one thing, stability is guaranteed as long as $|k_n| < 1$ for $n \le N$. In most linear predictive voice compression systems the k_n's are transmitted rather than the coefficients of $D_N(z)$ so that quantization and/or transmission errors cannot cause instability. For another, it is convenient to know a priori that the variables to be digitized are bounded in magnitude by unity.

Finally, it is worthwhile pointing out that we have shown, in fact, that any stable filter with a pulse transfer function of the form

$$1 \Big/ \Big(1 + \sum_{n=1}^{N} d_n z^{-n} \Big)$$

can be realized as a cascaded lattice network. The filter coefficients can be found using the backward recursion equations (7.6.41) and (7.6.42). Any numerator of degree N or less can be added to the pulse transfer function by forming a linear combination of the variables F_0 and $G_1, \ldots, G_N$ shown in Fig. 7.6.2 [40].

7.7 MIXED CONTINUOUS AND DISCRETE-TIME SYSTEMS

Actual systems frequently include both continuous and discrete-time subsystems. In this section we derive correlation and power spectral density relations for two common types of mixed systems.

First, let us consider the *type 1* mixed system shown in Fig. 7.7.1. We will assume that $a(t)$ and $b(t)$ are jointly wide-sense stationary random processes and that the sampling instants are $nT+X$ where X is a random variable uniformly distributed over $[0, T)$. Then from Fig. 7.7.1 we find that

$$a_1(t) = \sum_{k=-\infty}^{\infty} a(kT+X)h(t-kT-X)$$

Therefore

$$a_1(nT+X) = \sum_{k=-\infty}^{\infty} a(kT+X)h(nT-kT) \tag{7.7.1}$$

Also

$$c(t) = \sum_{n=-\infty}^{\infty} a_1(nT+X)g(t-nT-X)$$

$$= \sum_{n=-\infty}^{\infty} \sum_{k=-\infty}^{\infty} a(kT+X)h(nT-kT)g(t-nT-X) \tag{7.7.2}$$

Letting $n-k=m$ (7.7.2) becomes

$$c(t) = \sum_{m=-\infty}^{\infty} \sum_{k=-\infty}^{\infty} a(kT+X)h(mT)g(t-mT-kT-X) \tag{7.7.3}$$

The cross-correlation function for $c(t)$ and $b(t)$ is

$$R_{cb}(t+\tau, t) = E\{c(t+\tau)b(t)\} \tag{7.7.4}$$

Using the properties of conditional expectation [104] (7.7.4) can be expressed as

$$R_{cb}(t+\tau, t) = E\{E\{c(t+\tau)b(t)/X\}\} \tag{7.7.5}$$

Using (7.7.3) the conditional expectation for fixed X becomes

$$E\{c(t+\tau)b(t)/X=x\} = \sum_{m=-\infty}^{\infty} \sum_{k=-\infty}^{\infty} R_{ab}(kT+x-t)h(mT)g(t+\tau-mT-kT-x)$$

$$\tag{7.7.6}$$

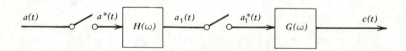

FIGURE 7.7.1. Type 1 mixed system.

Averaging (7.7.6) with respect to the probability density for X gives

$$R_{cb}(t+\tau,\,t)=\frac{1}{T}\int_0^T E\{c(t+\tau)b(t)/X=x\}\,dx$$

$$=\frac{1}{T}\sum_{m=-\infty}^{\infty}h(mT)\sum_{k=-\infty}^{\infty}\int_0^T R_{ab}(kT+x-t)g(t+\tau-mT-kT-x)\,dx$$

$$(7.7.7)$$

Letting $\mu=kT+x-t$ (7.7.7) becomes

$$R_{cb}(t+\tau,\,t)=\frac{1}{T}\sum_{m=-\infty}^{\infty}h(mT)\sum_{k=-\infty}^{\infty}\int_{kT-t}^{(k+1)T-t}R_{ab}(\mu)g(\tau-mT-\mu)\,d\mu$$

$$=\frac{1}{T}\sum_{m=-\infty}^{\infty}h(mT)\int_{-\infty}^{\infty}R_{ab}(\mu)g(\tau-mT-\mu)\,d\mu \qquad (7.7.8)$$

Notice that (7.7.8) is only a function of τ, so that $c(t)$ and $b(t)$ are jointly wide-sense stationary. In the future we will denote the cross-correlation function by $R_{cb}(\tau)$ instead of $R_{cb}(t+\tau,\,t)$. If the random variable X had not been added to the sampling instants, then $c(t)$ and $b(t)$ would not have been jointly wide-sense stationary. Taking the Fourier transform of (7.7.8) with respect to τ, we find that

$$S_{cb}(\omega)=\frac{1}{T}\sum_{m=-\infty}^{\infty}h(mT)S_{ab}(\omega)G(\omega)e^{-j\omega mT}$$

or

$$S_{cb}(\omega)=\frac{1}{T}H^*(\omega)G(\omega)S_{ab}(\omega) \qquad (7.7.9)$$

A *type 2* mixed system is shown in Fig. 7.7.2. For this system

$$c(t)=\sum_{k=-\infty}^{\infty}a(kT+X)g(t-kT-X) \qquad (7.7.10)$$

and

$$d(t)=\sum_{n=-\infty}^{\infty}b(nT+X)h(t-nT-X) \qquad (7.7.11)$$

As before

$$R_{cd}(\tau)=E\{c(t+\tau)d(t)\}=E\{E\{c(t+\tau)d(t)/X\}\} \qquad (7.7.12)$$

Using (7.7.10) and (7.7.11) we find that the conditional expectation in (7.7.12) for fixed X is

$$E\{c(t+\tau)d(t)/X=x\}=\sum_{k=-\infty}^{\infty}\sum_{n=-\infty}^{\infty}R_{ab}(kT-nT)g(t+\tau-kT-x)h(t-nT-x)$$

$$(7.7.13)$$

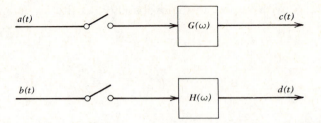

FIGURE 7.7.2. Type 2 mixed system.

Thus

$$R_{cd}(\tau) = \frac{1}{T} \int_0^T E\{c(t+\tau)d(t)/X = x\} \, dx$$

$$= \frac{1}{T} \sum_{k=-\infty}^{\infty} \sum_{n=-\infty}^{\infty} R_{ab}(kT-nT) \int_0^T g(t+\tau-kT-x)h(t-nT-x) \, dx$$

(7.7.14)

Changing the outer summation index so that $m = k - n$ and then making the change of variables $t - nT - x = \mu$ in the integral (7.7.14) becomes

$$R_{cd}(\tau) = \frac{1}{T} \sum_{m=-\infty}^{\infty} R_{ab}(mT) \sum_{n=-\infty}^{\infty} \int_{t-nT-T}^{t-nT} g(\mu+\tau-mT)h(\mu) \, d\mu$$

$$= \frac{1}{T} \sum_{m=-\infty}^{\infty} R_{ab}(mT) \int_{-\infty}^{\infty} g(\mu+\tau-mT)h(\mu) \, d\mu$$

(7.7.15)

Taking the Fourier transform of (7.7.15), it follows that

$$S_{cd}(\omega) = \frac{1}{T} \sum_{m=-\infty}^{\infty} R_{ab}(mT)G(\omega)H(-\omega)e^{-j\omega mT}$$

or

$$S_{cd}(\omega) = \frac{1}{T} G(\omega)H(-\omega)S_{ab}^*(\omega)$$

(7.7.16)

Example 7.7.1

Suppose that $G(\omega)$ in Fig. 7.7.2 is a zero-order hold so that

$$G(\omega) = \frac{1 - e^{-j\omega T}}{j\omega}$$

By letting $b(t) = a(t)$ and $H(\omega) = G(\omega)$, we find from (7.7.16) that

$$S_{cc}(\omega) = \frac{1}{T} G(\omega)G(-\omega)S_{aa}^*(\omega) = T \frac{\sin^2(\omega T/2)}{(\omega T/2)^2} S_{aa}^*(\omega)$$

(7.7.17)

$S_{cc}(\omega)$ can be interpreted as being the Fourier transform of the output of a filter with the transfer function $F(\omega) = (1/T)G(\omega)G(-\omega)$ driven by the input signal

$$R_{aa}^{*}(\tau) = \sum_{n=-\infty}^{\infty} R_{aa}(nT)\,\delta(\tau - nT)$$

The impulse response corresponding to $F(\omega)$ is

$$f(\tau) = \frac{1}{T}\int_{-\infty}^{\infty} g(\tau + \mu)g(\mu)\,d\mu \qquad (7.7.18)$$

Since

$$g(\tau) = \begin{cases} 1 & \text{for} \quad 0 \le \tau < T \\ 0 & \text{elsewhere} \end{cases}$$

it follows from (7.7.18) that

$$f(\tau) = \begin{cases} 1 - |\tau/T| & \text{for} \quad |\tau| < T \\ 0 & \text{elsewhere} \end{cases}$$

This is simply the impulse response of the linear point connector discussed in Section 3.8 with the one-sample delay removed. Therefore, $R_{cc}(\tau)$ consists of straight-line segments connecting the sample values $R_{aa}(nT)$. In equation form

$$R_{cc}(\tau) = \sum_{n=-\infty}^{\infty} R_{aa}(nT)\left(1 - \left|\frac{\tau - nT}{T}\right|\right) \qquad (7.7.19)$$

Equation 7.7.19 also follows directly from (7.7.15). ◀

7.8 OPTIMUM RECONSTRUCTION FILTERS

We will now examine the problem of finding the transfer function of the filter that is best in the minimum mean square error sense for reconstructing a continuous-time signal from a noise corrupted sampled signal. The problem is illustrated in Fig. 7.8.1. The signal $f(t)$ is a wide-sense stationary random process with power spectral density $S_{ff}(\omega)$. The signal $v(t)$ represents additive noise with power spectral density $S_{vv}(\omega)$. We will assume that $S_{fv}(\omega)$ is also known. The block labeled $L(\omega)$ represents a desired operation on $f(t)$. As in Section 7.7, we will assume that the noise corrupted signal $y(t)$ is sampled at time instants $nT + X$ where X is a random variable uniformly distributed over $[0, T)$. The problem is to choose $G(\omega)$ to minimize the mean square error between $c(t)$ and the desired signal $c_d(t)$.

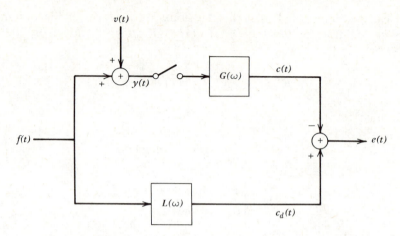

FIGURE 7.8.1. The signal reconstruction problem.

The mean square error can be calculated as

$$\Delta = E\{e^2(t)\} = R_{ee}(0) = \frac{1}{2\pi} \int_{-\infty}^{\infty} S_{ee}(\omega) \, d\omega \qquad (7.8.1)$$

Now

$$R_{ee}(\tau) = E\{[c_d(t+\tau) - c(t+\tau)][c_d(t) - c(t)]\}$$
$$= R_{c_d c_d}(\tau) - R_{cc_d}(\tau) - R_{c_d c}(\tau) + R_{cc}(\tau) \qquad (7.8.2)$$

so that

$$S_{ee}(\omega) = S_{c_d c_d}(\omega) - S_{cc_d}(\omega) - S_{c_d c}(\omega) + S_{cc}(\omega) \qquad (7.8.3)$$

From the input–output relationship for continuous-time systems, it follows that

$$S_{c_d c_d}(\omega) = L(\omega)L(-\omega)S_{ff}(\omega) \qquad (7.8.4)$$

According to (7.7.9) with $H^*(\omega) = 1$

$$S_{cc_d}(\omega) = \frac{1}{T} G(\omega)S_{yc_d}(\omega) = \frac{1}{T} G(\omega)L(-\omega)S_{yf}(\omega) \qquad (7.8.5)$$

Also

$$S_{c_d c}(\omega) = S_{cc_d}(-\omega) = \frac{1}{T} G(-\omega)L(\omega)S_{fy}(\omega) \qquad (7.8.6)$$

From (7.7.16) we find that

$$S_{cc}(\omega) = \frac{1}{T} G(\omega)G(-\omega)S_{yy}^*(\omega) \qquad (7.8.7)$$

Therefore

$$\Delta = \frac{1}{2\pi} \int_{-\infty}^{\infty} \left[L(\omega)L(-\omega)S_{ff}(\omega) - \frac{1}{T}L(\omega)G(-\omega)S_{fy}(\omega) \right.$$

$$\left. -\frac{1}{T}L(-\omega)G(\omega)S_{yf}(\omega) + \frac{1}{T}G(\omega)G(-\omega)S_{yy}^*(\omega) \right] d\omega \qquad (7.8.8)$$

Making the change of variables $w = -\omega$ in the integral over the second term on the right, we find that

$$\frac{1}{T}\int_{-\infty}^{\infty} L(\omega)G(-\omega)S_{fy}(\omega) \, d\omega = \frac{1}{T}\int_{-\infty}^{\infty} L(-w)G(w)S_{fy}(-w) \, dw$$

Since $S_{fy}(-w) = S_{yf}(w)$, we see that the integrals over the second and third terms on the right-hand side are identical. To simplify the notation we will denote $L(\omega)$ simply by L and $L(-\omega)$ by $\bar{L}$, etc. Combining the second and third terms and using the shorthand notation, the mean square error can be written as

$$\Delta = \frac{1}{2\pi} \int_{-\infty}^{\infty} \left[L\bar{L}S_{ff} - \frac{2}{T}L\bar{G}S_{fy} + \frac{1}{T}G\bar{G}S_{yy}^* \right] d\omega \qquad (7.8.9)$$

If $G(\omega)$ is replaced by the filter $G_1(\omega) = G(\omega) + \lambda(\omega)$ where $\lambda(\omega)$ is an arbitrary perturbation, the new mean square error becomes

$$\Delta_1 = \frac{1}{2\pi} \int_{-\infty}^{\infty} \left[L\bar{L}S_{ff} - \frac{2}{T}L(\bar{G} + \bar{\lambda})S_{fy} + \frac{1}{T}(G + \lambda)(\bar{G} + \bar{\lambda})S_{yy}^* \right] d\omega$$
$$(7.8.10)$$

With some manipulation (7.8.10) can be put in the form

$$\Delta_1 = \Delta + \frac{1}{2\pi T} \int_{-\infty}^{\infty} \lambda\bar{\lambda}S_{yy}^* \, d\omega + \frac{1}{\pi T} \int_{-\infty}^{\infty} \bar{\lambda}(GS_{yy}^* - LS_{fy}) \, d\omega \qquad (7.8.11)$$

For any real signal $a(t)$, $A(-\omega)$ is the complex conjugate of $A(\omega)$. Therefore

$$\lambda\bar{\lambda}S_{yy}^* = |\lambda(\omega)|^2 S_{yy}^*(\omega) \geq 0$$

so that the second term on the right-hand side of (7.8.11) must be nonnegative. Suppose G is chosen so that

$$\int_{-\infty}^{\infty} \bar{\lambda}(GS_{yy}^* - LS_{fy}) \, d\omega = 0 \qquad (7.8.12)$$

regardless of the perturbation λ. Then the third term on the right-hand side of (7.8.11) will be identically zero and $\Delta_1 \geq \Delta$ for any choice of λ. Therefore (7.8.12) is a necessary and sufficient condition that the optimum filter must satisfy.

If no realizability constraints are imposed, the obvious solution to (7.8.12) is

$$GS_{yy}^* - LS_{fy} = 0$$

or

$$G(\omega) = L(\omega)\frac{S_{fy}(\omega)}{S_{yy}^*(\omega)} \tag{7.8.13}$$

From (7.8.9) it follows that the corresponding mean square error is

$$\Delta = \frac{1}{2\pi}\int_{-\infty}^{\infty}\left[L\bar{L}S_{ff} - \frac{1}{T}G\bar{L}S_{fy}\right]d\omega \tag{7.8.14}$$

Let us now consider the special case of pure interpolation, that is, $v(t) = 0$, $L(\omega) = 1$, and no realizability constraint. Then $y(t) = f(t)$ so that

$$G(\omega) = \frac{S_{ff}(\omega)}{S_{ff}^*(\omega)} \tag{7.8.15}$$

and

$$\Delta = \frac{1}{2\pi}\int_{-\infty}^{\infty}S_{ff}(\omega)\left[1 - \frac{1}{T}G(\omega)\right]d\omega \tag{7.8.16}$$

The output of the reconstruction filter is

$$c(t) = \sum_{n=-\infty}^{\infty}f(nT+X)g(t-nT-X) \tag{7.8.17}$$

Notice that

$$G^*(\omega) = \frac{S_{ff}^*(\omega)}{S_{ff}^*(\omega)} = 1 \tag{7.8.18}$$

Therefore

$$g(nT) = \delta_{n0} \tag{7.8.19}$$

The output of the reconstruction filter at the sampling instant $kT + X$ is

$$c(kT+X) = \sum_{n=-\infty}^{\infty}f(nT+X)g(kT-nT) \tag{7.8.20}$$

Using (7.8.19), this reduces to

$$c(kT+X) = f(kT+X) \tag{7.8.21}$$

Thus the reconstructed signal passes directly through the input samples. This is what one would expect intuitively in the noiseless case.

Example 7.8.1 Sampling Theorem for Band-Limited Random Processes

Suppose that $f(t)$ is band-limited to $|\omega| < \omega_s/2$, i.e., $S_{ff}(\omega) \equiv 0$ for $|\omega| \geq \omega_s/2$. According to the aliasing formula

$$S_{ff}^*(\omega) = \frac{1}{T}\sum_{n=-\infty}^{\infty}S_{ff}(\omega - n\omega_s)$$

The aliases do not overlap because $S_{ff}(\omega)$ is sufficiently band-limited. Therefore

$$S_{ff}^*(\omega) = \frac{1}{T} S_{ff}(\omega) \qquad \text{for} \qquad |\omega| < \omega_s/2$$

In this case the optimum interpolation filter given by (7.8.15) reduces to

$$G(\omega) = \begin{cases} T & \text{for} \qquad |\omega| < \omega_s/2 \\ 0 & \text{elsewhere} \end{cases} \tag{7.8.22}$$

This is an ideal low-pass filter and has the impulse response

$$g(t) = \frac{\sin(\omega_s t/2)}{\omega_s t/2} \tag{7.8.23}$$

The reconstructed signal is

$$c(t) = \sum_{n=-\infty}^{\infty} f(nT - X) \frac{\sin[\omega_s(t - nT - X)/2]}{\omega_s(t - nT - X)/2} \tag{7.8.24}$$

It is easy to see from (7.8.16) that the mean square error is zero. Therefore, $c(t) = f(t)$ in the mean square error sense. Equation 7.8.24 is essentially identical to the cardinal series interpolation formula discussed in Section 2.4 for deterministic signals. ◀

Now let us require that the reconstruction filter be physically realizable. This is equivalent to requiring that the poles of $G(\omega)$ and $\lambda(\omega)$ lie in the upper half ω-plane. Assuming that $f(t)$ and $v(t)$ have finite power, it follows that the integrand in (7.8.12) must approach zero at least like $1/\omega^2$ as ω becomes infinite. Suppose that the path of integration is closed with an infinite radius semicircle around the upper half ω-plane. Then the integral along the semi-circle is zero and (7.8.12) can be satisfied by requiring that all poles of the integrand lie in the lower half plane. By assumption, all the poles of $\lambda(-\omega)$ must lie in the lower half plane so that it is only necessary to require that

$$G(\omega)S_{yy}^*(\omega) - L(\omega)S_{fy}(\omega) = Q(\omega) \tag{7.8.25}$$

where all the poles of $Q(\omega)$ lie in the lower half plane. It follows from Theorem 7.2.1, the Spectral Factorization Theorem, that $S_{yy}^*(\omega)$ can be represented as

$$S_{yy}^*(\omega) = \sigma^2 B^*(\omega) B^*(-\omega) \tag{7.8.26}$$

where all the poles and zeros of $B^*(\omega)$ lie in the upper half ω-plane so that all the poles and zeros of $B^*(-\omega)$ must lie in the lower half plane. Dividing (7.8.25) by $B^*(-\omega)$ gives

$$G(\omega)\sigma^2 B^*(\omega) - L(\omega)\frac{S_{fy}(\omega)}{B^*(-\omega)} = \frac{Q(\omega)}{B^*(-\omega)} \tag{7.8.27}$$

The second term on the left hand side of (7.8.27) can be partial fractioned and written as

$$L(\omega)\frac{S_{fy}(\omega)}{B^*(-\omega)} = \left[\frac{L(\omega)S_{fy}(\omega)}{B^*(-\omega)}\right]_+ + \left[\frac{L(\omega)S_{fy}(\omega)}{B^*(-\omega)}\right]_- \qquad (7.8.28)$$

where [ ]$_+$ indicates the sum of all partial fraction terms with poles in the upper half plane and [ ]$_-$ indicates the sum of all partial fraction terms with poles in the lower half plane. Using this decomposition (7.8.27) can be arranged in the form

$$G(\omega)\sigma^2 B^*(\omega) - \left[\frac{L(\omega)S_{fy}(\omega)}{B^*(-\omega)}\right]_+ = \frac{Q(\omega)}{B^*(-\omega)} + \left[\frac{L(\omega)S_{fy}(\omega)}{B^*(-\omega)}\right]_- \qquad (7.8.29)$$

Notice that all the poles of the left-hand side of (7.8.29) lie in the upper half plane while all the poles of the right-hand side lie in the lower half plane. Therefore, neither side can have any poles. According to Liouville's Theorem in complex variable theory, a rational function with no poles must be a constant. The factor $B^*(\omega)$ is bounded and periodic for real ω assuming that it has no poles along the real ω axis. Both $G(\omega)$ and $L(\omega)S_{fy}(\omega)$ must approach zero as ω becomes infinite in realistic situations. Therefore, this constant must be zero. Equating the left-hand side of (7.8.29) to zero, we find that the transfer function of the optimum realizable reconstruction filter is

$$G(\omega) = \frac{1}{\sigma^2 B^*(\omega)}\left[\frac{L(\omega)S_{fy}(\omega)}{B^*(-\omega)}\right]_+ \qquad (7.8.30)$$

Substituting (7.8.30) into (7.8.9) we find that the mean square error for the optimum realizable filter is

$$\Delta = \frac{1}{2\pi}\int_{-\infty}^{\infty}\left\{L\bar{L}S_{ff} - \frac{2}{T\sigma^2}\frac{LS_{fy}}{\bar{B}^*}\left[\frac{\overline{LS_{fy}}}{\bar{B}^*}\right]_+ + \frac{1}{T\sigma^2}\left[\frac{LS_{fy}}{\bar{B}^*}\right]_+\left[\frac{\overline{LS_{fy}}}{\bar{B}^*}\right]_+\right\}d\omega \qquad (7.8.31)$$

If we let

$$R(\omega) = \frac{L(\omega)S_{fy}(\omega)}{B^*(-\omega)} \qquad (7.8.32)$$

and

$$R_+(\omega) = [R(\omega)]_+ \qquad (7.8.33)$$

then (7.8.31) can be written as

$$\Delta = \frac{1}{2\pi}\int_{-\infty}^{\infty}\left[L\bar{L}S_{ff} - \frac{2}{T\sigma^2}R\bar{R}_+ + \frac{1}{T\sigma^2}R_+\bar{R}_+\right]d\omega \qquad (7.8.34)$$

In the time domain the transform $R_+(\omega)$ corresponds to $r_+(t) = r(t)u(t)$ where $u(t)$ is the unit step function. Using Parseval's Theorem we see that

$$\int_{-\infty}^{\infty}r(t)r_+(t)\,dt = \frac{1}{2\pi}\int_{-\infty}^{\infty}R\bar{R}_+\,d\omega = \frac{1}{2\pi}\int_{-\infty}^{\infty}\bar{R}R_+\,d\omega \qquad (7.8.35)$$

and

$$\int_{-\infty}^{\infty} r_+^2(t)\, dt = \frac{1}{2\pi} \int_{-\infty}^{\infty} R_+ \bar{R}_+ \, d\omega \tag{7.8.36}$$

Also

$$\int_{-\infty}^{\infty} r(t) r_+(t)\, dt = \int_{-\infty}^{\infty} r_+^2(t)\, dt = \int_{0}^{\infty} r^2(t)\, dt \tag{7.8.37}$$

Therefore, the integrals over the second and third terms in the integrand of (7.8.34) can be combined to give

$$\Delta = \frac{1}{2\pi} \int_{-\infty}^{\infty} \left[L\bar{L}S_{ff} - \frac{1}{T\sigma^2} R\bar{R}_+ \right] d\omega \tag{7.8.38}$$

Using (7.8.30), (7.8.32), and (7.8.33) we find that (7.8.38) can be written as

$$\Delta = \frac{1}{2\pi} \int_{-\infty}^{\infty} \left[L\bar{L}S_{ff} - \frac{1}{T} G\bar{L}S_{fy} \right] d\omega \tag{7.8.39}$$

Let us consider the special case of ideal extrapolation, that is, $v(t) = 0$, $L(\omega) = 1$, and $G(\omega)$ physically realizable. In this case $f(t) = y(t)$ so that

$$S_{yy}^*(\omega) = S_{ff}^*(\omega) = \sigma^2 B^*(\omega) B^*(-\omega)$$

According to (7.8.30), the optimum extrapolation filter has the transfer function

$$G(\omega) = \frac{1}{\sigma^2 B^*(\omega)} \left[\frac{S_{ff}(\omega)}{B^*(-\omega)} \right]_+ \tag{7.8.40}$$

According to (7.8.39), the corresponding mean square error is

$$\Delta = \frac{1}{2\pi} \int_{-\infty}^{\infty} S_{ff} \left[1 - \frac{1}{T} G \right] d\omega \tag{7.8.41}$$

The sampling operator $*$ and partial fractioning operator $[\ ]_+$ commute, so that

$$G^*(\omega) = \frac{1}{\sigma^2 B^*(\omega)} \left[\frac{S_{ff}^*(\omega)}{B^*(-\omega)} \right]_+ = \frac{1}{\sigma^2 B^*(\omega)} [\sigma^2 B^*(\omega)]_+ = 1$$

The last step follows because all the poles of $B^*(\omega)$ are in the upper half plane. Therefore

$$g(nT) = \delta_{n0} \tag{7.8.42}$$

Using the same argument as in the nonrealizable interpolation case, we find that the reconstructed signal must pass through the input samples.

Example 7.8.2

Suppose that $R_{ff}(\tau) = e^{-a|\tau|}$. Then

$$S_{ff}(\omega) = \frac{2a}{a^2 + \omega^2} = \frac{2a}{(\omega + ja)(\omega - ja)}$$

and

$$S_{ff}(z) = \frac{1 - e^{-2aT}}{(1 - e^{-aT}z^{-1})(1 - e^{-aT}z)}.$$

Therefore

$$\sigma^2 = 1 - e^{-2aT}$$

and

$$B(z) = 1/(1 - e^{-aT}z^{-1})$$

From (7.8.40) we find that the optimum extrapolation filter is

$$G(\omega) = \frac{1 - e^{-aT}z^{-1}}{1 - e^{-2aT}} \left[\frac{2a(1 - e^{-aT}z)}{a^2 + \omega^2} \right]_+$$

where $z = e^{j\omega T}$. The function inside the partial fraction operator has only one pole in the upper half plane. This is located at $\omega_0 = ja$. The residue at ja is $(1 - e^{-2aT})/j$. Therefore

$$G(\omega) = \frac{1 - e^{-aT}z^{-1}}{1 - e^{-2aT}} \frac{1 - e^{-2aT}}{j(\omega - ja)} = \frac{1 - e^{-aT}z^{-1}}{j\omega + a} \qquad (7.8.43)$$

The corresponding impulse response is

$$g(t) = e^{-at}[u(t) - u(t - T)] \qquad (7.8.44)$$

This is sketched in Fig. 7.8.2. The reconstructed signal passes through the input samples and decays exponentially between samples.

The mean square error can be easily evaluated by closing the integral in (7.8.41) around the lower half plane and using Cauchy's Residue Theorem. The result is

$$\Delta = 1 - \frac{1 - e^{-2aT}}{2aT} \qquad (7.8.45)$$

Notice that Δ converges to zero as T becomes zero. One would expect this result intuitively.

Equation 7.8.45 can be used to determine the required sampling frequency for a specified mean square error. For $|aT| \ll 1$, $\Delta \cong aT$. Suppose we require that $\Delta = 10^{-4}$. Then $f_s = 1/T \cong 10^4 a$. Often the 3 dB frequency $\omega_1 = a$ is called the signal bandwidth. This example points out that care must be taken in defining bandwidth when sampling is involved. ◄

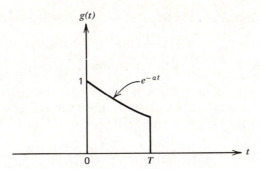

FIGURE 7.8.2. Impulse response of the optimum extrapolation filter for Example 7.8.2.

7.9 COMMON SUBOPTIMUM RECONSTRUCTION FILTERS: THE ZERO-ORDER HOLD, FIRST-ORDER HOLD, AND LINEAR POINT CONNECTOR

The mean square error can be calculated for any reconstruction filter using (7.8.2) or (7.8.9). We will now derive an expression for this error from (7.8.2) which is more convenient to use than (7.8.9) for filters that have short-duration impulse responses. The zero-order hold, first-order hold, and linear point connector are filters of this type. According to (7.8.1) and (7.8.2), the mean square error is

$$\Delta = R_{ee}(0) = R_{c_d c_d}(0) - 2R_{cc_d}(0) + R_{cc}(0) \qquad (7.9.1)$$

where the subscripts refer to the signals in Fig. 7.8.1. Using (7.7.8) with $h(mT) = \delta_{m0}$ and (7.7.15), we find that

$$\Delta = R_{c_d c_d}(0) - \frac{2}{T} \int_{-\infty}^{\infty} R_{yc_d}(\mu) g(-\mu) \, d\mu + \frac{1}{T} \sum_{m=-\infty}^{\infty} R_{yy}(mT) \int_{-\infty}^{\infty} g(\mu - mT) g(\mu) \, d\mu$$

$$(7.9.2)$$

In the noiseless reconstruction case (i.e., $v(t) = 0$) when the desired output is $c_d(t) = f(t)$, the mean square error becomes

$$\Delta = R_{ff}(0) - \frac{2}{T} \int_{-\infty}^{\infty} R_{ff}(\mu) g(\mu) \, d\mu + \frac{1}{T} \sum_{m=-\infty}^{\infty} R_{ff}(mT) \int_{-\infty}^{\infty} g(\mu - mT) g(\mu) \, d\mu$$

$$(7.9.3)$$

In Chapter 3 we defined the impulse response of the zero-order hold as

$$g_0(t) = \begin{cases} 1 & \text{for} \quad 0 \le t < T \\ 0 & \text{elsewhere} \end{cases} \tag{7.9.4}$$

Notice that $g_0(\mu - mT)g_0(\mu) \equiv 0$ for $m \ne 0$ so that the last summation on the right-hand side of (7.9.3) reduces to only the term for $m = 0$. Therefore, the mean square error for the zero-order hold is

$$\Delta_0 = 2R_{ff}(0) - \frac{2}{T} \int_0^T R_{ff}(\mu) \, d\mu \tag{7.9.5}$$

in the noiseless reconstruction case with $c_d(t) = f(t)$.

The impulse response of the first-order hold is

$$g_1(t) = \begin{cases} \dfrac{t+T}{T} & \text{for} \quad 0 \le t < T \\[2mm] \dfrac{T-t}{T} & \text{for} \quad T \le t < 2T \\[2mm] 0 & \text{elsewhere} \end{cases} \tag{7.9.6}$$

Since $g_1(t)$ has duration $2T$, the integral in the summation in (7.9.3) is nonzero only for $m = -1$, 0, *or* 1. After some calculation (7.9.3) reduces to

$$\Delta_1 = \tfrac{11}{3} R_{ff}(0) - \tfrac{5}{3} R_{ff}(T) + \frac{2}{T^2} \int_0^T \mu R_{ff}(\mu + T) \, d\mu - \frac{2}{T^2} \int_0^T (\mu + T) R_{ff}(\mu) \, d\mu \tag{7.9.7}$$

The impulse response of the linear point connector without delay is

$$g_L(t) = \begin{cases} 1 - |t/T| & \text{for} \quad |t| < T \\ 0 & \text{elsewhere} \end{cases} \tag{7.9.8}$$

This impulse response also has duration $2T$ so that the integral in the summation in (7.9.3) is nonzero only for $m = -1$, 0, and 1. In this case (7.9.3) can be reduced to

$$\Delta_L = \tfrac{5}{3} R_{ff}(0) + \tfrac{1}{3} R_{ff}(T) - \frac{4}{T^2} \int_0^T (T - \mu) R_{ff}(\mu) \, d\mu \tag{7.9.9}$$

The results derived here can be used to compare the performance of these suboptimum reconstruction filters with the optimum filters derived in Section 7.8. To be fair, the zero and first-order holds should be compared with the optimum realizable filter and the linear point connector should be compared with the optimum nonrealizable filter.

7.10 COMMENTS

In Sections 7.5 and 7.8 both the physically realizable and nonrealizable optimum filters were derived. The performance of the optimum nonrealizable filters can be achieved arbitrarily closely in practice by letting the desired operations $L(z)$ in Section 7.5 and $L(\omega)$ in Section 7.8 be sufficiently long ideal delays. Allowing a delay effectively lets the filter use some future data in making its estimate of the present signal value. From the mean square error formulas (7.5.24) and (7.8.39) for the optimum realizable filters, it can be shown that the mean square errors are nonincreasing with delay and converge to those for the optimum nonrealizable filters. A moderate delay is perfectly acceptable in many real-time filtering situations if it results in significantly better filtering. Of course, if previously recorded data is being processed off-line in nonreal time, delay is not really a factor. Thus nonrealizable filters often can be closely approximated in practice.

The optimum filters are rarely implemented in practice. For one thing, signal and noise power spectral densities are usually known only roughly and actually may vary with time. For another, the optimum filters may be costly to implement. Nevertheless, the results of the optimum filter problems are important because they provide standards against which to compare practical suboptimum filter designs.

8

THE DESIGN OF PULSE TRANSFER FUNCTIONS FOR DIGITAL FILTERING

8.1 INTRODUCTION

Linear, time-invariant filtering of signals using a digital computer is based on the fact that for zero initial conditions the input Z-transform $X(z)$ and output Z-transform $Y(z)$ of a discrete-time system with a pulse transfer function $G(z)$ are related by the equation $Y(z) = G(z)X(z)$. We have seen that

$$Y^*(\omega) = Y(z)\big|_{z=e^{j\omega T}} = G^*(\omega)X^*(\omega)$$

can be considered to be the spectrum of the discrete-time signal $y(nT)$. It was shown in Section 5.4 that $G^*(\omega)$ is the sinusoidal steady-state frequency response of the system. Thus the spectrum of the discrete-time input signal can be modified in some desired manner by choosing $G^*(\omega)$ appropriately. This modification is known as filtering.

If $x(nT)$ is obtained by sampling a continuous-time signal $x(t)$ whose Fourier transform $X(\omega)$ is band-limited to $|\omega| < \omega_s/2$, we have seen that

$$X(\omega) = TX^*(\omega) \qquad \text{for} \qquad |\omega| < \omega_s/2$$

The discrete-time signal $y(nT)$ can be considered to be the samples of a band-limited signal $y(t)$ with the Fourier transform

$$Y(\omega) = TY^*(\omega) = G^*(\omega)X(\omega) \qquad \text{for} \qquad |\omega| < \omega_s/2$$

Therefore, a band-limited continuous-time signal can be filtered by operating only on its samples. This is conventionally known as *digital filtering*. In practice, signals are only approximately band-limited. Care must be taken that the original continuous-time signal is sufficiently band-limited so that aliasing is negligible.

In Section 5.3 it was shown that a rational pulse transfer function of the form

$$G(z) = \frac{a_0 + a_1 z^{-1} + \cdots + a_M z^{-M}}{1 + b_1 z^{-1} + \cdots + b_N z^{-N}}$$

can be realized by the difference equation

$$y(nT) = \sum_{k=0}^{M} a_k x(nT - kT) - \sum_{k=1}^{N} b_k y(nT - kT)$$

It is customary to call a difference equation realization of a pulse transfer function a *digital filter*. If $b_k = 0$ for $k = 1, \ldots, N$ the filter is known as a *nonrecursive* filter. Otherwise, it is called a *recursive* filter. In this chapter we assume that M and N are finite. In the literature, some authors prefer to call nonrecursive filters with M finite, *finite-duration impulse response* (FIR) filters. They are also known as *moving average* or *transversal* filters. If $g(nT)$ has infinite duration, the filter is sometimes called an *infinite-duration impulse response* (IIR) filter.

Other difference equation realizations for $G(z)$ were presented in Sections 5.5 and 6.5. We will see in Chapter 9 that for practical reasons it is preferable, particularly in the case of recursive filters, to split $G(z)$ into low-order sections and use the parallel or cascade form realizations. The cascade form realization is most frequently used in practice since both the pole and zero locations can be accurately controlled.

In this chapter methods for finding pulse transfer functions with desired frequency responses are presented. The methods for designing recursive and nonrecursive filters are quite different. Most of the methods for designing recursive filters are based on transforming the design for an analog filter into the pulse transfer function for a digital filter. It will be assumed that the reader has some familiarity with standard filter terminology and types of analog filters like the Butterworth, Chebyshev, and elliptic filters. The design of analog filters is well-documented in many books [17,129,136]. The methods presented for designing nonrecursive filters are based on approximating the desired frequency response directly with a trigonometric polynomial.

8.2 DESIGN OF SIMPLE FILTERS BY TRIAL-AND-ERROR PLACEMENT OF POLES AND ZEROS IN THE z-PLANE

Simple filters can be rapidly designed by trial-and-error placement of poles and zeros in the z-plane. In Section 5.4 we saw how to graphically determine the frequency response of a digital filter with a rational pulse transfer function $G(z)$ from a pole-zero plot. The method is based on the observation that a factor of the form $z - z_0$ for $z = e^{j\omega T}$ corresponds to a vector drawn from z_0 to the point on the unit circle with argument ωT.

Let us assume that z_0 is close to the unit circle so that $|e^{j\omega T} - z_0|$ becomes small when $\omega T = \arg z_0$. Then, if z_0 is a zero of $G(z)$, the amplitude response $|G^*(\omega)| = |G(e^{j\omega T})|$ will have a null in the vicinity of $\omega_0 = \arg(z_0)/T$. The depth of the null will increase as z_0 is moved closer to the unit circle. When $G(z)$ has a pole at z_0, the factor $z - z_0$ will appear in the denominator so that the amplitude response will have a peak in the vicinity of ω_0. The height of the peak will increase as z_0 is moved closer to the unit circle.

The principles discussed in the previous paragraph provide a guide for placing poles and zeros in the z-plane to achieve simple low-pass, band-pass, high-pass, and band-stop amplitude responses. In addition, poles or zeros should be included in complex conjugate pairs so that $G(z)$ can be grouped into factors with real coefficients. The filter can then be implemented by difference equations using any of the forms discussed in Sections 5.3, 5.5, or 6.5. Because of practical considerations (see Chapter 9), the cascade or parallel form realizations with low-order sections should normally be used for recursive filters. If a recursive filter has zeros on or near the unit circle, then the cascade form realization should be used to precisely control the zero locations.

With a little practice, an intuitive feeling for how poles and zeros should be placed in order to obtain desired frequency responses can be developed. Of course, this trial-and-error procedure is most easily carried out with the aid of a digital computer and automatic plotter or interactive graphics terminal.

An example of a simple band-pass filter with a center frequency of $\omega_s/4$ is given in Example 5.4.1. Another example is presented below.

Example 8.2.1

Suppose that a three-pole, low-pass filter with an amplitude response that is relatively flat over the pass-band $|\omega T| < \pi/8$ is desired and that the phase response is not important. As a first trial, let us place poles at 0.8 and $0.8e^{\pm j\pi/8}$.

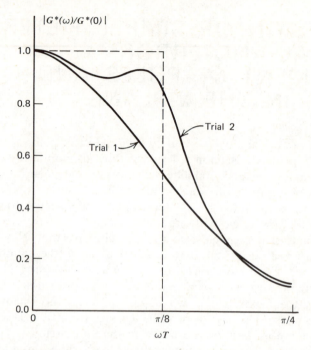

FIGURE 8.2.1.   Amplitude response for Example 8.2.1.

The resulting normalized amplitude response is shown in Fig. 8.2.1 for $0 <$ $\omega T < \pi/4$ as the curve labeled Trial 1. Clearly, the response drops too rapidly in the pass-band. To peak up the response toward the edge of the pass-band, let us move the complex poles closer to the unit circle to $0.88e^{\pm j\pi/8}$. The curve labeled Trial 2 in Fig. 8.2.1 shows the resulting amplitude response. This is moderately close to the desired response. The corresponding pulse transfer function is

$$G(z) = \frac{1}{(1 - 0.8z^{-1})(1 - 1.76z^{-1} \cos \pi/8 + 0.7744z^{-2})}$$

if a third-order zero is included at $z = 0$ to eliminate unnecessary delay. If $x(nT)$ is the input and $y(nT)$ is the output, the filter can be realized by the coupled difference equations

$$v(nT) = x(nT) + 0.8v(nT - T)$$

$$y(nT) = v(nT) + 1.76 \cos (\pi/8) y(nT - T) - 0.7744 y(nT - 2T)$$

This is a cascade form realization using a first-order section with a pole at 0.8 followed by a second-order section with poles at $0.88e^{\pm j\pi/8}$. To further improve the amplitude response, the pole at 0.8 might be moved away from the unit circle to 0.79 in an attempt to reduce the peak at zero. These iterations can be rapidly performed with the aid of a digital computer and automatic plotter. ◄

8.3 CONVERSION OF ANALOG FILTERS TO DIGITAL FILTERS BY USING GUARD FILTERS AND THE Z-TRANSFORM

When first presented with the problem of designing a digital filter with a prescribed frequency response, it is tempting to simply find a standard analog filter design with a rational transfer function $\mathcal{G}(s)$ that meets the requirements and use the corresponding Z-transform $G(z)$ as the pulse transfer function of the digital filter. The denominators of $\mathcal{G}(s)$ and, $G(z)$ will have the same degree so that the analog and digital filters will have, in a sense, the same complexity. However, this approach can lead to a digital filter frequency response $G^*(\omega)$ that, as a result of aliasing, is a poor approximation to the desired response.

From the aliasing formula (4.4.31) we can see that the magnitude of the aliasing error depends on the high-frequency asymptotic behavior of $\mathcal{G}(s)$ as well as the sampling frequency, ω_s. Suppose that $\mathcal{G}(s)$ behaves essentially as $1/(s/\omega_c)^m$ for $|s| > \omega_c$. If m is large and $\omega_c \ll \omega_s/2$, then the aliasing error will be small and simply using the Z-transform $G(z)$ corresponding to $\mathcal{G}(s)$ will result in a satisfactory digital filter. If m is small and $\omega_c \approx \omega_s/2$, then a poor or unusable digital filter design will be obtained. The popular elliptic and Chebyshev type II analog filters have transfer functions with m less than or equal to 1. If a wide-band, low-pass digital filter is required, that is, one with a bandwidth that is an appreciable fraction of $\omega_s/2$, then an analog filter with $\omega_c \approx \omega_s/2$ must be chosen.

To solve the aliasing problem, a low-pass filter $\mathcal{H}(s)$ that adequately band-limits $\mathcal{G}(s)$ to $|\omega| < \omega_s/2$ can be cascaded with $\mathcal{G}(s)$ to form the modified filter $\mathcal{G}_1(s) = \mathcal{G}(s)\mathcal{H}(s)$. The filter $\mathcal{H}(s)$ is often called a *guard filter*. The pulse transfer function $G_1(z)$ can then be used as the digital filter design. The amplitude response of the guard filter should be flat and its phase response linear over the frequency band where $G_1^*(\omega)$ must be a good approximation to $\mathcal{G}(j\omega)$. A Butterworth or Chebyshev type I all-pole low-pass filter cascaded with an all-pass phase equalizer can be used as the guard filter.

The guard filter and Z-transform method is a straightforward algebraic technique for converting analog filters into digital filters. However, this method is not often used because the resulting pulse transfer functions usually have more coefficients than "equally good" recursive and nonrecursive digital filters designed by other techniques that are described in this chapter. The extra complexity, particularly in the case of wide-band designs, is caused by the required complexity of the cascaded guard filter.

8.4 THE MATCHED Z-TRANSFORM

The matched Z-transform is another technique proposed for converting analog filters into digital filters. Suppose that an analog filter with the desired frequency response has the transfer function

$$\mathcal{G}(s) = \frac{\displaystyle\prod_{k=1}^{M} (s - a_k)}{\displaystyle\prod_{k=1}^{N} (s - b_k)} \tag{8.4.1}$$

The matched Z-transform of $\mathcal{G}(s)$ is defined to be

$$\hat{G}(z) = \frac{\displaystyle\prod_{k=1}^{M} (1 - e^{a_k T} z^{-1})}{\displaystyle\prod_{k=1}^{N} (1 - e^{b_k T} z^{-1})} \tag{8.4.2}$$

It is then assumed that $\hat{G}(z)$ is the pulse transfer function of a digital filter that adequately approximates the desired frequency response. Notice that the denominator of $\hat{G}(z)$ is the same as that of the ordinary Z-transform $G(z)$. However, the numerators differ. The philosophy is to map the zeros of $\mathcal{G}(s)$ into zeros of $\hat{G}(z)$ using the transformation $z_k = e^{a_k T}$.

The matched Z-transform method results in adequate digital filter designs for relatively simple, narrow-band filters. However, this method is not good for designing wide-band digital filters. The matched Z-transform does not preserve characteristics of typical prototype analog filters such as equal ripple amplitude response or linear phase response.

8.5 CONVERSION OF ANALOG FILTERS TO DIGITAL FILTERS WITH THE BILINEAR TRANSFORMATION

The conversion of analog filters to digital filters by a bilinear transformation is, perhaps, the easiest and most effective method for designing recursive digital filters that approximate piecewise constant amplitude responses. The well-known Butterworth, Chebyshev, and elliptic filters can be used as prototype analog designs. These all approximate ideal low-pass filters and can be easily converted into high-pass, band-pass, or band-stop analog filters by suitable substitutions for s.

The transformation

$$z = \frac{1+s}{1-s} \tag{8.5.1}$$

is known as a bilinear transformation. It is a one-to-one mapping of the extended s-plane onto the extended z-plane. Solving for s, we find that the inverse mapping is

$$s = \frac{z-1}{z+1} = \frac{1-z^{-1}}{1+z^{-1}} \tag{8.5.2}$$

The notation $\Omega = \text{Im } s$ and $\omega = \arg(z)/T$ will be used to distinguish between the real frequency variables for analog and digital filters. Letting $s = \sigma + j\Omega$, we find from (8.5.1) that

$$|z|^2 = \frac{(1+\sigma)^2 + \Omega^2}{(1-\sigma)^2 + \Omega^2} \tag{8.5.3}$$

Thus $|z| < 1$ for $\sigma < 0$, $|z| = 1$ for $\sigma = 0$, and $|z| > 1$ for $\sigma > 0$. In other words, the bilinear transformation maps the left half s-plane into the unit circle, the $j\Omega$ axis onto the unit circle, and the right half s-plane outside the unit circle in the z-plane.

Clearly, the infinite length $j\Omega$ axis must be compressed to fit uniquely onto the unit circle, or, equivalently, onto the Nyquist band $|\omega| < \omega_s/2$. This compression is known as *frequency warping*. The relationship between the frequency variables Ω and ω can be determined by letting $z = e^{j\omega T}$ in (8.5.2). This gives

$$s = \frac{e^{j\omega T} - 1}{e^{j\omega T} + 1} = j \tan(\omega T/2)$$

or

$$\Omega = \text{Im } s = \tan(\omega T/2) \tag{8.5.4}$$

Equation 8.5.4 is sketched in Fig. 8.5.1 for $|\omega| < \omega_s/2$. Notice that the frequency band $0 < \Omega < 1$ is mapped onto the interval $0 < \omega < \omega_s/4$ and the infinite length band $1 \leq \Omega$ is mapped onto $\omega_s/4 < \omega < \omega_s/2$. This illustrates the nonlinear nature of the frequency warping. The point $\Omega = 0$ is mapped to $\omega = 0$ and the points $\Omega = \pm\infty$ are mapped to $\omega = \pm\omega_s/2$.

Suppose that the rational function $\mathcal{G}(s)$ is the transfer function of an analog filter that has an approximately piecewise constant amplitude response. For example, $\mathcal{G}(s)$ might be a Butterworth low-pass filter with a 3 dB cutoff

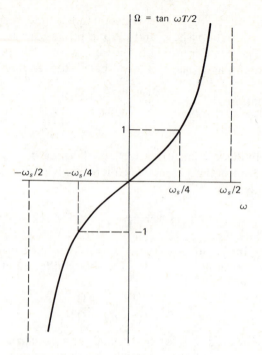

FIGURE 8.5.1. Frequency warping of the bilinear transformation.

frequency Ω_c. Basically, the bilinear transformation method of converting $\mathscr{G}(s)$ into a digital filter with a similar amplitude response is to replace s by $(1-z^{-1})/(1+z^{-1})$ to obtain the rational function of z^{-1}

$$G(z) = \mathscr{G}\left[\frac{1-z^{-1}}{1+z^{-1}}\right]$$

(8.5.5)

$G(z)$ will correspond to a stable digital filter since the bilinear transformation maps the left half s-plane into the unit circle. The frequency response of $G(z)$, according to (8.5.4), will be

$$G^*(\omega) = G(e^{j\omega T}) = \mathscr{G}\left(j \tan \frac{\omega T}{2}\right)$$

(8.5.6)

Thus the amplitude response of $G(z)$ will also be approximately piecewise constant since it is simply the response of $\mathscr{G}(s)$ compressed into the interval $|\omega| < \omega_s/2$. The equal ripple characteristics of Chebyshev and elliptic filters will be transformed into equal ripple characteristics. However, the critical frequencies such as cutoff frequencies, and pass-band and stop-band edges will appear at different points for the analog and digital filters as a result of the frequency warping. This must be taken into account when designing the prototype analog filter by prewarping the critical frequencies.

The design procedure can be summarized as follows:

1. Specify the set of critical frequencies $\{\omega_k\}$ for the desired digital filter.
2. Prewarp the critical frequencies to $\{\Omega_k = \tan(\omega_k T/2)\}$.
3. Design an analog filter with the transfer function $\mathcal{G}(s)$ using the pre-warped critical frequencies $\{\Omega_k\}$.
4. Calculate the desired digital filter pulse transfer function $G(z)$ using (8.5.5).
5. Realize $G(z)$ by difference equations, preferably using low-order sections.

In designing common analog filters, the required pole and zero locations are often calculated. Combining complex conjugate terms, it is then easy to express the transfer function in the factored form

$$\mathcal{G}(s) = \prod_{k=1}^{N} \mathcal{G}_k(s) \tag{8.5.7}$$

where the factors are low-order rational functions with real coefficients. Applying the bilinear transformation (8.5.7) becomes

$$G(z) = \prod_{k=1}^{N} \mathcal{G}\left[\frac{1-z^{-1}}{1+z^{-1}}\right] = \prod_{k=1}^{N} G_k(z) \tag{8.5.8}$$

Therefore, products are transformed into products. This is particularly convenient when $G(z)$ is to be realized in cascade form.

Example 8.5.1   Butterworth Low-Pass Filters

The squared amplitude response of an Nth order Butterworth low-pass filter is

$$A(\omega) = \frac{1}{1+(\omega/\Omega_c)^{2N}} \tag{8.5.9}$$

For $|\omega/\Omega_c| < 1$, $A(\omega)$ can be represented by the power series

$$A(\omega) = 1 - (\omega/\Omega_c)^{2N} + (\omega/\Omega_c)^{4N} - \cdots \tag{8.5.10}$$

It is easy to see from (8.5.10) that

$$\frac{d^k}{d\omega^k} A(\omega)\Big|_{\omega=0} = 0 \qquad \text{for} \qquad k = 1, \ldots, 2N-1$$

The Butterworth response is called *maximally flat* for this reason. Another property of the Butterworth response is that $A(\Omega_c) = \frac{1}{2}$ or $10 \log_{10} A(\Omega_c) \cong -3$

dB for all N. The frequency Ω_c is known as the 3 dB cutoff frequency for the filter. The Butterworth squared amplitude response converges to that of an ideal low-pass filter with a cutoff frequency Ω_c as N becomes large.

The transfer function $\mathcal{G}(s)$ of a Butterworth filter can be determined by observing that $A(\omega) = |\mathcal{G}(j\omega)|^2 = \mathcal{G}(j\omega)\mathcal{G}(-j\omega)$. Letting $\omega = s/j$, we find from (8.5.9) that

$$\mathcal{G}(s)\,\mathcal{G}(-s) = \frac{1}{1 + (-1)^N (s/\Omega_c)^{2N}} \qquad (8.5.11)$$

Therefore, the poles of $\mathcal{G}(s)$ must be the left half plane roots of $1 + (-1)^N (s/\Omega_c)^{2N}$. These roots are

$$s_k = \Omega_c \exp j\left[\frac{\pi}{N}\left(k + \frac{1}{2}\right) + \frac{\pi}{2}\right] \qquad \text{for} \qquad k = 0, \ldots, N-1 \qquad (8.5.12)$$

so that

$$\mathcal{G}(s) = \frac{\Omega_c^N}{\displaystyle\prod_{k=0}^{N-1}(s - s_k)} \qquad (8.5.13)$$

As a specific example, let us design a low-pass digital filter with a 3 dB cutoff frequency at $\omega_c = \omega_s/4$ using a third-order Butterworth prototype analog filter. The prewarped cutoff frequency for the analog filter must be $\Omega_c = \tan(\omega_c T/2) = 1$. According to (8.5.12), the poles of a third-order Butterworth filter with $\Omega_c = 1$ are at -1 and $e^{\pm j2\pi/3}$. Its transfer function is

$$\mathcal{G}(s) = 1/(s^3 + 2s^2 + 2s + 1) \qquad (8.5.14)$$

Making the substitution $s = (1 - z^{-1})/(1 + z^{-1})$ in (8.5.14), it follows that

$$G(z) = \frac{1}{6}\,\frac{1 + 3z^{-1} + 3z^{-2} + z^{-3}}{1 + \frac{1}{3}z^{-2}} \qquad (8.5.15)$$

The amplitude response of $G(z)$ is shown in Fig. 8.5.2. The filter can be realized by the difference equation

$$y(nT) = \tfrac{1}{6}x(nT) + \tfrac{1}{2}x(nT - T) + \tfrac{1}{2}x(nT - 2T) + \tfrac{1}{6}x(nT - 3T) - \tfrac{1}{3}y(nT - 2T) \quad \blacktriangleleft$$

We will now examine the effects of the bilinear transformation on envelope delay. Let us denote the phase responses of the analog and digital filters by $\varphi(\Omega) = \arg \mathcal{G}(j\Omega)$ and $\Phi(\omega) = \arg G^*(\omega)$, respectively. The *envelope* or *group delay* of the analog filter is defined as

$$\tau(\Omega) = -\frac{d}{d\Omega}\,\varphi(\Omega) \qquad (8.5.16)$$

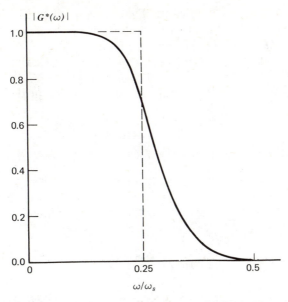

FIGURE 8.5.2. Amplitude response of $G(z)$ in Example 8.5.1.

Similarly, the envelope delay of the digital filter is

$$\Delta(\omega) = -\frac{d}{d\omega}\Phi(\omega) \tag{8.5.17}$$

It follows from (8.5.6) that

$$\Phi(\omega) = \varphi\left(\tan\frac{\omega T}{2}\right) \tag{8.5.18}$$

Taking the negative of the derivative of (8.5.18), we find that

$$\Delta(\omega) = \frac{T}{2}\sec^2(\omega T/2)\tau\left(\tan\frac{\omega T}{2}\right) \tag{8.5.19}$$

or

$$\Delta(\omega) = \frac{T}{2}(1+\Omega^2)\tau(\Omega), \qquad \Omega = \tan(\omega T/2) \tag{8.5.20}$$

When signal shapes must be preserved, linear phase and, consequently, constant envelope delay filters are required. From (8.5.19) we see that constant delay analog filters are not transformed into constant delay digital filters by the bilinear transformation. A solution to this problem is suggested by (8.5.20). Suppose that the function $\tau_0(\Omega)$ is constant over the band (Ω_a, Ω_b) and that

$\tau(\Omega) = \tau_0(\Omega)/(1+\Omega^2)$ over this band. Then, according to (8.5.20), the delay of the digital filter will be $\Delta(\omega) = (T/2)\tau_0(\Omega)$ and will be constant over the band (ω_a, ω_b) where $\Omega_a = \tan(\omega_a T/2)$ and $\Omega_b = \tan(\omega_b T/2)$. Thus the delay of the analog filter should be predistorted by the factor $1/(1+\Omega^2)$ over the band of interest.

Another transformation that maps the complex axis in the s-plane onto the unit circle in the z-plane is

$$s = \frac{z^2 - 2z \cos \omega_0 T + 1}{z^2 - 1} \tag{8.5.21}$$

This transformation can be used to convert a low-pass analog filter into a band-pass digital filter with a center frequency of ω_0. The roots of the numerator are $e^{\pm j\omega_0 T}$ so that the origin of the s-plane is mapped to these points. Letting $z = e^{j\omega T}$, it follows that Re $s = 0$ and

$$\Omega = \text{Im } s = \frac{\cos \omega_0 T - \cos \omega T}{\sin \omega T} \tag{8.5.22}$$

From a sketch of Ω versus ω it can be seen that the $j\Omega$ axis is mapped onto the top half and also onto the bottom half of the unit circle. If z is a real number r, then (8.5.21) can be written as

$$s = |r - e^{j\omega_0 T}|^2/(r^2 - 1) \tag{8.5.23}$$

In this case s is real. Also, $s < 0$ for $|r| < 1$. It follows from the Jordan mapping theorem in complex variable theory that the interior of the unit circle is mapped onto the left half s-plane and vice-versa. Therefore, stable analog filters are mapped into stable digital filters.

The procedure for designing a band-pass filter using (8.5.21) is similar to but slightly more complicated than the procedure for designing a low-pass filter using the bilinear transformation because the additional parameter ω_0 must be chosen. Suppose that a digital band-pass filter with upper and lower 3 dB cutoff frequencies of ω_2 and ω_1, respectively, is required. The amplitude response of the prototype analog filter must be an even function of Ω. Therefore, we must require that $\Omega_1 = -\Omega_2$ where Ω_1 and Ω_2 are the frequencies obtained by substituting ω_1 and ω_2 into (8.5.22). It follows that ω_0 should be chosen so that

$$\cos \omega_0 T = \frac{\sin (\omega_1 + \omega_2) T}{\sin \omega_1 T + \sin \omega_2 T} = \frac{\cos (\omega_1 + \omega_2) T/2}{\cos (\omega_1 - \omega_2) T/2} \tag{8.5.24}$$

Having calculated $\cos \omega_0 T$, Ω_2 can be calculated and the prototype analog filter designed. The digital filter is then obtained by replacing s by the right-hand side of (8.5.21).

Example 8.5.2

Suppose that a digital band-pass filter with upper and lower 3 dB cutoff frequencies of $\omega_2 = 3\omega_s/8$ and $\omega_1 = \omega_s/8$ is to be designed using a third-order Butterworth filter. From (8.5.24) we find that $\cos \omega_0 T = 0$ or $\omega_0 = \omega_s/4$ in this case. Then (8.5.21) becomes

$$s = \frac{z^2+1}{z^2-1} = \frac{1+z^{-2}}{1-z^{-2}} \qquad (8.5.25)$$

and (8.5.22) becomes

$$\Omega = -\cot \omega T \qquad (8.5.26)$$

Substituting ω_1 and ω_2 into (8.5.26), we find that $\Omega_2 = -\Omega_1 = 1$. Therefore, the required Butterworth filter is given by (8.5.14). Replacing s by the right-hand side of (8.5.25), we find that the pulse transfer function of the desired digital filter is

$$G(z) = \tfrac{1}{6} \frac{1-3z^{-2}+3z^{-4}-z^{-6}}{1+\tfrac{1}{3}z^{-4}} \qquad (8.5.27)$$

Notice that this is the same as (8.5.15) with z replaced by $-z^2$. ◀

8.6 DIGITAL ALL-PASS FILTERS

A filter with a flat amplitude response for all frequencies is known as an *all-pass* filter. All-pass filters are commonly used for phase and delay equalization. They are also used as phase splitters to generate output signals that are Hilbert transform pairs.

An Nth order digital all-pass filter has a pulse transfer function of the form

$$G(z) = K \frac{d_N + d_{N-1}z^{-1}+\cdots+d_1 z^{-(N-1)}+z^{-N}}{1+d_1 z^{-1}+\cdots+d_N z^{-N}}$$

$$= K \frac{z^{-N}D(z^{-1})}{D(z)} \qquad (8.6.1)$$

where K is a real constant and $D(z)$ is a polynomial with real coefficients and all its roots inside the unit circle. Notice that for $z = e^{j\omega T}$, $D(z)$ and $D(z^{-1})$ are complex conjugates. Therefore

$$|G^*(\omega)| = |G(e^{j\omega T})| = |K| \qquad (8.6.2)$$

Since the coefficients of $D(z)$ are real, the complex conjugate of each root must also be a root. The roots of $D(z^{-1})$ are the reciprocals of the roots of $D(z)$. Denoting the roots of $D(z)$ by $a_k e^{jb_k T}$ for $k = 1, \ldots, N$, $G(z)$ can be written in the factored form

$$G(z) = K \prod_{k=1}^{N} \frac{(1 - a_k e^{-jb_k T} z) z^{-1}}{1 - a_k e^{jb_k T} z^{-1}} \qquad (8.6.3)$$

We will now determine the envelope delay for a digital all-pass filter. The frequency response of any filter can be expressed in the polar form

$$G^*(\omega) = A(\omega) e^{j\Phi(\omega)} \qquad (8.6.4)$$

where $A(\omega) = |G^*(\omega)|$ and $\Phi(\omega) = \arg G^*(\omega)$. Thus

$$\ln G^*(\omega) = \ln A(\omega) + j\Phi(\omega)$$

so that

$$\Phi(\omega) = \operatorname{Im} \{\ln G^*(\omega)\} \qquad (8.6.5)$$

Therefore, the envelope delay can be expressed as

$$\Delta(\omega) = -\frac{d}{d\omega} \Phi(\omega) = -\operatorname{Im}\left[\frac{d}{d\omega} \ln G^*(\omega)\right] \qquad (8.6.6)$$

On the unit circle a typical term in the product in (8.6.3) has the form

$$G_k^*(\omega) = \frac{1 - a_k e^{j(\omega - b_k)T}}{1 - a_k e^{-j(\omega - b_k)T}} e^{-j\omega T} \qquad (8.6.7)$$

Substituting (8.6.7) into (8.6.6), we find that the envelope delay for this first-order all-pass section is

$$\Delta_k(\omega) = T \frac{1 - a_k^2}{|1 - a_k e^{j(\omega - b_k)T}|^2}$$

$$= T \frac{1 - a_k^2}{1 - 2a_k \cos(\omega - b_k)T + a_k^2} \qquad (8.6.8)$$

Since the log of a product is the sum of the logs, the total envelope delay is

$$\Delta(\omega) = \sum_{k=1}^{N} \Delta_k(\omega) \qquad (8.6.9)$$

The normalized envelope delay Δ_k/T versus normalized frequency $u = (\omega - b_k)/\omega_s$ for a first-order all-pass section is shown in Fig. 8.6.1 for several values of a_k. From (8.6.8) we can see that the delay has its maximum value, $(1 + a_k)/(1 - a_k)$, for $u = 0$ and its minimum value, $(1 - a_k)/(1 + a_k)$, for $u = 0.5$. Also, the delay "bump" becomes sharper as a_k approaches 1.

In various applications filters with linear phase or constant envelope delay over the input signal bandwidth are required. Linear phase analog filters are usually designed by cascading an all-pass delay equalizer with a filter having the desired amplitude response. Linear phase, recursive, digital filters can be

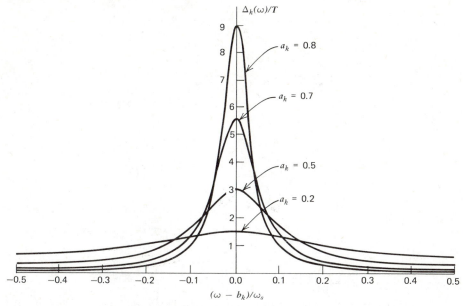

FIGURE 8.6.1.   Normalized delay vs. normalized frequency for first-order all-pass filter.

designed using the same technique. Basically, the method is to cascade second-order all-pass sections of the form

$$E_k(z) = \frac{a_k^2 - 2a_k z^{-1} \cos b_k T + z^{-2}}{1 - 2a_k z^{-1} \cos b_k T + a_k^2 z^{-2}} \tag{8.6.10}$$

with a filter $H(z)$ having the desired amplitude response to add delay bumps that flatten out the composite delay over the desired frequency band. This is illustrated in Fig. 8.6.2. The curve labeled "Original delay" represents the delay of $H(z)$. Second-order all-pass sections with resonant frequencies at b_1 and b_2 are added. The curve labeled "Equalized delay" is the sum of the delays of $H(z)$ and the two all-pass sections and is relatively flat over the band $\omega_L < \omega < \omega_H$.

An estimate of the number of second-order all-pass sections required can be obtained by dividing the area above the original delay curve that must be filled in by the area added by a single second-order section. The area added by a second-order all-pass section for positive frequencies is the area of $\Delta_k(\omega)$ in (8.6.8) over $-\omega_s/2 < \omega < \omega_s/2$. From (8.6.6) it follows that this area is simply the negative of the phase change in $G_k(z)$ as z makes one counterclockwise trip around the unit circle. The function $G_k(z)$ has one pole and no zeros inside the unit circle, so that, according to the argument principle discussed in Section 5.9, the phase change is -2π. Therefore, the estimate of the number of

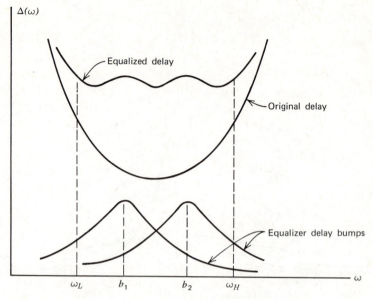

FIGURE 8.6.2. Envelope delay equalization using all-pass filters.

second-order all-pass sections required is

$$m = (\text{area to fill in})/2\pi \qquad (8.6.11)$$

Unfortunately, no simple analytical technique exists for optimally choosing the resonant frequencies b_k and pole radii a_k. With the aid of a digital computer and automatic plotter the design can be carried out reasonably quickly by trial and error. Automatic computer aided optimization procedures can also be used [28,37].

8.7 LINEAR PHASE NONRECURSIVE FILTERS

An appealing property of nonrecursive digital filters is that they can be designed with exactly linear phase over their pass-band. In this section we examine conditions on the pulse responses of nonrecursive filters that result in linear phase. Specific methods for designing nonrecursive filters with prescribed frequency responses are presented in following sections.

Suppose that the pulse transfer function of a nonrecursive filter with N taps is

$$G(z) = \sum_{k=0}^{N-1} g_k z^{-k} \tag{8.7.1}$$

For an odd number of taps, $N = 2r+1$ and the frequency response of the filter can be written as

$$G^*(\omega) = \sum_{k=0}^{2r} g_k e^{-j\omega kT} = e^{-j\omega rT} \sum_{k=-r}^{r} g_{k+r} e^{-j\omega kT}$$

$$= e^{-j\omega rT} \left[g_r + \sum_{k=1}^{r} (g_{r-k} e^{j\omega kT} + g_{r+k} e^{-j\omega kT}) \right] \tag{8.7.2}$$

The pulse response of the filter is said to have even symmetry if

$$g_k = g_{N-1-k} \qquad \text{for} \qquad k = 0, \ldots, N-1 \tag{8.7.3}$$

For N odd this is equivalent to

$$g_{r-k} = g_{r+k} \qquad \text{for} \qquad k = 0, \ldots, r \tag{8.7.4}$$

In this case (8.7.2) reduces to

$$G^*(\omega) = e^{-j\omega T(N-1)/2} G_1^*(\omega) \tag{8.7.5}$$

where

$$G_1^*(\omega) = g_r + 2 \sum_{k=1}^{r} g_{r-k} \cos \omega kT \tag{8.7.6}$$

Notice that $G_1^*(\omega)$ is a real and even function of ω. Thus the phase of $G^*(\omega)$ is linear except for jumps of π radians at the points where $G_1^*(\omega)$ changes sign. The complex exponential factor in (8.7.5) just corresponds to a delay of half the duration of the pulse response of the filter.

The pulse response is said to have odd symmetry if

$$-g_k = g_{N-1-k} \qquad \text{for} \qquad k = 0, \ldots, N-1 \tag{8.7.7}$$

For N odd this is equivalent to

$$-g_{r-k} = g_{r+k} \qquad \text{for} \qquad k = 0, \ldots, r \tag{8.7.8}$$

Equation 8.7.8 requires that $-g_r = g_r$ so that $g_r = 0$. In this case (8.7.2) reduces to

$$G^*(\omega) = e^{-j\omega T(N-1)/2} j G_2^*(\omega) \tag{8.7.9}$$

where

$$G_2^*(\omega) = 2 \sum_{k=1}^{r} g_{r-k} \sin \omega kT \tag{8.7.10}$$

Notice that $G_2^*(\omega)$ is a strictly real and odd function of ω. Therefore, $G^*(\omega)$

again has linear phase except for jumps of π radians at the points where $G_2^*(\omega)$ changes sign.

If N is even, it can be shown in a similar manner that for even symmetry

$$G^*(\omega) = e^{-j\omega T(N-1)/2} G_3^*(\omega) \qquad (8.7.11)$$

where

$$G_3^*(\omega) = 2 \sum_{k=1}^{r} g_{r-k} \cos \omega T(k - \tfrac{1}{2}) \qquad (8.7.12)$$

and that for odd symmetry

$$G^*(\omega) = e^{-j\omega T(N-1)/2} j G_4^*(\omega) \qquad (8.7.13)$$

where

$$G_4^*(\omega) = 2 \sum_{k=1}^{r} g_{r-k} \sin \omega T(k - \tfrac{1}{2}) \qquad (8.7.14)$$

Clearly, these filters also have piecewise linear phase.

8.8 THE FOURIER SERIES AND WINDOW FUNCTION METHOD OF DESIGNING NONRECURSIVE DIGITAL FILTERS

The guard filter and Z-transform method and the bilinear transformation method of designing recursive digital filters both required that analog filters with the desired characteristics be designed first. A straightforward analytical technique for designing nonrecursive digital filters directly from a desired frequency response specification $G^*(\omega)$ is now presented.

The frequency response $G^*(\omega)$ of any desired digital filter must be periodic with period ω_s. Therefore, $G^*(\omega)$ can be represented by the Fourier series

$$G^*(\omega) = \sum_{n=-\infty}^{\infty} g_n e^{-j\omega nT} \qquad (8.8.1)$$

where

$$g_n = \frac{1}{\omega_s} \int_{-\omega_s/2}^{\omega_s/2} G^*(\omega) e^{j\omega nT} \, d\omega \qquad (8.8.2)$$

The Fourier coefficient g_n is simply the pulse response sample $g(nT)$.

In most applications a real pulse response is required. According to (4.6.14), when g_n is real

$$G^*(-\omega) = \overline{G^*(\omega)} \qquad (8.8.3)$$

The converse can also be proved by observing from (8.8.2) that $\bar{g}_n = g_n$ when (8.8.3) holds. Thus the desired frequency response must be specified so that (8.8.3) is satisfied if a real pulse response is required.

Often, the desired frequency response is purely real or purely imaginary. An ideal low-pass filter and an ideal differentiator are examples of these two cases. If $G^*(\omega)$ is real, then (8.8.3) implies that $G^*(\omega)$ must also be an even function of ω. Making the change of variables $\omega' = -\omega$ in (8.8.2), it follows that $g_n = g_{-n}$, so that the pulse response has even symmetry when $G^*(\omega)$ is real and even. If $F^*(\omega)$ is real and $G^*(\omega) = jF^*(\omega)$, then (8.8.3) requires that $F^*(\omega)$ be an odd function of ω. It follows that $g_n = -g_{-n}$ in this case, so that, the pulse response has odd symmetry.

Theoretically, the output $y(nT)$ of the filter can be calculated from its input $x(nT)$ by the discrete-time convolution

$$y(nT) = \sum_{k=-\infty}^{\infty} g_k x(nT - kT) \qquad (8.8.4)$$

In practice, a nonrecursive filter can only approximate (8.8.4) by a finite summation. The first solution that usually comes to mind is to simply truncate the pulse response for $|n| > r$ and approximate the Fourier series for $G^*(\omega)$ by

$$G_r^*(\omega) = \sum_{n=-r}^{r} g_n e^{-j\omega nT} \qquad (8.8.5)$$

This corresponds to the pulse transfer function

$$G_r(z) = \sum_{n=-r}^{r} g_n z^{-n} \qquad (8.8.6)$$

$G_r(z)$ is not physically realizable, but this problem is easily solved by adding a delay of r samples. The resulting pulse transfer function is

$$H_r(z) = z^{-r} G_r(z) = \sum_{n=-r}^{r} g_n z^{-(n+r)} \qquad (8.8.7)$$

$H_r(z)$ is a nonrecursive filter with $N = 2r+1$ taps and can be realized by the discrete-time convolution

$$y(nT) = \sum_{k=-r}^{r} g_k x(nT - rT - kT) \qquad (8.8.8)$$

If g_n has even or odd symmetry, then we saw in Section 8.7 that $H_r^*(\omega)$ has a piecewise linear phase response.

A measure of the error introduced by truncating the Fourier series is the integral square error

$$e^2 = \frac{1}{\omega_s} \int_{-\omega_s/2}^{\omega_s/2} |E^*(\omega)|^2 \, d\omega \qquad (8.8.9)$$

where

$$E^*(\omega) = G^*(\omega) - G_r^*(\omega) = \sum_{|n|>r} g_n e^{-j\omega nT} \qquad (8.8.10)$$

Using the version of Parseval's Theorem given by (4.8.9), we find that

$$e^2 = \sum_{|n|>r} |g_n|^2 = \sum_{n=-\infty}^{\infty} |g_n|^2 - \sum_{n=-r}^{r} |g_n|^2$$

$$= \frac{1}{\omega_s} \int_{-\omega_s/2}^{\omega_s/2} |G^*(\omega)|^2 \, d\omega - \sum_{n=-r}^{r} |g_n|^2 \qquad (8.8.11)$$

If weights a_n different from g_n are used in the truncated series, then $e_n = g_n - a_n \neq 0$ for $|n| \leq r$ and the integral square error will be greater than e^2. Thus the truncated series $G_r^*(\omega)$ is best in the sense that of all the linear combinations of the same set of complex exponentials it minimizes the integral square error.

The number of terms in the truncated series necessary to reduce e^2 to a prescribed value can be determined from (8.8.11). The integral on the right-hand side is usually easy to calculate. New Fourier coefficients can be calculated until the sum on the right becomes sufficiently large to reduce e^2 to the prescribed value.

Example 8.8.1   A Hilbert Transform Filter

The samples of the Hilbert transform of a band-limited signal can be obtained by passing the signal samples through a digital filter with the frequency response

$$G^*(\omega) = -j \, \text{sign} \, \omega \qquad \text{for} \qquad |\omega| < \omega_s/2 \qquad (8.8.12)$$

where

$$\text{sign} \, \omega = \begin{cases} 1 & \text{for} & \omega > 0 \\ 0 & \text{for} & \omega = 0 \\ -1 & \text{for} & \omega < 0 \end{cases}$$

Using (8.8.2) we find that the Fourier coefficients of $G^*(\omega)$ are

$$g_n = \begin{cases} 0 & \text{for} & n \text{ even} \\ \dfrac{2}{n\pi} & \text{for} & n \text{ odd} \end{cases} \qquad (8.8.13)$$

Since the pulse response g_n has odd symmetry, it follows that

$$G^*(\omega) = -j \frac{4}{\pi} \sum_{\substack{n=1 \\ n \, \text{odd}}}^{\infty} \frac{1}{n} \sin n\omega T \qquad (8.8.14)$$

Equation 8.8.14 is just the Fourier series for a square wave with period ω_s. Notice that because of the $-j$ term in (8.8.14) each truncated approximation $G_r^*(\omega)$ will have exactly the desired phase response. However, the amplitude responses will not be perfect. As an example, $G_7^*(\omega)$ is shown in Fig. 8.8.1. The integral square error calculated by (8.8.11) is $e^2 = 5.04044 \times 10^{-2}$. This filter can be realized with a delay of seven samples by the pulse transfer function

$$H_7(z) = \frac{2}{\pi} \left(-\tfrac{1}{7} - \tfrac{1}{5}z^{-2} - \tfrac{1}{3}z^{-4} - z^{-6} + z^{-8} + \tfrac{1}{3}z^{-10} + \tfrac{1}{5}z^{-12} + \tfrac{1}{7}z^{-14} \right) \quad \blacktriangleleft$$

Example 8.8.1 illustrates the problem with simply truncating the Fourier series for $G^*(\omega)$ to obtain a finite term approximation. Notice that the frequency response shown in Fig. 8.8.1 has significant ripple. Normally, a flatter amplitude response is desirable. In general, a truncated Fourier series tends to have excessive ripple near the points where $G^*(\omega)$ has sharp jumps even though it is best in the least integral square error sense. This ripple is known as Gibbs' phenomenon.

Before proposing a cure, let us examine the cause of Gibbs' phenomenon more carefully. We can consider $G_r^*(\omega)$ to be the transform of the product of g_n and the rectangular "window"

$$p_{n,r} = \begin{cases} 1 & \text{for} \quad |n| \le r \\ 0 & \text{elsewhere} \end{cases} \qquad (8.8.15)$$

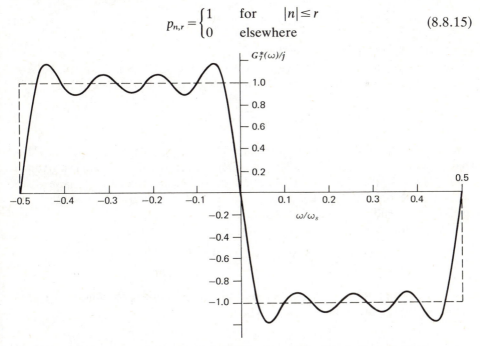

FIGURE 8.8.1. Frequency response of $r = 7$ Hilbert transform filter in Example 8.8.1.

According to (4.7.18), the transform of $g_n p_{n,r}$ can be calculated by the frequency domain convolution

$$G_r^*(\omega) = \frac{1}{\omega_s} \int_{-\omega_s/2}^{\omega_s/2} P_r^*(\lambda) G^*(\omega - \lambda) \, d\lambda \qquad (8.8.16)$$

where

$$P_r^*(\omega) = \sum_{n=-r}^{r} e^{-j\omega nT} \qquad (8.8.17)$$

$P_r^*(\omega)$ can also be written as

$$P_r^*(\omega) = e^{j\omega rT}(1 + e^{-j\omega T} + \cdots + e^{-j\omega 2rT}) \qquad (8.8.18)$$

The term in parenthesis is a geometric series, so that

$$P_r^*(\omega) = e^{j\omega rT} \frac{1 - e^{-j\omega T(2r+1)}}{1 - e^{-j\omega T}} \qquad (8.8.19)$$

Now

$$e^{j\omega rT} = e^{j\omega T(2r+1)/2} / e^{j\omega T/2} \qquad (8.8.20)$$

Substituting (8.8.20) into (8.8.19), multiplying out the numerator and denominator, and then dividing both by $2j$, we obtain

$$P_r^*(\omega) = \frac{\sin(2r+1)\omega T/2}{\sin \omega T/2} \qquad (8.8.21)$$

$P_r^*(\omega)$ is known as the *Fourier kernel.*

The ripples in $G_r^*(\omega)$ are a result of the ripples in $P_r^*(\omega)$. From (8.8.17) we can see that $P_r^*(0) = 2r + 1 = N$. The numerator of (8.8.21) is zero for

$$\omega = k \frac{2\pi}{T(2r+1)} = k\omega_s/N \qquad (8.8.22)$$

The denominator is zero when $\omega = k\omega_s$. Thus in the interval $|\omega| < \omega_s/2$, $P_r^*(\omega)$ has a peak of N at $\omega = 0$ and has zero crossings at the frequencies given by (8.8.22) for k not zero. In this interval $P_r^*(\omega)$ approximates the shape of the $\sin x/x$ type function plotted in Fig. 2.4.1. The function $P_r^*(\omega)$ repeats with period ω_s outside the Nyquist band.

To examine what happens at a jump, let us assume that

$$G^*(\omega) = \begin{cases} 1 & \text{for} \quad 0 \le \omega < \omega_s/2 \\ 0 & \text{for} \quad -\omega_s/2 \le \omega < 0 \end{cases} \qquad (8.8.23)$$

Since the integrand in (8.8.16) has period ω_s, the limits of integration can be replaced by $\omega - \omega_s/2$ and $\omega + \omega_s/2$. Therefore

$$G_r^*(\omega) = \frac{1}{\omega_s} \int_{\omega - \omega_s/2}^{\omega + \omega_s/2} P_r^*(\lambda) G^*(\omega - \lambda) \, d\lambda = \frac{1}{\omega_s} \int_{\omega - \omega_s/2}^{\omega} P_r^*(\lambda) \, d\lambda \quad (8.8.24)$$

The ripples in $G_r^*(\omega)$ occur as different ripples in $P_r^*(\omega)$ are included in the region of integration in (8.8.24). The largest change in $G_r^*(\omega)$ occurs as the central lobe of $P_r^*(\omega)$, that is, the positive lobe between the zeros at $\pm\omega_s/N$, is included.

According to (8.8.24)

$$G_r^*(0) = \frac{1}{\omega_s} \int_{-\omega_s/2}^{0} P_r^*(\lambda)\,d\lambda \tag{8.8.25}$$

Since $P_r^*(\omega)$ is even

$$G_r^*(0) = \frac{1}{2\omega_s} \int_{-\omega_s/2}^{\omega_s/2} P_r^*(\omega)\,d\omega = \tfrac{1}{2}p_{0,r} = \tfrac{1}{2} \tag{8.8.26}$$

In other words, all the truncated series pass through the point halfway up the jump.

Now let us assume that N is large and that $0 \le \omega \le \omega_s/N$. Then the lower limit in (8.8.24) can essentially be replaced by $-\omega_s/2$. With this approximation, we can see that the first peak in $G_r^*(\omega)$ for $\omega > 0$ occurs for $\omega = \omega_s/N$ because the entire area of the central lobe is then included. This peak value is

$$G_r^*(\omega_s/N) \cong \frac{1}{\omega_s} \int_{-\omega_s/2}^{\omega_s/N} P_r^*(\lambda)\,d\lambda = \frac{1}{2} + \frac{1}{\omega_s} \int_{0}^{\omega_s/N} P_r^*(\lambda)\,d\lambda \tag{8.8.27}$$

Substituting (8.8.21) for $P_r^*(\lambda)$ and making the change of variables $x = \lambda NT/2$, we find that

$$G_r^*(\omega_s/N) = \frac{1}{2} + \frac{1}{\pi} \int_{0}^{\pi} \frac{\sin x}{N \sin x/N}\,dx \tag{8.8.28}$$

Now

$$\lim_{N \to \infty} G_r^*(\omega_s/N) = \frac{1}{2} + \frac{1}{\pi} \int_{0}^{\pi} \lim_{N \to \infty} \frac{\sin x}{N \sin x/N}\,dx$$

$$= \frac{1}{2} + \frac{1}{\pi} \int_{0}^{\pi} \frac{\sin x}{x}\,dx \tag{8.8.29}$$

The integral in (8.8.29) must be numerically evaluated and approximately has the value 0.5895. Therefore

$$\lim_{N \to \infty} G_r^*(\omega_s/N) = 1.0895 \tag{8.8.30}$$

Surprisingly, even for N infinite the series has an 8.95 percent overshoot at the jump. The term $\sin x/N$ in (8.8.28) can be closely approximated by x/N for $x/N < \pi/N < 0.1$ or, equivalently, $N > 10\pi \cong 31$. Thus even for moderate N an 8.95 percent overshoot can be expected. If succeeding ripples are examined, it is found that their amplitudes decrease essentially as $1/(\omega N)$ in the vicinity of the jump. As N becomes large, all the ripples effectively move over to the jump.

Since $P_r^*(\omega)$ is even, the ripples before and after a jump have the same characteristics. Therefore, in the case of the jump we have been investigating, the first negative peak to the left of $\omega = 0$ undershoots zero to -0.0895 for moderate N. This large undershoot and relatively slow ripple decay causes difficulty when stop-band rejections of a little more than 21 dB are required. In most filtering applications stop-band rejections of 40 dB or more are required.

For practical filters that approximate ideal filters with piecewise constant frequency responses, the allowable widths of the transition bands between the constant characteristic regions are usually specified. An estimate of the number of coefficients required in a nonrecursive filter to obtain a specified transition bandwidth can be determined by observing that the distance between the positive and negative peaks immediately adjacent to a jump is the width of the main lobe of $P_r^*(\omega)$. Thus an estimate of the transition bandwidth for an N tap filter is

$$\omega_t = 2\omega_s/N \qquad (8.8.31)$$

If ω_t is specified, then the required value of N is

$$N = 2\omega_s/\omega_t \qquad (8.8.32)$$

Example 8.8.2

Suppose that a low-pass nonrecursive filter is required with $\omega_s = 2\pi \times 10^4$ and $\omega_t \leq 2\pi \times 10^2$. According to (8.8.32), the required number of coefficients is $N \geq 200$. This illustrates one drawback of nonrecursive filters: a large number of coefficients are required for a small transition bandwidth. ◄

The excess ripple problem can be solved by replacing the rectangular truncation window $p_{n,r}$ by a truncation window with a frequency response that has small amplitude side lobes in relation to the central lobe. In order to obtain smaller side lobes, the central lobe must be broadened. In other words, the ripple can be decreased at the expense of a wider transition band.

One good, easy-to-use window is the Hamming window

$$h_{n,r} = (0.54 + 0.46 \cos 2\pi n/N)p_{n,r} \qquad N = 2r+1 \qquad (8.8.33)$$

Using the frequency translation property in Section 4.6, we can see that

$$H_r^*(\omega) = 0.54 P_r^*(\omega) + 0.23 P_r^*(\omega - \omega_s/N) + 0.23 P_r^*(\omega + \omega_s/N) \qquad (8.8.34)$$

We have already observed that in the Nyquist band $P_r^*(\omega)$ has a central lobe extending from $-\omega_s/N$ to ω_s/N and has zeros at $k\omega_s/N$ except for a peak of height N for $k = 0$. From a sketch of the three components in (8.8.34) we can see that the central lobe of $H_r^*(\omega)$ extends from $-2\omega_s/N$ to $2\omega_s/N$ and that outside this interval the side lobes of the three components tend to cancel. For N sufficiently large, $H_r^*(\omega)$ has 99.963 percent of its energy over the Nyquist band in its central lobe, and sidelobes that are down by more than 40 dB from the central peak. The width of the central lobe is $4\omega_s/N$, so that to achieve a transition bandwidth of ω_t at a jump

$$N = 4\omega_s/\omega_t \qquad (8.8.35)$$

filter coefficients are required.

In summary, the design procedure is to first calculate the Fourier coefficients g_n of the desired frequency response and then use the modified coefficients $\hat{g}_n = g_n h_{n,r}$ as the actual filter coefficients. The resulting frequency response can be calculated directly from $\hat{G}_r(z)$. The effect of the window can be visualized from (8.8.16) with $P_r^*(\lambda)$ replaced by $H_r^*(\lambda)$.

Example 8.8.3

The frequency response of the $r = 7$ or $N = 15$ Hilbert transform filter discussed in Example 8.8.1 with coefficients modified by the Hamming window is shown in Fig. 8.8.2. Notice that the transition bandwidth has been doubled and that the ripples cannot be seen on the scale of the drawing.   ◄

Another easy-to-use window is the Blackman window

$$b_{n,r} = (0.42 + 0.5 \cos 2\pi n/N + 0.08 \cos 4\pi n/N) p_{n,r} \qquad (8.8.36)$$

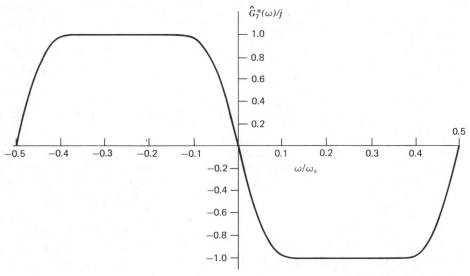

FIGURE 8.8.2. $r = 7$ Hilbert transform filter of Example 8.8.1 modified by the Hamming window.

The corresponding frequency response is

$$B_r^*(\omega) = 0.42 P_r^*(\omega) + 0.25[P_r^*(\omega + \omega_s/N) + P_r^*(\omega - \omega_s/N)]$$
$$+ 0.04[P_r^*(\omega + 2\omega_s/N) + P_r^*(\omega - 2\omega_s/N)] \tag{8.8.37}$$

The width of the central lobe is $6\omega_s/N$ so that

$$N = 6\omega_s/\omega_t \tag{8.8.38}$$

coefficients are required to obtain a transition bandwidth of ω_t at a jump. The side lobes are down by more than 80 dB from the central peak.

Kaiser has proposed an excellent window. The weighting sequence is

$$k_{n,r} = \frac{I_0(\omega_a T \sqrt{r^2 - n^2})}{I_0(r\omega_a T)} p_{n,r} \tag{8.8.39}$$

where $I_0(x)$ is the modified Bessel function of the first kind and zero order. If the function $k(t)$, time limited to $|t| < rT$, is formed by letting $t = nT$ in (8.8.39), it can be shown that

$$K(\omega) = \frac{2}{I_0(r\omega_a T)} \frac{\sin (rT\sqrt{\omega^2 - \omega_a^2})}{\sqrt{\omega^2 - \omega_a^2}} \tag{8.8.40}$$

For reasonable values of $r\omega_a T$ aliasing is negligible and $K_r^*(\omega)$ is essentially equal to $K(\omega)$ in the Nyquist band. The Kaiser window closely approximates the zero-order prolate spheroidal wave functions which have optimum band-limiting properties [123]. A trade-off between the width of the central lobe and amplitudes of the side lobes can be obtained by varying $r\omega_a T$. Typically $r\omega_a T$ is chosen between 4 and 9. From (8.8.40) we can see that the zeros of $K(\omega)$ occur at

$$\omega = \pm \sqrt{\omega_a^2 + \left(\frac{k\omega_s}{2r}\right)^2} \qquad \text{for} \qquad k = \pm 1, \pm 2, \ldots \tag{8.8.41}$$

For r large, $N = 2r + 1 \cong 2r$. Using this approximation, the first zero in the frequency response of the Hamming window occurs at ω_s/r. According to (8.8.41), the first zero in $K(\omega)$ occurs at ω_s/r when $r\omega_a T = \pi\sqrt{3} = 5.4414$. The first zero of the frequency response of the Blackman window occurs at approximately $1.5\omega_s/r$ and the corresponding value of $r\omega_a T$ is $2\sqrt{2}\,\pi = 8.885$. When $r\omega_a T$ is appropriately chosen, the Kaiser window is superior to both the Hamming and Blackman windows. Some design curves and tables are given in Kuo and Kaiser [81]. An estimate of the number of taps required for a 0.3 percent peak overshoot is

$$N = 4\omega_s/\omega_t \tag{8.8.42}$$

For a 0.01% or $-80\,dB$ peak overshoot the estimate becomes

$$N = 5.75\omega_s/\omega_t \tag{8.8.43}$$

Helms [43] has proposed the Dolph-Chebyshev window. This window is optimum in the sense that the central lobe width is the smallest possible for a given peak side lobe amplitude. However, it does not follow that the integrated frequency response has optimum properties. In fact, the Kaiser window is slightly more efficient than the Dolph-Chebyshev window because it requires a slightly smaller number of taps to achieve a specified transition bandwidth and peak overshoot. As is usually the case in practical situations, the choice of window functions is somewhat arbitrary and depends on the designer's preferences as well as the specific design requirements.

8.9 WEIGHTED LEAST INTEGRAL SQUARE ERROR DESIGN OF NONRECURSIVE DIGITAL FILTERS

A truncated Fourier series minimizes the magnitude squared error averaged over the entire band $|\omega| < \omega_s/2$. The excess ripple in the vicinity of a jump allows the truncated series to have a rapid transition that minimizes large errors near the jump at the expense of smaller errors elsewhere. This suggests solving the excess ripple problem by choosing the coefficients of the finite series to minimize a weighted integral square error with the error in a transition band around a jump given zero weighting.

Let us assume that $G^*(\omega)$ is a desired frequency response that is to be approximated by the finite trigonometric series

$$F^*(\omega) = \sum_{n=-r}^{r} f_n e^{-j\omega nT} \tag{8.9.1}$$

The coefficients are to be chosen to minimize the weighted integral square error

$$v^2 = \frac{1}{\omega_s} \int_{-\omega_s/2}^{\omega_s/2} |G^*(\omega) - F^*(\omega)|^2 Q(\omega)\, d\omega \tag{8.9.2}$$

where $Q(\omega)$ is a nonnegative, real, even function. Now

$$|G^*(\omega) - F^*(\omega)|^2 = [G^*(\omega) - F^*(\omega)][\bar{G}^*(\omega) - \bar{F}^*(\omega)]$$
$$= |G^*(\omega)|^2 - F^*(\omega)\bar{G}^*(\omega) - \bar{F}^*(\omega)G^*(\omega) + |F^*(\omega)|^2 \tag{8.9.3}$$

Substituting (8.9.1) and (8.9.3) into (8.9.2), we find that

$$v^2 = \frac{1}{\omega_s} \int_{-\omega_s/2}^{\omega_s/2} |G^*(\omega)|^2 Q(\omega)\, d\omega - \sum_{n=-r}^{r} \bar{f}_n a_n - \sum_{n=-r}^{r} f_n \bar{a}_n + \sum_{n=-r}^{r} \sum_{m=-r}^{r} f_n \bar{f}_m q_{m-n} \tag{8.9.4}$$

where

$$a_n = \frac{1}{\omega_s} \int_{-\omega_s/2}^{\omega_s/2} G^*(\omega) e^{j\omega n T} Q(\omega)\, d\omega \tag{8.9.5}$$

and

$$q_n = \frac{1}{\omega_s} \int_{-\omega_s/2}^{\omega_s/2} e^{j\omega n T} Q(\omega)\, d\omega \tag{8.9.6}$$

Since $Q(\omega)$ is even, $q_n = q_{-n}$. In terms of the matrices

$$\mathbf{F} = \begin{bmatrix} f_r \\ f_{r-1} \\ \cdot \\ \cdot \\ \cdot \\ f_{-r} \end{bmatrix} \qquad \mathbf{A} = \begin{bmatrix} a_r \\ a_{r-1} \\ \cdot \\ \cdot \\ \cdot \\ a_{-r} \end{bmatrix} \qquad \mathbf{Q} = \begin{bmatrix} q_0 & q_1 & q_2 & \cdots & q_{2r} \\ q_1 & q_0 & q_1 & \cdots & q_{2r-1} \\ \cdot & & & & \cdot \\ \cdot & & & & \cdot \\ \cdot & & & & \cdot \\ q_{2r} & & & \cdots & q_0 \end{bmatrix}$$

the integral square error is

$$v^2 = \frac{1}{\omega_s} \int_{-\omega_s/2}^{\omega_s/2} |G^*(\omega)|^2 Q(\omega)\, d\omega - \mathbf{F}^\dagger \mathbf{A} - \mathbf{A}^\dagger \mathbf{F} + \mathbf{F}^\dagger \mathbf{Q} \mathbf{F} \tag{8.9.7}$$

where † denotes conjugate transpose. Equation 8.9.7 can also be written as

$$v^2 = \frac{1}{\omega_s} \int_{-\omega_s/2}^{\omega_s/2} |G^*(\omega)|^2 Q(\omega)\, d\omega + (\mathbf{F} - \mathbf{Q}^{-1}\mathbf{A})^\dagger \mathbf{Q} (\mathbf{F} - \mathbf{Q}^{-1}\mathbf{A}) - \mathbf{A}^\dagger \mathbf{Q}^{-1} \mathbf{A} \tag{8.9.8}$$

The last term on the right-hand side of (8.9.4) is positive since it is the integral of $|F^*(\omega)|^2 Q(\omega)$. Therefore, $\mathbf{Q}$ is a positive definite matrix and the second term on the right of (8.9.8) is nonnegative for any choice of $\mathbf{F}$. Consequently, the $\mathbf{F}$ that minimizes v^2 is

$$\mathbf{F} = \mathbf{Q}^{-1}\mathbf{A} \tag{8.9.9}$$

and the corresponding integral square error is

$$v^2 = \frac{1}{\omega_s} \int_{-\omega_s/2}^{\omega_s/2} |G^*(\omega)|^2 Q(\omega)\, d\omega - \mathbf{A}^\dagger \mathbf{Q}^{-1} \mathbf{A} \tag{8.9.10}$$

Excellent designs can be achieved by properly choosing the weighting function $Q(\omega)$. Nevertheless, the weighted least integral square error method has not been used to any great extent because the computation required is more complex than for the window function method, particularly for large r.

However, the required computation is not as great as it appears since $\mathbf{Q}$ is a Toeplitz matrix and can be inverted efficiently [145].

8.10 FREQUENCY SAMPLING FILTERS

Nonrecursive digital filters can be designed by polynomial interpolation on the unit circle. There is a unique polynomial of degree $N-1$ in z^{-1} that passes through the samples of a desired pulse transfer function at N points distributed around the unit circle. This polynomial can be found by using the Lagrange interpolation formula discussed in Chapter 3. By continuity, the polynomial must approximate the desired response reasonably well at points on the unit circle between the sampling points for N sufficiently large. The resulting filter is called a *frequency sampling* filter.

We will only consider the case where the sampling points are uniformly distributed around the unit circle. In this case, the interpolation formula simplifies considerably. Suppose that the desired sampling points are

$$z_k = e^{jk2\pi/N} \qquad \text{for} \qquad k=0,\ldots,N-1 \qquad (8.10.1)$$

These correspond to the frequencies

$$\omega_k = k\omega_s/N \qquad \text{for} \qquad k=0,\ldots,N-1 \qquad (8.10.2)$$

The z_k's are the N roots of the polynomial $1-z^{-N}$. Therefore, the functions

$$F_k(z) = \frac{1-z^{-N}}{N(1-e^{jk2\pi/N}z^{-1})} \qquad \text{for} \qquad k=0,\ldots,N-1 \qquad (8.10.3)$$

are equivalent to polynomials of degree $N-1$. At the sampling points, $F_k(z)$ has the values

$$F_k(e^{jr2\pi/N}) = \begin{cases} 0 & \text{for} \quad r \neq k \\ 1 & \text{for} \quad r = k \end{cases} \qquad (8.10.4)$$

The zeros for $r \neq k$ are obvious. The value 1 for $r=k$ can be easily verified by using l'Hospital's rule. At other points on the unit circle

$$F_k^*(\omega) = F_k(e^{j\omega T}) = \frac{1}{N} \frac{1-e^{-j\omega NT}}{1-e^{-j(\omega-k\omega_s/N)T}}$$

$$= \frac{1}{N} \frac{1-e^{-jN(\omega-k\omega_s/N)T}}{1-e^{-j(\omega-k\omega_s/N)T}}$$

$$= e^{-j(N-1)(\omega-k\omega_s/N)T/2} \frac{\sin N(\omega-k\omega_s/N)T/2}{N \sin(\omega-k\omega_s/N)T/2} \qquad (8.10.5)$$

If we denote the desired frequency response samples by

$$G_k = G^*(k\omega_s/N) \qquad \text{for} \qquad k = 0, \ldots, N-1 \qquad (8.10.6)$$

then the function

$$G(z) = \sum_{k=0}^{N-1} G_k F_k(z) = \frac{1 - z^{-N}}{N} \sum_{k=0}^{N-1} \frac{G_k}{1 - e^{jk2\pi/N} z^{-1}} \qquad (8.10.7)$$

has this same set of samples due to (8.10.4). Since the F_k's are equivalent to polynomials of degree $N-1$, $G(z)$ must be equivalent to the desired interpolation polynomial of degree $N-1$. According to (8.10.5) and (8.10.7), the entire interpolated frequency response on the unit circle is

$$G^*(\omega) = e^{-j(N-1)\omega T/2} \sum_{k=0}^{N-1} G_k e^{j(N-1)k\pi/N} \frac{\sin N(\omega - k\omega_s/N)T/2}{N \sin (\omega - k\omega_s/N)T/2} \qquad (8.10.8)$$

The factor $e^{-j(N-1)\omega T/2}$ corresponds to a delay of half the duration of the pulse response of the filter.

The polynomial representation for $G(z)$ can be found by using the fact that

$$\frac{1 - z^{-N}}{1 - e^{jk2\pi/N} z^{-1}} = \sum_{n=0}^{N-1} (e^{jk2\pi/N} z^{-1})^n \qquad (8.10.9)$$

Substituting (8.10.9) into (8.10.7), we find that

$$G(z) = \frac{1}{N} \sum_{k=0}^{N-1} G_k \sum_{n=0}^{N-1} (e^{jk2\pi/N} z^{-1})^n$$

$$= \sum_{n=0}^{N-1} \left[\frac{1}{N} \sum_{k=0}^{N-1} G_k e^{j(2\pi/N)nk} \right] z^{-n} \qquad (8.10.10)$$

Therefore

$$G(z) = \sum_{n=0}^{N-1} g_n z^{-n} \qquad (8.10.11)$$

where

$$g_n = \frac{1}{N} \sum_{k=0}^{N-1} G_k e^{j(2\pi/N)nk} \qquad \text{for} \qquad n = 0, \ldots, N-1 \qquad (8.10.12)$$

The frequency sampling filter can be implemented by the convolution

$$y(nT) = \sum_{k=0}^{N-1} g_k x(nT - kT) \qquad (8.10.13)$$

where $y(nT)$ is the output and $x(nT)$ is the input.

The sequence $g_0, \ldots, g_{N-1}$ is known as the *inverse discrete Fourier transform* (IDFT) of the sequence $G_0, \ldots, G_{N-1}$. From (8.10.11) we find that the frequency response samples can be calculated as

$$G_k = G(e^{jk\omega_s T/N}) = \sum_{n=0}^{N-1} g_n e^{-j(2\pi/N)nk} \qquad \text{for} \qquad k = 0, \ldots, N-1$$

$$(8.10.14)$$

The sequence $G_0, \ldots, G_{N-1}$ is known as the *discrete Fourier transform* (DFT) of the sequence $g_0, \ldots, g_{N-1}$. The DFT and IDFT have found many applications in signal processing and will be discussed in detail in Chapter 10.

Normally, a real pulse response is required, so the frequency response samples cannot be arbitrarily specified. When g_n is real, we can see from (8.10.11) that $G^*(-\omega) = \overline{G^*(\omega)}$ and so $G^*(-k\omega_s/N) = \overline{G^*(k\omega_s/N)}$ or

$$G_{-k} = \bar{G}_k \qquad (8.10.15)$$

Since $G^*(\omega)$ has period ω_s, G_k must have period N. Therefore (8.10.15) is equivalent to

$$G_{N-k} = \bar{G}_k \qquad \text{for} \qquad k = 0, \ldots, N-1 \qquad (8.10.16)$$

when g_n is real for all n. The converse is also true. According to (8.10.12)

$$\bar{g}_n = \frac{1}{N} \sum_{k=0}^{N-1} \bar{G}_k e^{-j(2\pi/N)nk} \qquad (8.10.17)$$

Changing the index to $r = N - k$ (8.10.17) becomes

$$\bar{g}_n = \frac{1}{N} \sum_{r=1}^{N} \bar{G}_{N-r} e^{-j(2\pi/N)n(N-r)} = \frac{1}{N} \sum_{r=0}^{N-1} \bar{G}_{N-r} e^{j(2\pi/N)nr} \qquad (8.10.18)$$

Comparing (8.10.18) and (8.10.12), we find that $g_n = \bar{g}_n$ when (8.10.16) is true. Therefore, the pulse response will be real if and only if the frequency response samples satisfy (8.10.16).

A linear phase response is often desirable in addition to a real pulse response. From (8.10.8) we can see that one way to obtain a linear phase response is to require that the quantities

$$G'_k = G_k e^{j(N-1)k\pi/N} \qquad \text{for} \qquad k = 0, \ldots, N-1 \qquad (8.10.19)$$

be real. This choice results in a real frequency response with samples G'_k if the linear phase factor is ignored. In this case, the original frequency response samples should be chosen as

$$G_k = G'_k e^{-j(N-1)k\pi/N} \qquad \text{for} \qquad k = 0, \ldots, N-1 \qquad (8.10.20)$$

with G'_k real. To obtain a real pulse response (8.10.16) also requires that

$$G_{N-k} = G'_{N-k} e^{-j(N-1)\pi} e^{j(N-1)k\pi/N} = \bar{G}_k \qquad \text{for} \qquad k = 0, \ldots, N-1$$

$$(8.10.21)$$

For N odd (8.10.21) is satisfied if

$$G'_{N-k} = G'_k \qquad \text{for} \qquad k = 0, \ldots, N-1 \tag{8.10.22}$$

For N even (8.10.21) is satisfied if

$$G'_{N-k} = -G'_k \qquad \text{for} \qquad k = 0, \ldots, N-1 \tag{8.10.23}$$

Substituting (8.10.20) into (8.10.12), we find that

$$g_n = \frac{1}{N} \sum_{k=0}^{N-1} G'_k \exp j \frac{2\pi}{N} k \left(n - \frac{N-1}{2} \right) \tag{8.10.24}$$

If N is odd, $(N-1)/2$ is an integer and (8.10.24) can be written as

$$g_n = g'_{n-(N-1)/2} \qquad \text{for} \qquad n = 0, \ldots, N-1 \tag{8.10.25}$$

where $g'_n = \text{IDFT}\{G'_k\}$. We can see from (8.10.12) that an IDFT has period N if n is allowed to take on any integer value. Therefore

$$(g_0, \ldots, g_{N-1}) = (g'_{(N+1)/2}, \ldots, g'_{N-1}, g'_0, g'_1, \ldots, g'_{(N-1)/2})$$

That is, the pulse response g_n is the sequence $(g'_0, \ldots, g'_{N-1})$ cyclically shifted $(N-1)/2$ samples so that g'_0 is in the center of the sequence. If N is even, $(N-1)/2$ is a point halfway between the two center samples. If G'_k is nonnegative and N odd, we find from (8.10.24) that

$$|g_n| \leq \frac{1}{N} \sum_{k=0}^{N-1} G'_k = g_{(N-1)/2}$$

so that the peak of the pulse response occurs at its center. This is a desirable property when the frequency sampling filter is being used to simulate, for example, the impulse response of a telephone line channel.

From the discussion in Section 8.7, we would expect that g_n should have even symmetry about its center when G'_k is real and chosen so that g_n is real. This can be shown to be the case by observing from (8.10.24) that

$$g_{N-1-n} = \frac{1}{N} \sum_{k=0}^{N-1} G'_k \exp -j \frac{2\pi}{N} k \left(n - \frac{N-1}{2} \right) = \bar{g}_n = g_n$$

$$\text{for} \qquad n = 0, \ldots, N-1 \tag{8.10.26}$$

If, neglecting the factor $e^{-j(N-1)\omega T/2}$, a purely imaginary frequency response is required, we can let

$$G_k e^{j(N-1)k\pi/N} = jG'_k$$

or

$$G_k = jG'_k e^{-j(N-1)k\pi/N} \qquad \text{for} \qquad k = 0, \ldots, N-1 \tag{8.10.27}$$

with G'_k real. Applying the condition $G_{N-k} = \bar{G}_k$ necessary for a real pulse response, we find that G'_k must be chosen so that

$$G'_{N-k} = (-1)^N G'_k \qquad \text{for} \qquad k = 0, \ldots, N-1 \qquad (8.10.28)$$

Substituting (8.10.27) into (8.10.12) yields

$$g_n = \frac{1}{N} \sum_{k=0}^{N-1} jG'_k \exp j\frac{2\pi}{N} k\left(n - \frac{N-1}{2}\right) \qquad (8.10.29)$$

Therefore,

$$g_{N-1-n} = \frac{1}{N} \sum_{k=0}^{N-1} jG'_k \exp -j\frac{2\pi}{N} k\left(n - \frac{N-1}{2}\right) = -\bar{g}_n$$

$$= -g_n \qquad \text{for} \qquad n = 0, \ldots, N-1 \qquad (8.10.30)$$

Thus the pulse response has odd symmetry in this case.

Example 8.10.1

Suppose that an $N = 15$, linear phase, low-pass filter with a cutoff frequency $\omega_c = 4\omega_s/15$ is required. Notice that ω_c corresponds to the frequency sampling point $k = 4$. According to (8.10.22), we can choose the modified frequency samples as

$$G'_k = \begin{cases} 1 & \text{for} \quad k = 0, \ldots, 4 \text{ and } 11, \ldots, 14 \\ 0 & \text{for} \quad k = 5, \ldots, 10 \end{cases}$$

According to 8.10.20, the actual frequency samples are

$$G_k = G'_k e^{-j14k\pi/15} \qquad \text{for} \qquad k = 0, \ldots, N-1$$

The pulse response g_n can be calculated by (8.10.12) and the filter implemented as a discrete-time convolution. The resulting frequency response, neglecting the linear phase factor $e^{-j7\omega T}$, is shown in Fig. 8.10.1. ◀

Figure 8.10.1 illustrates the problem with frequency sampling filters. As in the case of truncated Fourier series, frequency sampling filters have excessive ripple in the vicinity of a jump. Again, the problem can be solved by trading off the width of the transition band for reduced ripple amplitude. The ripple can be significantly reduced by allowing a transition band with several frequency response samples that are distributed to give a smoother transition. Figure 8.10.2 shows the frequency response obtained when the filter in Example 8.10.1 is modified by allowing one transition band sample, $G'_4 = G'_{11} = 0.5$. Notice that the ripple has been significantly reduced but at the expense of a doubled transition bandwidth. The ripple could be further reduced by decreasing $G'_3 = G'_{12}$ and increasing $G'_5 = G'_{10}$. With the aid of a computer and automatic plotter, the transition band samples can be quickly optimized by trial and error. A computer-aided design technique using linear programming has

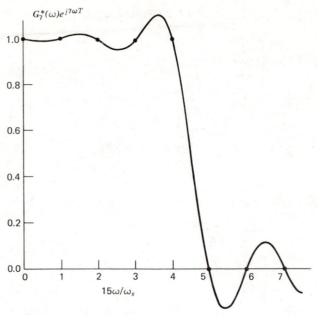

FIGURE 8.10.1. Frequency response of $N = 15$ frequency sampling filter in Example 8.10.1.

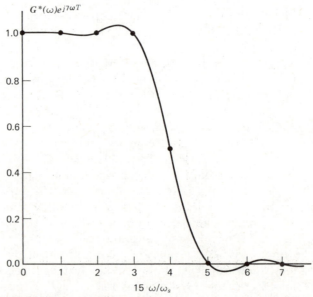

FIGURE 8.10.2. Filter of Example 8.10.1 modified by allowing one transition band sample, $G'_4 = G'_{11} = 0.5$.

been developed for finding the transition band samples that minimize the peak stop-band ripple subject to a constraint on the peak pass-band ripple [109]. Since the optimization is directly on the frequency response, these filters are usually somewhat more efficient in terms of the number of coefficients required than filters designed by the Fourier series and window function method.

Finally, it is interesting to observe that frequency sampling filters can be implemented in recursive form. From (8.10.7) we see that $G(z)$ can be realized as a cascade of the "comb" filter $1 - z^{-N}$ and a parallel bank of resonators of the form $(G_k/N)/(1 - e^{jk2\pi/N}z^{-1})$. Resonators with complex conjugate poles can be combined into second-order sections with real coefficients. This structure is appealing in that the frequency response samples appear directly. Notice that the resonator poles are on the unit circle. This can cause stability problems. With the availability of a digital computer, $g_n = \text{IDFT}\{G_k\}$ can be calculated easily and the filter implemented in nonrecursive form without any stability problem.

8.11 TRANSFORMATION OF A PROTOTYPE LOW-PASS DIGITAL FILTER INTO OTHER DIGITAL FILTERS

A standard method for designing low-pass, high-pass, band-pass, and band-stop analog filters is to transform a prototype low-pass filter with a cutoff frequency of unity into the desired filter by a substitution for s that maps the $j\omega$ axis onto itself. A similar procedure can be used for digital filters by making a substitution for z that maps the unit circle onto itself.

We will now use the variable Z as the argument of the pulse transfer function of the prototype low-pass filter and Ω as the real frequency variable, so that, $Z = e^{j\Omega T}$ on the unit circle. $G_0(Z)$ will represent the pulse transfer function and $G_0^*(\Omega) = G_0(e^{j\Omega T})$ the frequency response of the prototype filter. We will use the variable z as the argument of the pulse transfer function of the desired filter and let $z = e^{j\omega T}$ on the unit circle.

Any transformation generated by a rational function of z with real coefficients that maps the unit circle in the z-plane onto the unit circle in the Z-plane must have the form

$$Z^{-1} = f(z) = \pm \frac{d_N + d_{N-1}z^{-1} + \cdots + z^{-N}}{1 + d_1 z^{-1} + \cdots + d_N z^{-N}} = \pm \frac{z^{-N}D(z^{-1})}{D(z)} \qquad (8.11.1)$$

where $D(z)$ is a polynomial with real coefficients and all its roots inside the unit circle. The roots of $D(z^{-1})$ are the reciprocals of the roots of $D(z)$ and must lie outside the unit circle. Obviously, $|f(z)|=1$ for $z=e^{j\omega T}$ since $D(z)$ and $D(z^{-1})$ are then complex conjugates. Notice that $f(z)$ is simply the pulse transfer function of an all-pass filter with $K=\pm 1$ as discussed in Section 8.6. Since $D(z)$ has real coefficients, the conjugate of any complex root must also be a root. Therefore, if $p_1,\ldots,p_N$ are the roots of $D(z)$, (8.11.1) can be written in the factored form

$$Z^{-1} = \pm \prod_{k=1}^{N} \frac{z^{-1}-\bar{p}_k}{1-p_k z^{-1}} \tag{8.11.2}$$

Let us consider the single factor

$$f_1(z) = (z^{-1}-\bar{p}_k)/(1-p_k z^{-1}) = (1-\bar{p}_k z)/(z-p_k) \tag{8.11.3}$$

Then

$$|f_1(z)|^2 = f_1(z)\overline{f_1(z)} = \frac{1-2\,\mathrm{Re}\,(p_k\bar{z})+|p_k|^2|z|^2}{|z|^2-2\,\mathrm{Re}\,(p_k\bar{z})+|p_k|^2} \tag{8.11.4}$$

Subtracting the denominator from the numerator of (8.11.4) yields

$$v = (1-|z|^2)(1-|p_k|^2) \tag{8.11.5}$$

The factor $1-|p_k|^2$ is positive since $|p_k|<1$. Therefore

$$|f_1(z)| \begin{cases} <1 & \text{for} \quad v<0 \text{ or } |z|>1 \\ =1 & \text{for} \quad v=0 \text{ or } |z|=1 \\ >1 & \text{for} \quad v>0 \text{ or } |z|<1 \end{cases} \tag{8.11.6}$$

The product in (8.11.2) must exhibit the same behavior. Therefore

$$|Z| \begin{cases} <1 & \text{for} \quad |z|<1 \\ =1 & \text{for} \quad |z|=1 \\ >1 & \text{for} \quad |z|>1 \end{cases} \tag{8.11.7}$$

Thus, the transformation $Z^{-1}=f(z)$ maps the unit circle onto itself.

The problem now is to find all-pass functions that generate the desired frequency transformations. The order N of the required all-pass function can be determined by observing that as z makes one counterclockwise rotation around the unit circle, Z makes N counterclockwise rotations according to the argument principle. To map the prototype low-pass filter $G_0(Z)$ into another low-pass or a high-pass filter, the unit circle must be mapped onto itself once requiring that $N=1$. To map the prototype filter into a band-pass or band-stop filter, the unit circle must be mapped onto itself twice requiring that $N=2$. The $N=1$ and $N=2$ cases are of primary concern since more complicated multi-band filters can be realized by combinations of these simpler filters.

Let us now determine the substitution required for a low-pass to low-pass transformation. Suppose that we wish to transform a prototype low-pass filter $G_0(Z)$ with cutoff frequency Ω_c into a low-pass filter $G(z)$ with the cutoff frequency ω_c. According to the previous paragraph, a first-order all-pass function is required. In addition, the point $z = 1$ corresponding to $\omega = 0$ should be mapped to the point $Z = 1$ corresponding to $\Omega = 0$ and the point $z = -1$ corresponding to $\omega = \pm\omega_s/2$ should be mapped to the point $Z = -1$ corresponding to $\Omega = \pm\omega_s/2$. The transformation

$$Z^{-1} = \frac{z^{-1} - p}{1 - pz^{-1}} \tag{8.11.8}$$

satisfies these requirements. We can determine p from the requirement

$$e^{-j\Omega_c T} = \frac{e^{-j\omega_c T} - p}{1 - pe^{-j\omega_c T}} \tag{8.11.9}$$

which gives

$$p = \sin\left[(\Omega_c - \omega_c)T/2\right]/\sin\left[(\Omega_c + \omega_c)T/2\right] \tag{8.11.10}$$

The pulse transfer function of the desired filter is

$$G(z) = G_0(Z)$$

where Z is given by (8.11.8).

A low-pass filter can be transformed into a high-pass filter by rotating the unit circle 180°. Therefore, in addition to $N = 1$, we must require that $z = 1$ maps to $Z = -1$ and $z = -1$ maps to $Z = 1$. The transformation

$$Z^{-1} = -\frac{z^{-1} - p}{1 - pz^{-1}} \tag{8.11.11}$$

has these properties. If the prototype low-pass filter has the cutoff frequency Ω_c and the desired cutoff frequency for the high-pass filter is ω_c, we must require that $-\Omega_c$ transforms to ω_c or

$$e^{j\Omega_c T} = -\frac{e^{-j\omega_c T} - p}{1 - pe^{-j\omega_c T}} \tag{8.11.12}$$

or

$$p = \cos\left[(\Omega_c + \omega_c)T/2\right]/\cos\left[(\Omega_c - \omega_c)T/2\right] \tag{8.11.13}$$

The required transformation for mapping the prototype low-pass filter into a band-pass filter with center frequency ω_0, upper cutoff frequency ω_2, and lower cutoff frequency ω_1 can be shown to be [23]

$$Z^{-1} = -\frac{d_2 + d_1 z^{-1} + z^{-2}}{1 + d_1 z^{-1} + d_2 z^{-2}} \tag{8.11.14}$$

where

$$d_1 = -2ak/(k+1) \tag{8.11.15}$$

$$d_2 = (k-1)/(k+1) \tag{8.11.16}$$

$$k = \cot\left[(\omega_2 - \omega_1)T/2\right] \tan(\Omega_c T/2) \tag{8.11.17}$$

and

$$a = \cos\left[(\omega_2 + \omega_1)T/2\right]/\cos\left[(\omega_2 - \omega_1)T/2\right] = \cos \omega_0 T \tag{8.11.18}$$

The center frequency ω_0 corresponds to the frequency $\Omega = 0$ or point $Z = 1$. The upper cutoff frequency $\omega = \omega_2$ is mapped to $\Omega = \Omega_c$ and the lower cutoff frequency $\omega = \omega_1$ is mapped to $\Omega = -\Omega_c$. The points $z = \pm 1$ are mapped to $Z = -1$.

The required transformation for mapping the prototype low-pass filter into a band-stop filter with center frequency ω_0, upper cutoff frequency ω_2, and lower cutoff frequency ω_1 can be shown to be [23]

$$Z^{-1} = \frac{d_2 + d_1 z^{-1} + z^{-2}}{1 + d_1 z^{-1} + d_2 z^{-2}} \tag{8.11.19}$$

where

$$d_1 = -2a/(1+k) \tag{8.11.20}$$

$$d_2 = (1-k)/(1+k) \tag{8.11.21}$$

$$k = \tan\left[(\omega_2 - \omega_1)T/2\right] \tan(\Omega_c T/2) \tag{8.11.22}$$

and

$$a = \cos\left[(\omega_2 + \omega_1)T/2\right]/\cos\left[(\omega_2 - \omega_1)T/2\right] = \cos \omega_0 T \tag{8.11.23}$$

The frequency $\omega = \omega_0$ is mapped to $\Omega = \omega_s/2$, $\omega = \omega_1$ is mapped to $\Omega = \Omega_c$, and $\omega = \omega_2$ is mapped to $\Omega = -\Omega_c$. The points $z = \pm 1$ are mapped to $Z = 1$.

Example 8.11.1

Let us use the filter given by (8.5.15) as the prototype low-pass filter. This filter has a 3 dB cutoff frequency of $\Omega_c = \omega_s/4$. Suppose that a band-stop filter with upper and lower cutoff frequencies $\omega_2 = 3\omega_s/8$ and $\omega_1 = \omega_s/8$ is required. According to (8.11.19) through (8.11.23), the necessary transformation is

$$Z^{-1} = z^{-2} \tag{8.11.24}$$

From (8.11.23) we find that $\cos \omega_0 T = 0$, so that, the center frequency is $\omega_0 = \omega_s/4$. It follows from (8.11.24) that $\Omega = 2\omega$. Substituting (8.11.24) into (8.5.15), we find that the required pulse transfer function is

$$G(z) = \tfrac{1}{6} \frac{1 + 3z^{-2} + 3z^{-4} + z^{-6}}{1 + \tfrac{1}{3}z^{-4}} \tag{8.11.25}$$

The amplitude response of $G(z)$ is shown in Fig. 8.11.1   ◄

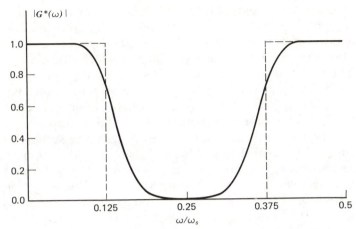

FIGURE 8.11.1. Amplitude response of band-stop filter in Example 8.11.1.

In general, these all-pass substitutions are not suitable for transforming nonrecursive filters into nonrecursive filters. Only in the unusual case where the required substitution has the form $Z^{-1} = \pm z^{-N}$, as in Example 8.11.1, is a nonrecursive filter transformed into a nonrecursive filter. However, the standard techniques for designing nonrecursive filters allow the desired frequency response to be approximated directly and no transformations are required. In some cases, the design for a low-pass filter $G_0(z)$ with good stop-band rejection may be available, and we would like to obtain a band-pass filter quickly without any extensive redesign. The sum

$$G(z) = G_0(ze^{j\omega_0 T}) + G_0(ze^{-j\omega_0 T}) \qquad (8.11.26)$$

has the frequency response

$$G^*(\omega) = G_0^*(\omega + \omega_0) + G_0^*(\omega - \omega_0) \qquad (8.11.27)$$

which is a band-pass characteristic with center frequency ω_0. Care must be taken to make sure that the distortion from the overlapping tails is not too great.

8.12 A BRIEF STUDY OF COMPUTER-AIDED DESIGN TECHNIQUES

Most of the methods described in this chapter for designing digital filters have been well-defined analytical procedures requiring only simple computations to calculate the filter coefficients. The resulting filters may be very good but are not usually optimum according to various criteria. Minimization of the Chebyshev norm, that is, the peak error over the Nyquist band, is a popular criterion. Iterative numerical methods must be normally used to find the optimum filters.

Computer-aided optimization techniques have been most effectively applied to the design of nonrecursive filters. In this case, the frequency response at any given frequency is a linear combination of the pulse response samples or, equivalently, the frequency response samples discussed in Section 8.10. Linear programming can be applied to problems of this type. Rabiner [109] has used linear programming to optimize frequency sampling filters. In his technique, the pass-band samples are set to 1 and the stop-band samples to zero. Several transition band samples are adjusted to minimize the peak stop-band ripple subject to a constraint on the peak pass-band ripple. Theoretically, all the samples could be adjusted, but the constraint of reasonable computation time has limited the number of free variables to about four. The filters designed by Rabiner's method are not optimum in the minimax or Chebyshev sense because most of the frequency samples are fixed. However, the resulting designs are very good.

It can be shown that nonrecursive filters that are optimum in the Chebyshev sense must have equal ripple behavior. Herrmann [46] developed a method for designing linear phase, equal ripple, nonrecursive filters by numerically solving a set of nonlinear equations involving the stop and pass-band ripple tolerances, the extremal frequencies, and the filter coefficients. Because of its computational complexity, his method is limited to designing filters with about 60 taps or less. Hofstetter, Oppenheim, and Siegel [48] discovered a method for designing linear phase, equal ripple filters equivalent to Herrmann's filters that can be used to design filters with essentially any number of taps. In their method, tolerances on the pass and stop-band ripple, as well as the number of ripples in each band, are specified. An initial guess at the extremal frequencies is made and a polynomial is fit through these points by Lagrange interpolation. The extremal frequencies of this polynomial are used in the next iteration. One drawback of this method is that the edges of the stop and pass-bands cannot be specified beforehand. McClellan and Parks [106,90] have found a solution by formulating the problem as a weighted Chebyshev approximation problem, which can be solved using the Remez exchange algorithm. Their method is similar in many respects to Hofstetter's method. The filter order, pass and stop-band edges, and an error weighting function can be specified. A FORTRAN computer program for designing these optimum filters has been published [91]. The program is efficient and high-order filters can be designed in only a few seconds. The McClellan and Parks technique is the best computer-aided design method for linear phase nonrecursive filters presently available.

It should be pointed out that nonrecursive filters designed using the Fourier series and Kaiser window function are not as efficient but compare reasonably well with the optimum equal ripple designs.

Computer-aided design of recursive filters is more difficult since the filter coefficients enter into the cost functions nonlinearly. Steiglitz [126] has investigated a method for choosing the coefficients in a cascade of second-order sections to minimize the integral square error between the actual and desired amplitude response. Deczky [28] has investigated the more general problem of finding the coefficients in a cascade of second-order sections to minimize a weighted sum of weighted integrals of the pth powers of the absolute amplitude response and envelope delay errors. If only the amplitude response or envelope delay is included and p is allowed to become large, the solutions approach the Chebyshev or minimax solutions. Both Steiglitz and Deczky used the Fletcher-Powell optimization algorithm. Both methods require a significant amount of computation time. Because of the nonlinear nature of the problem, convergence only to a local minima can be guaranteed. However, computational experience has indicated that good designs can be obtained.

8.13 COMMENTS AND CONCLUSIONS

Techniques for designing both recursive and nonrecursive digital filters have been presented in this chapter. The methods for designing these two classes of filters are essentially disjoint. The simplest and most effective method for designing recursive filters that approximate ideal low-pass, high-pass, band-pass, and band-stop amplitude characteristics is to transform an analog filter with the desired characteristics into a digital filter using the bilinear transformation. Excellent methods for designing analog filters of these types exist. The all-pass substitutions discussed in Section 8.11 can also be used to transform a prototype low-pass digital filter into these four types of digital filters. As a result of nonlinear frequency warping, the bilinear transformation cannot be used to transform an analog filter with an amplitude response that is not piecewise constant into a digital filter with proper characteristics. The same is true for linear phase analog filters. In this case the guard filter and Z-transform method can be used. If only a desired frequency response is specified, it can be approximated directly by computer optimization of the parameters in a recursive filter structure.

The methods for designing nonrecursive digital filters approximate the desired frequency response directly. They are not based on transforming an analog filter into a digital filter. Filters that closely approximate any reasonable frequency response can be quickly designed using the Fourier series and window function method. Frequency sampling filters are more difficult to design since trial-and-error or computer-aided design procedures must be used to select the transition band samples that minimize ripples. Linear phase, nonrecursive filters that are optimum in the Chebyshev sense can now be efficiently designed using the method of McClellan and Parks discussed in Section 8.12.

The choice between recursive and nonrecursive filters depends on the particular application. When phase response is not important and an amplitude response with steep skirts is required, recursive filters require significantly fewer coefficients than nonrecursive filters. When a linear phase response is required, the picture can change. Nonrecursive filters can be designed with exactly linear phase. The phase response of a recursive filter can be equalized with all-pass sections, but the number of coefficients in the combined recursive filter and equalizer may approach that of the linear phase nonrecursive filter. Finally, filtering using the discrete Fourier transform has become popular and is equivalent to nonrecursive filtering. This application will be discussed in Chapters 10 and 11.

9

EFFECTS OF QUANTIZATION AND FINITE WORD LENGTH ARITHMETIC IN DIGITAL FILTERS

9.1 INTRODUCTION

Until now we have assumed perfect implementation for pulse transfer functions. We will now investigate some of the practical problems that must be taken into account when a pulse transfer function is actually implemented digitally. These problems are all a result of the fact that numbers must be quantized and represented as finite bit binary words in digital machines. The quantization process is an irreversible nonlinear operation.

Quantization errors are initially introduced when the analog input signal is sampled and converted into a sequence of binary numbers. We will model this effect by simply adding noise to the ideal samples.

None of the digital filter design techniques presented in Chapter 8 incorporated the constraint of quantized coefficients. When the coefficients are quantized for implementation, the resulting filter must be checked to insure that its frequency response is still acceptable. We will find that small changes in the coefficients of a polynomial can cause large changes in the location of its roots when the roots are clustered near the unit circle. The changes are larger for higher order polynomials. This effect is particularly important in recursive filters since their frequency responses and stability are very sensitive to the position of poles near the unit circle. We will conclude that practical recursive digital filters of order greater than two should rarely be implemented in direct form. A solution is to implement them using a cascade or parallel form realization with first and second-order sections.

A third type of quantization error is introduced by the rounding of products or sums of products to the original machine word length. This is known as finite word length arithmetic. We will see how to estimate the noise introduced by finite word length arithmetic when the input signal has a large dynamic range relative to the quantization step size. When the input to a recursive filter is zero or a constant, we will find that fixed point finite word length arithmetic causes unwanted self sustaining oscillations known as limit cycles. Again, we will conclude that the direct form realization of recursive filters of order greater than two should usually be avoided.

9.2 QUANTIZATION NOISE INTRODUCED BY ANALOG-TO-DIGITAL CONVERSION

The process of approximating a sample of a continuous-time signal by a finite digit binary number is known as analog-to-digital conversion. The binary number generated by an analog-to-digital converter (ADC) is almost always in a fixed point format. Signed magnitude, one's complement, or two's complement fixed point formats can be used [20]. The two's complement format is frequently chosen since subtraction can be performed by adding the two's complement of the subtrahend to the minuend eliminating the need for a separate subtracter. The normalized two's complement representation of any number x with $-A \leq x < A$ has the form

$$x/A = -b_0 + \sum_{n=1}^{\infty} b_n 2^{-n} \tag{9.2.1}$$

where the b_n's can have only the values 0 or 1. For x positive $b_0 = 0$ and for x negative $b_0 = 1$. Therefore, b_0 is called the sign bit.

In an actual digital machine only a finite number of bits can be used to represent any number. Let us assume that numbers are to be represented in the two's complement format. Then, by simply truncating the series in (9.2.1) we can obtain the $K+1$ bit approximation

$$[x]_t = A\left(-b_0 + \sum_{n=1}^{K} b_n 2^{-n}\right) \tag{9.2.2}$$

which can be represented by the binary word $(b_0, b_1, \ldots, b_K)$ in the machine.

Any number of this form must be an integral multiple of

$$q = A2^{-K} \tag{9.2.3}$$

The quantity q is called the *quantization step size*. The relationship between $[x]_t$ and x is illustrated in Fig. 9.2.1. We can see that the truncation error, $e_t = x - [x]_t$, must lie in the semi-open interval $[0, q)$. Since the truncation error has a positive bias that can accumulate in a sequence of arithmetic operations, truncation is usually avoided.

Rounding x to the nearest integral multiple of q is a better method of approximating x by a $K+1$ bit binary number. The rounded number can be represented as

$$[x]_r = A\left(-b_0 + \sum_{n=1}^{K} b_n 2^{-n} + b_{K+1}2^{-K}\right) \tag{9.2.4}$$

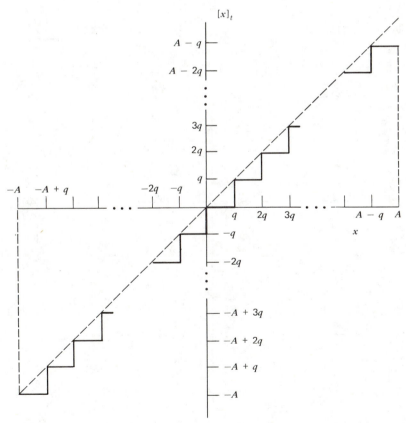

FIGURE 9.2.1. Truncation of two's complement numbers.

The relationship between $[x]_r$ and x is shown in Fig. 9.2.2. For $A(1 - 2^{-K-1}) = A - q/2 \le x < A$, $b_0 = 0$ and $b_1 = b_2 = \cdots = b_{K+1} = 1$. In this case, the last term $b_{K+1}2^{-K}$ on the right-hand side of (9.2.4) will cause an overflow into the sign bit if no overflow detection is used. The overflow causes a jump to the value $-A$ as shown on the bottom right of Fig. 9.2.2. An advantage of using two's complement arithmetic is that if the total sum of a set of normalized numbers is in the range $[-1, 1)$, then, even though partial sums overflow or underflow, the correct total sum will be obtained. Therefore, overflow and underflow detection is commonly omitted. The roundoff error $e_r = x - [x]_r$ is confined to the interval $[-q/2, q/2)$ except in the small overflow region.

In both roundoff and truncation, numbers are quantized to a set of uniformly spaced levels. In some special applications like pulse code modulation voice transmission, signals are quantized to nonuniformly spaced levels to more accurately represent the signal amplitudes that occur most frequently.

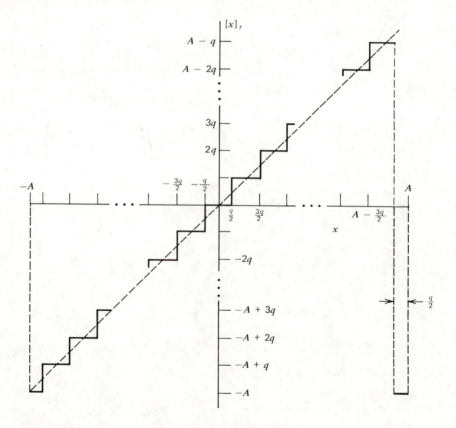

FIGURE 9.2.2. Rounding of two's complement numbers.

Nonuniform quantization can be achieved by first passing the signal through an instantaneous nonlinearity and then into a uniform quantizer. The instantaneous nonlinearity is often called a compander. In the remainder of this chapter we assume that uniform quantization is used.

Let us now assume that the input to the quantizer (i.e., analog-to-digital converter) is a random variable X with the probability density function $f_X(x)$. In addition, let us assume that X can be quantized to any integral multiple of the quantization step size q so that overflow and saturation do not occur. Then it follows that the probability density function for the roundoff error is

$$f_E(e) = \begin{cases} \sum_{n=-\infty}^{\infty} f_X(e + nq) & \text{for} \quad -q/2 \le e < q/2 \\ 0 & \text{elsewhere} \end{cases} \tag{9.2.5}$$

If N is an integer and X is uniformly distributed over $[-Nq, Nq)$, then we find from (9.2.5) that the roundoff error is uniformly distributed over $[-q/2, q/2)$, that is

$$f_E(e) = \begin{cases} 1/q & \text{for} \quad -q/2 \le e < q/2 \\ 0 & \text{elsewhere} \end{cases} \tag{9.2.6}$$

In this case, the roundoff error has zero mean and variance $q^2/12$. If $f_X(x)$ is moderately broad relative to q, then the roundoff error is still almost uniformly distributed over $[-q/2, q/2)$. We can argue similarly that the truncation error is almost uniformly distributed over $[0, q)$. In the remainder of this chapter we will always assume that numbers are quantized by rounding.

If a signal $x(t)$ is sampled and quantized, then

$$[x(nT)]_r = x(nT) - e(nT) \tag{9.2.7}$$

where $e(nT)$ is the roundoff error sequence. Theoretical analyses and numerous simulations have shown that, when the probability density function for $x(t)$ is moderately broad relative to q and the frequency spectrum of $x(t)$ is sufficiently broad so that a number of quantization levels are normally crossed from sample to sample, $e(nT)$ can be closely approximated by a white noise sequence uncorrelated with $x(nT)$ and uniformly distributed over $[-q/2, q/2)$ [5,139,140]. With these assumptions, $e(nT)$ has zero mean, variance $q^2/12$, and the sampled power spectral density

$$S_{ee}(z) = q^2/12 \tag{9.2.8}$$

In summary, the effect of analog-to-digital conversion can usually be modeled by simply adding a zero mean white noise sequence of variance $q^2/12$ to the original unquantized discrete-time signal.

The additive noise model can be used to obtain an estimate of the number of bits needed to represent a signal $x(nT)$, limited to the dynamic range $(-A, A)$, with a specified signal-to-quantization noise ratio. If $K+1$ bits are used, then the quantization step size is $q = A2^{-K}$. Representing the quantizer output by (9.2.7), we define the signal-to-quantization noise ratio as

$$P = E\{x^2(nT)\}/E\{e^2(nT)\} \qquad (9.2.9)$$

or

$$P_0 = 10 \log_{10} P \text{ dB} \qquad (9.2.10)$$

We will assume that $e(nT)$ is a zero mean white noise sequence with variance $q^2/12 = E\{e^2(nT)\}$. Clearly, P depends on the statistics of $x(nT)$. If we assume that $x(nT)$ is uniformly distributed over $(-A, A)$, then $E\{x^2(nT)\} = A^2/3$ and $P = 2^{2(K+1)}$. Therefore, the required number of bits is

$$K+1 = \lceil \tfrac{1}{2} \log_2 P \rceil = \left\lceil \frac{P_0}{20} \log_2 10 \right\rceil \qquad (9.2.11)$$

where $\lceil y \rceil$ is the smallest integer greater than or equal to y.

9.3 A BOUND ON THE ERROR IN THE FREQUENCY RESPONSE OF A NONRECURSIVE FILTER CAUSED BY COEFFICIENT QUANTIZATION

In Chapter 8 several methods for designing a nonrecursive filter with a desired frequency response were presented. We implicitly assumed that all numbers were represented exactly. When the filter is implemented, the coefficients must be quantized and represented by a finite number of bits. When hardware simplicity is required and for high-speed real-time applications, the coefficients are usually represented in fixed point form and fixed point arithmetic is used. Suppose that the desired pulse transfer function is

$$G_0(z) = \sum_{n=0}^{N-1} g_n z^{-n} \qquad (9.3.1)$$

The coefficients rounded to the nearest multiple of q can be written as

$$[g_n]_r = g_n - e_n \qquad (9.3.2)$$

where

$$|e_n| \leq q/2 \qquad (9.3.3)$$

Therefore, the pulse transfer function actually realized is

$$G(z) = \sum_{n=0}^{N-1} [g_n]_r z^{-n} = G_0(z) - \sum_{n=0}^{N-1} e_n z^{-n} \qquad (9.3.4)$$

The error introduced by the coefficient quantization is

$$E(z) = G_0(z) - G(z) = \sum_{n=0}^{N-1} e_n z^{-n} \qquad (9.3.5)$$

On the unit circle, we find from (9.3.5) and (9.3.3) that

$$|E^*(\omega)| = |E(e^{j\omega T})| \leq \sum_{n=0}^{N-1} |e_n| \leq Nq/2 \qquad (9.3.6)$$

While the right-hand side of (9.3.6) is a worst case bound, it still gives a ball park estimate of the error.

The error for nonrecursive filters was easy to bound because the coefficients enter into the frequency response linearly. The same approach cannot be used for recursive filters because the coefficients enter nonlinearly. However, in either case the error can be easily calculated exactly, for all practical purposes, using a general purpose digital computer. The frequency response of a filter with quantized coefficients should always be checked to be sure that it is close enough to the desired response.

9.4 THE EFFECT OF COEFFICIENT QUANTIZATION ON THE LOCATION OF THE POLES AND ZEROS OF DIGITAL FILTERS

In this section we will indirectly investigate the effect of coefficient quantization on the frequency response of a digital filter by examining its effect on the location of the poles and zeros of the filter. Suppose that the pulse transfer function of the desired filter has the form

$$G(z) = A(z)/B(z) \qquad (9.4.1)$$

where

$$A(z) = \sum_{k=0}^{M} a_k z^{-k} \qquad (9.4.2)$$

and

$$B(z) = 1 + \sum_{k=1}^{N} b_k z^{-k} = \prod_{k=1}^{N} (1 - p_k z^{-1}) \qquad (9.4.3)$$

If the filter is realized using one of the direct forms discussed in Chapters 5 and 6, then the denominator coefficients $1, b_1, \ldots, b_N$ will appear directly in the required difference equations. To obtain a rough estimate of the accuracy with which these coefficients must be represented to maintain stability, let us assume that $G(z)$ is a narrow-band low-pass filter. Then the poles of $G(z)$, $p_1, \ldots, p_N$, will be clustered inside the unit circle close to the point $z = 1$. Therefore, we can write that

$$p_k = 1 + e_k \qquad \text{with } |e_k| \ll 1 \qquad \text{for } k = 1, \ldots, N \qquad (9.4.4)$$

If a single coefficient b_r is changed to $b_r' = b_r + \delta$, then the new denominator will be

$$B'(z) = 1 + \sum_{k=1}^{N} b_k z^{-k} + \delta z^{-r} = B(z) + \delta z^{-r} \qquad (9.4.5)$$

As δ is increased in magnitude, roots of $B'(z)$ will eventually move outside the unit circle. In general, the roots will cross the unit circle at different points. It is particularly easy to check for roots crossing at $z = 1$. From (9.4.5) we can see that if

$$\delta = -B(1) = -\left(1 + \sum_{k=1}^{N} b_k\right) = -\prod_{k=1}^{N} (1 - p_k) \qquad (9.4.6)$$

then $B'(z)$ will have a zero at $z = 1$. Substituting (9.4.4) into (9.4.6), we find that

$$|\delta| = \prod_{k=1}^{N} |e_k| \ll 1 \qquad (9.4.7)$$

Thus only a small coefficient perturbation is required to cause instability. From (9.4.7), we can see that the accuracy requirements are more severe when the filter order N is large. Similar results can be obtained for other common types of filters that have poles clustered near points on the unit circle.

Example 9.4.1

Suppose that the desired pulse transfer function is

$$G(z) = \frac{1}{(1 - 0.99z^{-1})^3} = \frac{1}{1 - 2.97z^{-1} + 2.9403z^{-2} - 0.970299z^{-3}}$$

According to (9.4.6), incrementing any denominator coefficient by only $-B(1) = -10^{-6}$ will cause a pole at $z = 1$ when a direct form realization is used. Since $2^{-20} < 10^{-6} < 2^{-19}$, at least 19 binary digits to the right of the binary point are required to achieve an error of less than 10^{-6} in the rounded binary fixed point representation of any coefficient. On the other hand, if the filter is implemented by cascading three first-order sections of the form $1/(1 - 0.99z^{-1})$, then in each section the coefficient 0.99 can be increased by as much as 10^{-2} before instability occurs. An accuracy of 10^{-2} can be obtained by using six bits or more to the right of the binary point. ◀

In addition to maintaining stability, we must insure that the poles and zeros of the implemented filter are sufficiently close to those of the desired filter so that its frequency response is acceptable. Changes in the zeros of $B(z)$ for incremental changes in its coefficients can be examined by using the total differential rule

$$dp_i = \sum_{n=1}^{N} \frac{\partial p_i}{\partial b_n}\bigg|_{z=p_i} db_n \qquad (9.4.8)$$

The partial derivatives can be calculated from the polynomial and factored forms for $B(z)$ in (9.4.3) using the rule

$$\frac{\partial p_i}{\partial b_n} = \frac{\partial B/\partial b_n}{\partial B/\partial p_i} \qquad (9.4.9)$$

When $B(z)$ has only first-order zeros, we find from (9.4.8) and (9.4.9) that

$$dp_i = - \sum_{n=1}^{N} \frac{p_i^{1-n}}{\displaystyle\prod_{\substack{k=1 \\ k\neq i}}^{N} (1 - p_k p_i^{-1})} db_n \qquad (9.4.10)$$

If $B(z)$ has a tightly clustered set of zeros and p_k and p_i are in this set, then $p_k p_i^{-1}$ is close to 1 so that the product in (9.4.10) will be small and its reciprocal large. In this case, small changes in the coefficients of $B(z)$ will cause large changes in its zeros. This effect becomes more pronounced as the number of zeros in the cluster increases. The same argument applies to the numerator $A(z)$. However, the frequency response of a filter is significantly more sensitive to changes in poles near the unit circle than to changes in zeros.

Changes in the zeros of $B(z)$ when a single coefficient is varied can also be examined by using the root locus method discussed in Section 5.10 since the right-hand side of (9.4.5) can be considered to be the characteristic polynomial for a single loop negative feedback system with the open loop gain $\delta z^{-r}/B(z)$.

We have discovered that the accuracy requirements become greater as the filter poles cluster closer together. For low-pass filters the poles cluster near $z = 1$, and for high-pass filters they cluster near $z = -1$. For band-pass filters they cluster near $z = e^{\pm j\omega_0 T}$ where ω_0 is the center frequency of the filter. If the bandwidth of the filter is W, then a measure of the tightness of the clustering is W/ω_s. The clustering becomes tighter as this ratio decreases. If the filter bandwidth W is held fixed and the sampling rate ω_s is increased, then we see that the poles become more tightly clustered and the accuracy requirements increase.

The results of this section can be summed up by saying that a direct form implementation of a practical recursive digital filter of order greater than two should usually be avoided. In Example 9.4.1 we saw that even for a third-order filter the accuracy requirements for a direct form realization can be significant. A solution to the problem is to realize the filter by paralleling or cascading first- and second-order sections. The cascade form is most often chosen so that the zeros as well as the poles can be directly controlled. Recent studies have also shown that the cascaded lattice structure discussed at the end of Section 7.6 is also insensitive to coefficient quantization.

9.5 FIXED POINT FINITE WORD LENGTH ARITHMETIC

Suppose that in a particular digital computer numbers are stored in a fixed point format using words of $K+1$ bits including the sign bit. When two of these numbers are added, the sum can be represented by $K+1$ bits except when an overflow occurs. When two of the numbers are multiplied using a fixed point algorithm, the full accuracy product contains $2K+1$ bits. The typical operation performed in implementing a digital filter is a sum of products. The sum can be carried out using the full $2K+1$ bit products or products rounded to a fewer number of bits. The total sum must then be rounded to $K+1$ bits for storage. This process is known as *finite word length arithmetic*.

The accuracy of the stored total sum depends on the number of bits retained in the products for addition as well as the number of bits used for storage. Suppose that products are rounded to less than $2K+1$ but more than $K+1$ bits and that the resulting numbers correspond to multiples of the arithmetic quantization step size q_a. Let us assume that the stored numbers correspond to multiples of the storage quantization step size q_s. Suppose that we wish to calculate the sum of products

$$S = \sum_{n=1}^{N} a_n b_n = \sum_{n=1}^{N} c_n \qquad (9.5.1)$$

where a_n and b_n are $K+1$ bit numbers. The rounded products can be written as

$$[c_n]_r = c_n + e_n \qquad (9.5.2)$$

where $|e_n| \le q_a/2$. Thus, the computed sum can be written as

$$S_1 = \sum_{n=1}^{N} [c_n]_r = S + \sum_{n=1}^{N} e_n \qquad (9.5.3)$$

We will assume that numbers have been scaled so that the probability of overflow is negligible. The computed sum rounded to $K+1$ bits for storage can be written as

$$S_2 = S_1 + v = S + \sum_{n=1}^{N} e_n + v \qquad (9.5.4)$$

where $|v| < q_s/2$. From (9.5.4) we can see that

$$|S_2 - S| \le N q_a/2 + q_s/2 \qquad (9.5.5)$$

When the full accuracy $2K+1$ bit products are used, $e_n = 0$ for $n = 1, \ldots, N$ so that

$$|S_2 - S| \le q_s/2 \qquad (9.5.6)$$

If the products are rounded to the storage accuracy of $K+1$ bits before addition, the resulting sum has $K+1$ bits and can be stored directly, so $v = 0$ and

$$|S_2 - S| \le N q_s/2 \qquad (9.5.7)$$

The bounds given by (9.5.5), (9.5.6), and (9.5.7) are achievable worst case bounds. A little thought will show that when N is greater than or equal to two, the composite bound decreases from $N q_s/2$ to $q_s/2$ as the number of bits retained in products increases from $K+1$ to $2K+1$.

A less conservative estimate of the noise introduced by finite word length arithmetic can be obtained by an approximate statistical approach. When products are rounded to more than $K+1$ and less than $2K+1$ bits in such a way that the quantization errors $e_1, \ldots, e_N$ and v in (9.5.4) can take on 16 or more values, simulations have verified that these errors can be adequately modeled as zero mean, uncorrelated random variables with e_n uniformly distributed over $(-q_a/2, q_a/2)$ and v uniformly distributed over $(-q_s/2, q_s/2)$. Under these assumptions, the variance of e_n is $q_a^2/12$, the variance of v is $q_s^2/12$, and it follows from (9.5.4) that the total quantization noise variance is

$$E\{(S_2 - S)^2\} = N q_a^2/12 + q_s^2/12 \qquad (9.5.8)$$

When the full accuracy $2K+1$ bit products are used, $e_1 = \cdots = e_n = 0$, so that

$$E\{(S_2 - S)^2\} = q_s^2/12 \qquad (9.5.9)$$

If products are rounded to the storage accuracy of $K+1$ bits, then $v = 0$ and

$$E\{(S_2 - S)^2\} = N q_s^2/12 \qquad (9.5.10)$$

9.6 NOISE IN THE OUTPUT OF A RECURSIVE FILTER CAUSED BY FIXED POINT FINITE WORD LENGTH ARITHMETIC

The noise introduced by finite word length arithmetic can be analyzed by replacing each rounded term by its original value plus an error term limited in magnitude to half the quantization step size. In this section we will use the approximate statistical approach discussed in Section 9.5 and assume that the different rounding errors are zero mean, uncorrelated random variables each having variance $q^2/12$ where q is the appropriate quantization step size.

The output of a finite tap nonrecursive filter is a weighted sum of inputs. Therefore, the error in the calculation of the present output introduced by finite word length arithmetic does not propagate into the calculation of future outputs. The resulting output noise can be characterized by the appropriate equation in Section 9.5.

The output of a recursive filter is a weighted sum of present and past inputs and past outputs. In this case, the rounding errors propagate into the calculation of successive outputs. To illustrate this effect, let us assume that the pulse transfer function given in (9.4.1) is implemented using a type 0 direct form realization. If $x(nT)$ is the filter input and $y(nT)$ is its output, then the ideal input output relation is

$$y(nT) = \sum_{k=0}^{M} a_k x(nT - kT) - \sum_{k=1}^{N} b_k y(nT - kT) \qquad (9.6.1)$$

We will assume that a_k, b_k, and $x(nT)$ have already been quantized to the required word lengths and that these effects can be analyzed separately. We will assume that overflows do not occur. To simplify the analysis slightly, we will assume that products and the total sum are both rounded to multiples of q. Then the computed and stored output $y_1(nT)$ is

$$y_1(nT) = \sum_{k=0}^{M} [a_k x(nT - kT)]_r - \sum_{k=1}^{N} [b_k y_1(nT - kT)]_r \qquad (9.6.2)$$

The rounded products can be written as

$$[a_k x(nT - kT)]_r = a_k x(nT - kT) + e_k(nT)$$

and

$$[b_k y_1(nT - kT)]_r = b_k y_1(nT - kT) + f_k(nT)$$

Therefore,

$$y_1(nT) = \sum_{k=0}^{M} a_k x(nT-kT) - \sum_{k=1}^{N} b_k y_1(nT-kT) + e(nT) \qquad (9.6.3)$$

where

$$e(nT) = \sum_{k=0}^{M} e_k(nT) - \sum_{k=1}^{N} f_k(nT) \qquad (9.6.4)$$

The filter with the roundoff errors is illustrated in Fig. 9.6.1. Assuming that the roundoff errors are zero mean, uncorrelated random variables each with variance $q^2/12$, we find that

$$E\{e^2(nT)\} = (M+N+1)q^2/12 \qquad (9.6.5)$$

Taking the Z-transform of (9.6.3) yields

$$Y_1(z) = Y(z) + V(z) \qquad (9.6.6)$$

where

$$Y(z) = X(z)A(z)/B(z) \qquad (9.6.7)$$

and

$$V(z) = E(z)/B(z) \qquad (9.6.8)$$

Thus the computed output $y_1(nT)$ is the sum of the desired output $y(nT)$ and a noise signal $v(nT)$. Assuming that $e(nT)$ is a white noise sequence, then it follows from (7.2.4), (7.4.8), (9.6.5), and (9.6.8) that the output noise power is

$$E\{v^2(nT)\} = \frac{q^2}{12}(M+N+1)\frac{1}{2\pi j}\oint \frac{1}{B(z)B(z^{-1})}\frac{dz}{z} \qquad (9.6.9)$$

where the unit circle can be taken as the contour of integration.

The output noise power in parallel and cascade form realizations can be determined using the same approach [74]. The output noise power is usually less for parallel and cascade form realizations than for direct form realizations. In cascade form realizations the ordering of sections should be selected to minimize the output noise.

Numerous simulations have verified that the approximate statistical approach gives results that are quite good for design purposes.

Example 9.6.1

Let the desired pulse transfer function be

$$G(z) = \frac{0.4}{(1-0.9z^{-1})(1-0.8z^{-1})}$$

From (9.6.9) we find that the output noise power in a type 0 direct form realization is $22.4q^2$.

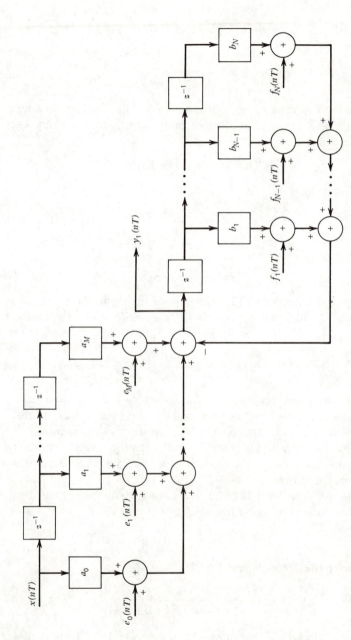

FIGURE 9.6.1. Noise in a type 0 direct form recursive filter realization caused by fixed point finite word length arithmetic.

The desired filter can be realized by the coupled difference equations

$$a(nT) = 0.4x(nT) + 0.9a(nT - T) \tag{9.6.10}$$

$$y(nT) = a(nT) + 0.8y(nT - T) \tag{9.6.11}$$

which correspond to the cascade of $0.4/(1 - 0.9z^{-1})$ followed by $1/(1 - 0.8z^{-1})$. When products are rounded (9.6.10) can be written as

$$a_1(nT) = 0.4x(nT) + e_1(nT) + 0.9a_1(nT - T) + f_1(nT)$$

$$= 0.4x(nT) + 0.9a_1(nT - T) + e(nT) \tag{9.6.12}$$

where

$$e(nT) = e_1(nT) + f_1(nT)$$

Equation 9.6.11 becomes

$$y_1(nT) = a_1(nT) + 0.8y_1(nT - T) + f(nT) \tag{9.6.13}$$

Taking the Z-transform of (9.6.12) gives

$$A_1(z) = \frac{0.4}{1 - 0.9z^{-1}} X(z) + \frac{1}{1 - 0.9z^{-1}} E(z) \tag{9.6.14}$$

Taking the Z-transform of (9.6.13) and using (9.6.14), we find that

$$Y_1(z) = G(z)X(z) + \frac{E(z)}{(1 - 0.9z^{-1})(1 - 0.8z^{-1})} + \frac{F(z)}{1 - 0.8z^{-1}}$$

The term $G(z)X(z)$ is the desired output and the remaining terms are noise. Assuming that the noise sequences are mutually uncorrelated and white, we find that the sampled power spectral density for the total output noise is

$$S(z) = \frac{q^2/6}{(1 - 0.9z^{-1})(1 - 0.8z^{-1})(1 - 0.9z)(1 - 0.8z)} + \frac{q^2/12}{(1 - 0.8z^{-1})(1 - 0.8z)}$$

Integrating $S(z)$ around the unit circle we find that the total output noise power is $15.13q^2$.

In partial fraction form

$$G(z) = \frac{3.6}{1 - 0.9z^{-1}} - \frac{3.2}{1 - 0.8z^{-1}}$$

$G(z)$ can be realized in parallel form by the equations

$$a(nT) = 3.6x(nT) + 0.9a(nT - T)$$

$$b(nT) = -3.2x(nT) + 0.8b(nT - T)$$

and

$$y(nT) = a(nT) + b(nT)$$

Using (9.6.9) we find that the variances of the noises introduced into $a(nT)$ and $b(nT)$ by rounding are respectively

$$\sigma_1^2 = 2\left(\frac{q^2}{12}\right)\left(\frac{1}{0.19}\right) \quad \text{and} \quad \sigma_2^2 = 2\left(\frac{q^2}{12}\right)\left(\frac{1}{0.36}\right)$$

Assuming that the noises are uncorrelated, the total output noise power is $\sigma_1^2 + \sigma_2^2 = 1.34q^2$. Thus, the parallel form realization has significantly less output noise than the type 0 direct form and cascade form realizations for this example. ◄

9.7 THE DEADBAND EFFECT IN RECURSIVE FILTERS

When the input to a digital filter is a constant, the rounding errors resulting from fixed point finite word length arithmetic cannot be modeled as uncorrelated random variables. If the input to a nonrecursive filter with the pulse response $g_0, \ldots, g_{N-1}$ is the constant A, then its computed output will be

$$y_1(nT) = \sum_{k=0}^{N-1} [g_k A]_r$$

which is a unique constant. We will find in this section that for recursive filters there may be no unique steady-state output; that in fact there may be a "deadband" containing many possible steady-state outputs for a given constant input. This is not too surprising when we consider that a recursive filter with finite word length arithmetic is a nonlinear feedback system. It is well known that unforced nonlinear systems can have stable periodic oscillations known as limit cycles. The steady-state outputs in the deadband are zero frequency limit cycles. More general limit cycles will be discussed in Section 9.8.

As usual, let us assume that the desired pulse transfer function has the form $G(z) = A(z)/B(z)$ where $A(z)$ and $B(z)$ are polynomials defined by (9.4.2) and (9.4.3). The difference equation describing the type 0 direct form realization of $G(z)$ is

$$y(nT) = \sum_{k=0}^{M} a_k x(nT - kT) - \sum_{k=1}^{N} b_k y(nT - kT) \tag{9.7.1}$$

If all the poles of $G(z)$ are strictly inside the unit circle, $y(nT)$ will converge to some final value y when the input is $x(nT) = c$ and the arithmetic is perfect. The final value must satisfy (9.7.1), so that

$$y = c \sum_{k=0}^{M} a_k - y \sum_{k=1}^{N} b_k \tag{9.7.2}$$

and

$$y = c\,\frac{\displaystyle\sum_{k=0}^{M} a_k}{1 + \displaystyle\sum_{k=1}^{N} b_k} = cG(1) \qquad (9.7.3)$$

Equation 9.7.3 could have been derived using the Final Value Theorem. However, we will need to recognize y in the form of (9.7.2) in the following analysis.

Let us assume that a_k, b_k, and $x(nT) = c$ have already been quantized to finite word length. We will assume that (9.7.1) is implemented by summing the full accuracy products to give the computed output $y_1(nT)$ which is then rounded to the nearest multiple of q and stored as $y_2(nT)$. With these assumptions, the computed output is

$$y_1(nT) = c\sum_{k=0}^{M} a_k - \sum_{k=1}^{N} b_k y_2(nT - kT) \qquad (9.7.4)$$

Let us assume that the stored output $y_2(nT)$ reaches the steady-state value

$$y_2 = y - e \qquad (9.7.5)$$

which must be a multiple of q. From (9.7.4) we find that the computed output must be

$$y_1(nT) = \left(c\sum_{k=0}^{M} a_k - y\sum_{k=1}^{N} b_k\right) + e\sum_{k=1}^{N} b_k \qquad (9.7.6)$$

Using (9.7.2) and adding and subtracting 1 from the sum on the far right-hand side of (9.7.6), we find that

$$y_1(nT) = y - e + e\left(1 + \sum_{k=1}^{N} b_k\right) = y_2 + eB(1) \qquad (9.7.7)$$

The computed output $y_1(nT)$ will be rounded to y_2 if

$$-q/2 \le eB(1) < q/2 \qquad (9.7.8)$$

When all the zeros of $B(z)$ are inside the unit circle, $B(1)$ is positive and (9.7.8) can be written as

$$-\frac{q}{2B(1)} \le e < \frac{q}{2B(1)} \qquad (9.7.9)$$

Therefore, the stored output will remain at the value y_2 if

$$y - \frac{q}{2B(1)} < y_2 \le y + \frac{q}{2B(1)} \qquad (9.7.10)$$

and N consecutive stored outputs are y_2. This interval is known as the deadband [9].

The width of the deadband is inversely proportional to $B(1)$. In Section 9.4 we observed that $B(1)$ is also a measure of the coefficient accuracy required for stability. We saw that $B(1)$ is small for narrow band low-pass filters and decreases as the filter order N increases. The deadband can be many quantization steps wide. Once again, we must conclude that a direct form realization of a recursive digital filter of order greater than two should usually be avoided.

Example 9.7.1

Suppose that $G(z) = 1/(1 - 0.9z^{-1})$ and $x(nT) = 10$. According to (9.7.3) the final value for the output of the ideal filter is $y = 100$. We will assume that outputs are rounded to the nearest integer, so that $q = 1$. The calculated output is

$$y_1(nT) = 10 + 0.9y_2(nT - T) \qquad (9.7.11)$$

According to (9.7.10), the deadband is $95 < y_2 \leq 105$. To illustrate the behavior of the implemented filter, let us assume that the initial stored output is above the deadband at $y_2(0) = 107$. Using (9.7.11), the sequence of calculated and stored outputs shown in Table 9.7.1 is obtained.

TABLE 9.7.1 Output Started above the Deadband				TABLE 9.7.2 Output Started below the Deadband		
n	$y_1(nT)$	$y_2(nT)$		n	$y_1(nT)$	$y_2(nT)$
0	—	107		0	—	94
1	106.3	106		1	94.6	95
2	105.4	105		2	95.5	96
3	104.5	105		3	96.4	96
.	.			.	.	.
.	.			.	.	.
.	.			.	.	.

When the output is started below the deadband at $y_2(0) = 94$, the output sequence shown in Table 9.7.2 is obtained.

If any integer in the deadband is chosen for $y_2(0)$, then $y_2(nT)$ will remain at this value. ◄

The output of the filter described in Example 9.7.1 "hangs up" at the edge of the deadband when it is started outside the deadband. For higher order filters the behavior is not this simple. Depending on the filter and the initial conditions, the output may penetrate into the deadband or may even jump through to the other side of the deadband. Also, stable nonzero frequency limit cycles are likely.

The width of the deadband can be decreased by decreasing q. If the deadband is too wide and q cannot be decreased for practical reasons, the problem can be remedied by adding a small dither signal of the form $(-1)^n d$ with $d \approx q/2$ to the right-hand side of (9.7.4). The dither signal helps the output jump across the quantization thresholds and reach a small steady-state oscillation about the desired final output value. Random noise on the input signal can have the effect of a dither signal when filtering actual data.

9.8 BOUNDS ON LIMIT CYCLES IN RECURSIVE DIGITAL FILTERS USING FIXED POINT FINITE WORD LENGTH ARITHMETIC

The limit cycles in implemented recursive digital filters rarely can be calculated by simple analytical methods. The zero frequency limit cycles associated with the deadband discussed in Section 9.7 are an unusual exception. Normally, the limit cycles can be determined only by directly evaluating the nonlinear difference equation describing the actual filter. Since these filters are finite state machines, limit cycles must be reached in a finite number of steps.

Although limit cycles are difficult to predict exactly, we can show that a limit cycle will exist in a recursive filter implemented in direct form if the poles of the desired filter are moderately close to the unit circle and the filter is started in any nonzero state. This is an important observation because many practical filters have their poles near the unit circle. We will assume that the desired pulse transfer function has the form $G(z) = A(z)/B(z)$ as in Section 9.7 and is implemented as a type 0 direct form realization with products summed with full accuracy or rounded before summation. In the remainder of this section we will always assume that the input is identically zero. Suppose that the rounded and stored output $y_2(nT)$ is identically zero for $n > k - N$ and that $y_2(kT - NT)$ is not zero. Then, from (9.7.4) we find that the output at time kT before rounding is

$$y_1(kT) = -b_N y_2(kT - NT) \qquad (9.8.1)$$

By hypothesis, $|y_2(kT - NT)| \geq q$. Therefore, $|y_1(kT)| > q/2$ if $|b_N| > 0.5$, and the rounded output $y_2(kT)$ will not be zero. Consequently, when $|b_N| > 0.5$ the zero state cannot be reached from a nonzero state and a limit cycle must exist. The magnitude of b_N is equal to the magnitude of the product of the zeros of $B(z)$ and will be significantly greater than 0.5 in many filters.

A variety of bounds on limit cycles caused by roundoff errors have been developed [56,86,105,120]. These bounds are useful design equations since exact determination of all possible limit cycles is usually exceedingly tedious. We will derive the bounds of Long and Trick. With zero input the implemented type 0 direct form difference equation can be written as

$$y_2(nT) = -\sum_{k=1}^{N} b_k y_2(nT - kT) + e(nT) \tag{9.8.2}$$

where $y_2(nT)$ is the quantized and stored output and $e(nT)$ is the total roundoff error. We will assume that no overflows occur. Then $|e(nT)| \leq \delta$ where

$$\delta = \begin{cases} q/2 & \text{products summed with full accuracy} \\ Nq/2 & \text{products rounded before summation} \end{cases} \tag{9.8.3}$$

We will assume that the filter is asymptotically stable, that is, has all its poles strictly inside the unit circle. We observed in Section 5.7 that the output of a stable filter is bounded for any bounded input. Therefore, $y_2(nT)$ must be bounded. Taking the Z-transform of (9.8.2) we see that

$$Y_2(z) = E(z)/B(z) \tag{9.8.4}$$

If $h(n)$ is the pulse response of $H(z) = 1/B(z)$, then $y_2(nT)$ can theoretically be calculated as

$$y_2(nT) = \sum_{k=0}^{\infty} h(k)e(nT - kT) \tag{9.8.5}$$

Now let us assume that a limit cycle of period L samples exists, so that $e(nT) = e(nT + LT)$ for all n. Using the periodicity of $e(nT)$ (9.8.5) can be written as

$$y_2(nT) = \sum_{r=0}^{L-1} e(nT - rT) \sum_{k=0}^{\infty} h(r + kL) \tag{9.8.6}$$

Therefore

$$|y_2(nT)| \leq \delta \sum_{r=0}^{L-1} \left| \sum_{k=0}^{\infty} h(r + kL) \right| \tag{9.8.7}$$

Equation 9.8.7 provides an absolute bound on the peak value of any limit cycle with period L.

The sum over k in (9.8.7) can be calculated by Theorem 4.8.1. If

$$p(n) = \begin{cases} 1 & \text{for} \quad n = 0, L, 2L, \ldots \\ 0 & \text{elsewhere} \end{cases} \tag{9.8.8}$$

then

$$P(z) = 1/(1 - z^{-L}) \qquad \text{for} \qquad |z| > 1 \tag{9.8.9}$$

Therefore

$$S_r = \sum_{k=0}^{\infty} h(r + kL) = \sum_{k=0}^{\infty} h(r + k) p(k)$$

$$= \frac{1}{2\pi j} \oint \frac{H(z) z^{r-1}}{1 - z^L} \, dz \qquad \text{for} \qquad r = 0, \ldots, L - 1 \tag{9.8.10}$$

where the contour of integration encloses all the poles of $H(z)$ but none of the zeros of $1 - z^L$. Evaluating the integral by taking the sum of the residues inside the contour we obtain

$$S_r = \sum_{z_k} \text{Res} \left[\frac{H(z) z^{r-1}}{1 - z^L}, z_k \right] \qquad \text{for} \qquad r = 0, \ldots, L - 1 \tag{9.8.11}$$

where $\{z_k\}$ is the set of poles of $H(z)$. The integral can also be evaluated by taking the negative of the sum of the residues at the zeros of $1 - z^L$. This yields

$$S_r = \frac{1}{L} \sum_{k=0}^{L-1} H(e^{jk2\pi/L}) e^{jrk2\pi/L} \qquad \text{for} \qquad r = 0, \ldots, L - 1 \tag{9.8.12}$$

For zero frequency limit cycles $L = 1$ and (9.8.12) reduces to $S_0 = H(1) = 1/B(1)$. Since $B(1)$ is positive for stable filters, the bound given by (9.8.7) is

$$|y_2(nT)| \le \delta/B(1) \qquad \text{for} \qquad L = 1 \tag{9.8.13}$$

Notice that (9.8.13) agrees with the deadband result derived in Section 9.7 when products are summed with full accuracy.

In general, the limit cycle periods for different initial conditions cannot be determined except by exact simulation. For design purposes, a bound independent of L can be determined by observing from (9.8.5) or (9.8.7) that

$$|y_2(nT)| \le \delta \sum_{k=0}^{\infty} |h(k)| \tag{9.8.14}$$

Let us apply these bounds to second-order filters. Suppose that $B(z) = 1 + b_1 z^{-1} + b_2 z^{-2}$. According to the Schur-Cohn Test discussed in Section 5.8, the necessary and sufficient conditions for stability are

$$|b_2| < 1 \quad \text{and} \quad 1 - |b_1| + b_2 > 0 \qquad (9.8.15)$$

From (9.8.7), (9.8.12), and (9.8.15) we find that

$$|y_2(nT)| \le \delta/(1 - |b_1| + b_2) \quad \text{for} \quad L = 2 \qquad (9.8.15)$$

Equation 9.8.7 can be evaluated for other L in a straightforward manner using (9.8.11) or (9.8.12).

For second-order filters with two real poles the absolute bound (9.8.14) has the closed form

$$|y_2(nT)| \le \delta/(1 - |b_1| + b_2) \quad \text{for} \quad b_1^2 \ge 4b_2 \qquad (9.8.16)$$

For complex poles (9.8.14) cannot generally be put in closed form. In this case

$$h(k) = \frac{a^k}{\sin \theta} \sin(k+1)\theta \qquad (9.8.17)$$

where $a = \sqrt{b_2}$, $\cos \theta = -b_1/(2\sqrt{b_2})$, and $4b_2 > b_1^2$. One upper bound on the sum in (9.8.14) is

$$\sum_{k=0}^{\infty} |h(k)| \le \sum_{k=0}^{\infty} \left| \frac{a^k}{\sin \theta} \right| = \frac{1 + \sqrt{b_2}}{(1 - b_2)\sqrt{1 - \dfrac{b_1^2}{4b_2}}} \qquad (9.8.18)$$

It can be shown that $|\sin[(k+1)\theta]/\sin \theta| \le k+1$, so another bound is

$$\sum_{k=0}^{\infty} |h(k)| \le \sum_{k=0}^{\infty} (k+1)a^k = \frac{1}{(1 - \sqrt{b_2})^2} \qquad (9.8.19)$$

Selecting the tightest bound from (9.8.16), (9.8.18), and (9.8.19) for each b_1

and b_2 gives the composite absolute bound

$$|y_2(nT)| \leq \begin{cases} \dfrac{\delta}{1-|b_1|+b_2} & \text{for} \quad b_2 \leq 0, \text{ or } b_2 > 0 \text{ and } 2\sqrt{b_2} \leq |b_1| \\[3ex] \dfrac{\delta}{(1-\sqrt{b_2})^2} & \text{for} \quad b_2 > 0 \text{ and } 2b_2\sqrt{\dfrac{2}{\sqrt{b_2}}-1} \leq |b_1| \leq 2\sqrt{b_2} \\[3ex] \dfrac{\delta(1+\sqrt{b_2})}{(1-b_2)\sqrt{1-\dfrac{b_1^2}{4b_2}}} & \text{for} \quad b_2 > 0 \text{ and } |b_1| \leq 2b_2\sqrt{\dfrac{2}{\sqrt{b_2}}-1} \end{cases}$$

$$(9.8.20)$$

The zero input limit cycle behavior of a digital filter realized as a cascade of first- and second-order sections is more difficult to analyze since only the initial section in the cascade has zero input. A crude bound on the output limit cycle could be developed by extending an absolute bound down the cascade section by section.

We have assumed in the derivations in this section that the total sums never overflow. If the overflow characteristics of two's complement arithmetic are taken into account, it can be shown that large periodic oscillations may be possible [32]. This behavior can be avoided by scaling the input so that overflows do not occur.

9.9 FLOATING POINT FINITE WORD LENGTH ARITHMETIC

The best signal to roundoff noise ratios are achieved in filters using fixed-point arithmetic when signal levels are as large as possible without causing overflow. Signal levels at points within a filter can be estimated for an input with a given power spectral density by calculating the signal variances at these points. Fixed internal scaling can often be used to optimize the signal levels.

Floating point arithmetic is a method for providing automatic scaling. An arbitrary finite number x can be represented exactly using the floating point binary representation

$$x = \text{sign}\,(x)c2^b \qquad (9.9.1)$$

where c, the mantissa, is an infinite precision binary number contained in $\{0\} \cup [\frac{1}{2}, 1)$ and b, the exponent, is an integer. We have arbitrarily chosen to represent $x2^{-b}$ as the signed magnitude number $\text{sign}\,(x)c$. We could have just as easily chosen the two's complement or any other standard representation.

In an actual machine both b and c must be represented by finite bit words. We will assume that enough bits have been allocated to b to cover the required dynamic range. If c is rounded to K bits, then the ideal and actual machine mantissas can be related by the equation

$$[c]_r = c + \delta \qquad (9.9.2)$$

where

$$|\delta| \leq 2^{-K}/2 \qquad (9.9.3)$$

Therefore, when x is not zero, the actual machine floating point number has the form

$$x_1 = \text{sign}\,(x)(c + \delta)2^b = x(1 + e) \qquad (9.9.4)$$

where

$$e = \delta/c \qquad (9.9.5)$$

From (9.9.3) and the fact that the smallest nonzero value for c is $\tfrac{1}{2}$, we find that

$$|e| \leq 2^{-K} \qquad (9.9.6)$$

Thus the error introduced in forming the floating point machine number satisfies the bound

$$|x - x_1| \leq |x|\,2^{-K} \qquad (9.9.7)$$

This is quite different than the fixed point case where the magnitude of the error is bounded by a constant independent of x.

Equation 9.9.4 forms the basis for analyzing the errors introduced by floating point finite word length arithmetic. We will use $[\cdot]_{\text{fl}}$ to denote a computed floating point machine number. The computed floating point sum of two numbers x and y satisfies

$$[x + y]_{\text{fl}} = (x + y)(1 + e) \qquad (9.9.8)$$

and the computed product satisfies

$$[xy]_{\text{fl}} = xy(1 + v) \qquad (9.9.9)$$

where $|e| \leq 2^{-K}$ and $|v| \leq 2^{-K}$.

The typical computation required for calculating a digital filter output is a sum of products. Suppose

$$S = x_1 y_1 + x_2 y_2 + \cdots + x_N y_N \qquad (9.9.10)$$

where x_n and y_n for $n = 1, \ldots, N$ are floating point machine numbers. According to (9.9.8) and (9.9.9), the computed sum, assuming that products are added from left to right, can be written as

$$[S]_{\text{fl}} = [\cdots [[x_1 y_1 (1+v_1) + x_2 y_2 (1+v_2)](1+e_2) + x_3 y_3 (1+v_3)](1+e_3)$$
$$+ \cdots + x_N y_N (1+v_N)](1+e_N)$$

$$= x_1 y_1 (1+v_1) \prod_{k=2}^{N} (1+e_k) + \sum_{n=2}^{N} x_n y_n (1+v_n) \prod_{k=n}^{N} (1+e_k) \qquad (9.9.11)$$

where $|e_k| \le 2^{-K}$ for $k = 2, \ldots, N$ and $|v_n| \le 2^{-K}$ for $n = 1, \ldots, N$. Notice that when products are expanded, the right-hand side of (9.9.11) will contain a term equal to the desired sum S. The error analysis can proceed from this point using a deterministic bounding approach [119,142] or an approximate statistical approach [69].

In the approximate statistical approach the e's and v's are considered to be mutually uncorrelated random variables uniformly distributed over $(-2^{-K}, 2^{-K})$. In addition, they are assumed to be uncorrelated with the desired products. Linear terms in the e's and v's are retained and higher order error products are neglected. Analytical results obtained using these assumptions agree reasonably well with simulation results.

Example 9.9.1 First-Order Floating Point Filter

Let the desired difference equation be

$$y(nT) = x(nT) + \beta y(nT - T) \qquad (9.9.12)$$

which corresponds to the pulse transfer function $G(z) = 1/(1 - \beta z^{-1})$. For simplicity, we will assume that β and $x(nT)$ are floating point machine numbers. The actually computed floating point output can be written as

$$y_1(nT) = [x(nT) + \beta y_1(nT - T)]_{\text{fl}}$$
$$= \{x(nT) + \beta y_1(nT - T)[1 + v(nT)]\}[1 + e(nT)] \qquad (9.9.13)$$

We will define the output error sequence to be $w(nT) = y_1(nT) - y(nT)$. Subtracting (9.9.12) from (9.9.13) and neglecting second-order terms in e, v, and w, we find that

$$w(nT) - \beta w(nT - T) = r(nT) \qquad (9.9.14)$$

where

$$r(nT) = x(nT)e(nT) + [v(nT) + e(nT)]\beta y(nT - T) \qquad (9.9.15)$$

Assuming that $e(nT)$ and $v(nT)$ are white noise sequences and that $e(nT)$, $v(nT)$, and $x(nT)$ are mutually uncorrelated signals, it follows that $r(nT)$ is a white noise sequence with variance

$$R_{rr}(0) = \sigma^2 R_{xx}(0) + 2\sigma^2 \beta^2 R_{yy}(0) + \sigma^2 \beta R_{xy}(T) \qquad (9.9.16)$$

where

$$\sigma^2 = E\{e^2(nT)\} = E\{v^2(n)\} = 2^{-2K}/3 \tag{9.9.17}$$

Using (7.4.7) and (7.2.4), it follows that the output noise power is

$$E\{w^2(nT)\} = R_{ww}(0) = R_{rr}(0)/(1-\beta^2) \tag{9.9.18}$$

As a specific example, suppose that $x(nT)$ is a white noise sequence with variance σ_x^2. Then (9.9.18) becomes

$$R_{ww}(0) = \sigma_x^2 \sigma^2 (1+\beta^2)/(1-\beta^2)^2 \tag{9.9.19}$$

◄

The output noise in fixed and floating point filter realizations can be compared using the statistical approaches. If the floating point mantissa and fixed point word have the same number of bits, then the floating point implementation is superior. If the total floating point word including the exponent is constrained to have the same number of bits as the fixed point word, then the relative ranking depends on the input probability density function, input power spectral density, and filter frequency response. Conditions can be found where fixed point arithmetic is better than floating point arithmetic and vice-versa. Generally, when a large dynamic range is required, as in a high gain resonator, floating point arithmetic generates less output noise.

Limit cycles in recursive filters are usually not as large with floating point arithmetic as with fixed point arithmetic. In many cases they cannot even exist [119].

Floating point arithmetic is significantly more complicated than fixed point arithmetic. When cost and/or speed become important factors, fixed point arithmetic is usually chosen. Recently, some high-speed programmable signal processing computers and a variety of relatively low-cost microprocessors have appeared on the market. Since programmable devices have the appeal of flexibility, many future digital filtering operations will be implemented using these devices with the appropriate software programs. Using the software approach, floating point arithmetic is no more costly to implement than fixed point arithmetic if speed is not a factor.

10

THE DISCRETE FOURIER TRANSFORM: THEORY, EFFICIENT COMPUTATION, APPLICATIONS TO DIGITAL FILTERING AND CORRELATION

10.1 INTRODUCTION

In the discussion of frequency sampling filters in Section 8.10, we found that the pulse response of a nonrecursive filter with N taps can be uniquely determined from samples of its frequency response taken at N points uniformly distributed around the unit circle. The rules for finding the frequency response samples from the pulse response and vice-versa were called the *discrete Fourier transform* (DFT) and *inverse discrete Fourier transform* (IDFT), respectively. In this chapter, we study the DFT in detail and find that it has properties analogous to those of the Fourier, Laplace, and Z-transforms.

In particular, we will find that the product of the DFT's of two N-point sequences corresponds to what we will call the cyclic convolution of the two sequences. We will see how this result can be used to perform ordinary discrete-time convolution and hence digital filtering by the indirect frequency domain approach rather than by the direct-time-domain difference equation approach. The frequency domain approach to digital filtering using the DFT was considered to be impractical until 1965 because it was generally believed that the computation required to process an N-point sequence by this method was proportional to N^2. At that time, Cooley and Tukey [26] published a paper presenting an algorithm for finding a DFT with computation proportional to $N \log_2 N$. This algorithm is now called the fast Fourier transform (FFT) and two variations are presented in Section 10.7. The FFT made the frequency domain approach to filtering competitive with and even more efficient than the time-domain approach in many instances. The FFT also greatly reduced the time required to perform correlation and spectral analysis. The DFT is now an important, flexible, and popular tool in digital signal processing.

10.2 THE DISCRETE AND INVERSE DISCRETE FOURIER TRANSFORMS

The discrete Fourier transform or DFT of the N-point sequence $f_0, f_1, \ldots, f_{N-1}$ is defined as the N-point sequence

$$F_k = \sum_{n=0}^{N-1} f_n e^{-j2\pi nk/N} \qquad \text{for} \qquad k = 0, \ldots, N-1 \qquad (10.2.1)$$

Frequently, we will use the notation $F_k = \text{DFT}\{f_n\}$ to denote this sequence. When k is allowed to take on values outside the set $\{0, \ldots, N-1\}$, F_k repeats with period N, that is

$$F_k = F_{N+k} \qquad \text{for all } k \qquad (10.2.2)$$

In the discussion of frequency sampling filters in Section 8.10, we observed that F_k is the sequence obtained by sampling

$$F(z) = \sum_{n=0}^{N-1} f_n z^{-n} \qquad (10.2.3)$$

at the equally spaced points on the unit circle $e^{jk\omega_s T/N} = e^{jk2\pi/N}$ for $k = 0, \ldots, N-1$. Thus the index k corresponds to the frequency $k\omega_s/N$.

The sequence $f_0, \ldots, f_{N-1}$ can be uniquely determined from its DFT. Consider the sum

$$S_n = \frac{1}{N} \sum_{k=0}^{N-1} F_k e^{j2\pi nk/N} \qquad \text{for} \qquad 0 \le n \le N-1 \qquad (10.2.4)$$

Replacing F_k by the right-hand side of (10.2.1) with n changed to r, we obtain

$$S_n = \frac{1}{N} \sum_{k=0}^{N-1} \left[\sum_{r=0}^{N-1} f_r e^{-j2\pi rk/N} \right] e^{j2\pi nk/N}$$

$$= \sum_{r=0}^{N-1} f_r \left[\frac{1}{N} \sum_{k=0}^{N-1} e^{j2\pi(n-r)k/N} \right] \qquad (10.2.5)$$

The term in brackets on the right-hand side of (10.2.5) is a geometric series and sums to

$$\frac{1}{N} \frac{1-e^{j2\pi(n-r)}}{1-e^{j2\pi(n-r)/N}} = \begin{cases} 1 & \text{for} \quad n-r \text{ a multiple of } N \\ 0 & \text{elsewhere} \end{cases} \qquad (10.2.6)$$

Therefore (10.2.5) reduces to $S_n = f_n$ and we find that the inverse discrete Fourier transform formula or IDFT is

$$f_n = \text{IDFT}\{F_k\} = \frac{1}{N} \sum_{k=0}^{N-1} F_k e^{j2\pi nk/N} \qquad \text{for} \quad n = 0, \ldots, N-1 \quad (10.2.7)$$

It will be convenient in some cases to let n take on values outside the set $\{0, \ldots, N-1\}$. It follows from (10.2.7) that

$$f_{n+N} = f_n \qquad \text{for all } n \qquad (10.2.8)$$

Thus the basic sequence $f_0, \ldots, f_{N-1}$ repeats with period N. We use this convention in the remainder of the book.

Equation 10.2.7 can also be written as

$$f_n = \frac{1}{N} \overline{\sum_{k=0}^{N-1} \bar{F}_k e^{-j2\pi nk/N}} \qquad (10.2.9)$$

The significance of this alternate inversion formula is that IDFT's can be obtained using a DFT algorithm by adding the scale factor $1/N$ and taking the indicated conjugates.

In our discussion of frequency sampling filters in Section 8.10, we initially derived a formula for $F(z)$ in terms of F_k. This formula can also be derived by replacing f_n in (10.2.3) by the IDFT inversion formula to obtain

$$F(z) = \sum_{n=0}^{N-1} \frac{1}{N} \sum_{k=0}^{N-1} F_k e^{j2\pi nk/N} z^{-n}$$

$$= \frac{1}{N} \sum_{k=0}^{N-1} F_k \sum_{n=0}^{N-1} (e^{j2\pi k/N} z^{-1})^n$$

$$= \frac{1-z^{-N}}{N} \sum_{k=0}^{N-1} \frac{F_k}{1-e^{j2\pi k/N} z^{-1}} \qquad (10.2.10)$$

On the unit circle, $F(z)$ can be put in the form of (8.10.8).

10.3 FREQUENCY SELECTIVITY OF THE DFT

The Fourier transform of $f(t) = e^{j\omega_0 t}$ is the impulse $2\pi\delta(\omega - \omega_0)$. We might expect the DFT of $f(nT) = e^{j\omega_0 nT}$ to have an analogous property. This DFT is

$$F_k = \sum_{n=0}^{N-1} e^{j\omega_0 nT} e^{-j2\pi nk/N}$$

$$= e^{-j(N-1)vT/2} \frac{\sin NvT/2}{\sin vT/2} \qquad \text{for} \qquad k = 0, \ldots, N-1 \qquad (10.3.1)$$

where $v = k\omega_s/N - \omega_0$. We encountered a similar function in our discussion of frequency sampling filters in Section 8.10.

If k and m are integers in the set $\{0, \ldots, N-1\}$ and $\omega_0 = m\omega_s/N$, we find that

$$F_k = \begin{cases} N & \text{for} \quad k = m \\ 0 & \text{elsewhere} \end{cases} \qquad (10.3.2)$$

This is the DFT analogy of the Fourier transform of the complex exponential being an impulse function. The frequency selective property of the DFT has recently been used to implement HF modems. In these systems, data is transmitted by digitally phase modulating a set of tones separated by the frequency increment ω_s/N. An IDFT is used for modulation and a DFT is used for demodulation.

If ω_0 is not one of the frequency sampling points $m\omega_s/N$, then none of the DFT values are zero. This phenomenon is called leakage. By sketching F_k it is easy to see that $|F_k|$ has its peak value when $k\omega_s/N$ is closest to ω_0.

10.4 SOME USEFUL TRANSFORM RELATIONSHIPS

In this section, a number of relatively simple but useful DFT relationships are presented. Since the DFT and IDFT formulas are similar in form, each DFT property has an analogous IDFT property.

1. *Linearity*

    If c_1 and c_2 are two constants, then

$$\text{DFT}\{c_1 f_n + c_2 g_n\} = c_1 F_k + c_2 G_k \qquad (10.4.1)$$

Proof This result follows directly from (10.2.1)

Q.E.D.

2. *Time Reversal*

$$\text{DFT}\{f_{-n}\} = F_{-k} = F_{N-k} \qquad (10.4.2)$$

Proof   According to (10.2.7)

$$f_{-n} = \frac{1}{N} \sum_{k=0}^{N-1} F_k e^{-j2\pi nk/N} \qquad (10.4.3)$$

Letting $r = N - k$ (10.4.3) becomes

$$f_{-n} = \frac{1}{N} \sum_{r=1}^{N} F_{N-r} e^{-j2\pi n(N-r)/N} = \frac{1}{N} \sum_{r=1}^{N} F_{N-r} e^{j2\pi nr/N} \qquad (10.4.4)$$

Since $F_N = F_0$ and $e^{j2\pi n} = 1$ (10.4.4) can be written as

$$f_{-n} = \frac{1}{N} \sum_{k=0}^{N-1} F_{N-k} e^{j2\pi nk/N} \qquad (10.4.5)$$

Thus, $f_{-n} = \text{IDFT}\{F_{N-k}\}$. According to (10.2.2), $F_{N-k} = F_{-k}$.

Q.E.D.

3. *Cyclic Time and Frequency Shifts*

$$\text{DFT}\{f_{n-r}\} = F_k e^{-j2\pi rk/N} \qquad (10.4.6)$$

and

$$\text{IDFT}\{F_{k-r}\} = f_n e^{j2\pi nr/N} \qquad (10.4.7)$$

Proof   Substituting the right-hand side of (10.4.6) into the IDFT formula gives

$$\frac{1}{N} \sum_{k=0}^{N-1} F_k e^{j2\pi(n-r)k/N} = f_{n-r}$$

Thus (10.4.6) must be true because of the uniqueness of transform pairs. Equation 10.4.7 can be proved in a similar manner.

Q.E.D.

The shifts in the relationship just presented are called cyclic because the N-point sequences $\{f_{n-r}\}_{n=0}^{N-1}$ and $\{F_{k-r}\}_{k=0}^{N-1}$ can be written, since f_n and F_k repeat with period N, as

$$f_{N-p}, \ldots, f_{N-1}, f_0, f_1, \ldots, f_{N-p-1}$$

and

$$F_{N-p}, \ldots, F_{N-1}, F_0, F_1, \ldots, F_{N-p-1}$$

where $r = mN + p$ with m and p integers and $0 \le p \le N-1$. These are just the sequences $f_0, \ldots, f_{N-1}$ and $F_0, \ldots, F_{N-1}$ shifted cyclically to the right r positions.

4. *DFT of Complex Conjugate Sequence*
   Let $\text{DFT}\{f_n\} = F_k$. Then

$$\text{DFT}\{\bar{f}_n\} = \bar{F}_{-k} = \bar{F}_{N-k} \qquad (10.4.8)$$

Proof Taking the conjugate of both sides of (10.2.1) yields

$$\bar{F}_k = \sum_{n=0}^{N-1} \bar{f}_n e^{j2\pi nk/N}$$

Replacing k by $-k$ we obtain

$$\bar{F}_{-k} = \sum_{n=0}^{N-1} \bar{f}_n e^{-j2\pi nk/N}$$

Q.E.D.

5. *Conditions on F_k for Real f_n*
    The sequence f_n is real if and only if

$$\bar{F}_{N-k} = F_k \qquad \text{for} \qquad k = 0, \ldots, N-1 \tag{10.4.9}$$

Proof See Section 8.10, equations (8.10.15)–(8.10.18).

Q.E.D.

10.5 CYCLIC CONVOLUTION AND CORRELATION

We have frequently used the fact that the Z-transform of the convolution of two sequences is the product of their Z-transforms. A similar property holds for DFTs and is presented in the following theorem. This property is the basis for the DFT frequency domain approach to digital filtering that is presented in Sections 10.9 and 10.10.

THEOREM 10.5.1. Cyclic Convolution Theorem
Let the cyclic convolution of two sequences f_n and g_n with $f_{n+N} = f_n$ and $g_{n+N} = g_n$ for all n be defined as

$$f_n \circledast g_n = \sum_{r=0}^{N-1} f_r g_{n-r} = \sum_{r=0}^{N-1} g_r f_{n-r} \tag{10.5.1}$$

Then

$$\text{DFT}\{f_n \circledast g_n\} = F_k G_k \tag{10.5.2}$$

where $F_k = \text{DFT}\{f_n\}$ and $G_k = \text{DFT}\{g_n\}$. Also

$$\frac{1}{N} F_k \circledast G_k = \frac{1}{N} \sum_{r=0}^{N-1} F_r G_{k-r} = \frac{1}{N} \sum_{r=0}^{N-1} G_r F_{k-r}$$

$$= \text{DFT}\{f_n g_n\} \tag{10.5.3}$$

Proof   Using (10.4.1) and (10.4.6), we find that

$$\text{DFT}\left\{\sum_{r=0}^{N-1} f_r g_{n-r}\right\} = \sum_{r=0}^{N-1} f_r \text{DFT}\{g_{n-r}\}$$

$$= \sum_{r=0}^{N-1} f_r e^{-j2\pi rk/N} G_k = F_k G_k$$

The equality between the two forms for the cyclic convolution in (10.5.1) can be proved directly using the periodicity of f_n and g_n or by observing that F_k and G_k can be interchanged without altering the product $F_k G_k$. Equation 10.5.3 can be proved in a similar manner by applying the IDFT to the cyclic convolution of F_k and G_k.

Q.E.D.

Cyclic convolution is similar to ordinary discrete-time convolution in that one sequence is reversed in time, delayed by n units, multiplied by the other sequence on a term-by-term basis and then the products are summed. It differs because the two sequences are required to repeat evey N samples and only N consecutive products are summed. Because f_n and g_n are periodic, the sum can be taken over any N consecutive products. Also, the sequence resulting from the cyclic convolution must repeat every N samples.

As r varies from 0 to $N-1$ in (10.5.1), g_{n-r} takes on the values

$$g_n, g_{n-1}, \ldots, g_1, g_0, g_{-1}, \ldots, g_{n-N+1}$$

assuming $0 \le n \le N-1$. Since $g_{n+N} = g_n$, this sequence can be written as

$$g_n, g_{n-1}, \ldots, g_1, g_0, g_{N-1}, \ldots, g_{n+1}$$

and is simply the reversed N-point sequence

$$g_0, g_{N-1}, g_{N-2}, \ldots, g_2, g_1$$

shifted cyclically to the right n positions. The convolution defined in Theorem 10.5.1 is called cyclic convolution for this reason. It is also known as periodic or circular convolution.

Example 10.5.1

Let $N = 8$, $f_0 = f_1 = f_2 = f_3 = f_4 = f_5 = 1$, $f_6 = f_7 = 0$, $g_0 = g_1 = g_2 = g_6 = g_7 = 1$, and $g_3 = g_4 = g_5 = 0$. The sequences f_r, g_{-r}, g_{1-r}, and g_{2-r} for $r = 0, \ldots, N-1$ are shown in Fig. 10.5.1. Notice that g_{1-r} is g_{-r} cyclically shifted one position to the right and g_{2-r} is g_{1-r} cyclically shifted one position to the right. Calculating the sum of products required in (10.5.1) for each cyclic shift, the sequence $f_n \circledast g_n$ shown at the bottom of Fig. 10.5.1 is obtained for $0 \le n \le N-1$.   ◄

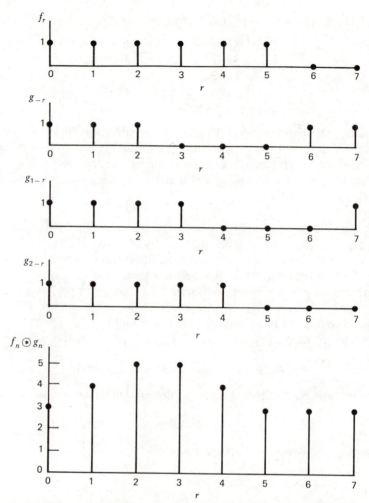

FIGURE 10.5.1. Cyclic convolution of f_n and g_n in Example 10.5.1.

Correlation or forming lagged products is an operation that is required in a variety of signal processing applications. The following theorem forms the basis for using the DFT for correlation.

THEOREM 10.5.2.   Cyclic Correlation Theorem

Let f_n and g_n be defined as in Theorem 10.5.1. Then

$$\text{DFT}\left\{\sum_{r=0}^{N-1} f_{r+n}\bar{g}_r\right\} = F_k\bar{G}_k \qquad (10.5.4)$$

and

$$\text{IDFT}\left\{\frac{1}{N}\sum_{r=0}^{N-1} F_{r+k}\bar{G}_r\right\} = f_n\bar{g}_n \tag{10.5.5}$$

Proof Using (10.4.1) and (10.4.6), we find that

$$\text{DFT}\left\{\sum_{r=0}^{N-1} f_{r+n}\bar{g}_r\right\} = \sum_{r=0}^{N-1}\text{DFT}\{f_{r+n}\}\bar{g}_r$$

$$= F_k\sum_{r=0}^{N-1}\bar{g}_r e^{j2\pi rk/N} = F_k\bar{G}_k$$

The proof of (10.5.5) is similar.

Q.E.D.

Correlation is similar to convolution. However, convolution is always commutative, that is, f_n and g_n can be interchanged but correlation is not commutative in general.

10.6 SUMS OF PRODUCTS

The sum of term-by-term products of two N-point sequences can be calculated directly or by summing term-by-term products of their appropriately modified DFTs. Three variations of this result are presented in the following theorem.

THEOREM 10.6.1. Parseval's Theorem
Let $\text{DFT}\{f_n\} = F_k$ and $\text{DFT}\{g_n\} = G_k$. Then

$$\sum_{n=0}^{N-1} f_n g_n = \frac{1}{N}\sum_{k=0}^{N-1} F_k G_{-k} \tag{10.6.1}$$

$$\sum_{n=0}^{N-1} f_n \bar{g}_n = \frac{1}{N}\sum_{k=0}^{N-1} F_k \bar{G}_k \tag{10.6.2}$$

and

$$\sum_{n=0}^{N-1} |f_n|^2 = \frac{1}{N}\sum_{k=0}^{N-1} |F_k|^2 \tag{10.6.3}$$

Proof According to (10.5.3)

$$\sum_{n=0}^{N-1} f_n g_n e^{-j2\pi nk/N} = \frac{1}{N}\sum_{r=0}^{N-1} F_r G_{k-r}$$

Letting $k = 0$ yields

$$\sum_{n=0}^{N-1} f_n g_n = \frac{1}{N}\sum_{r=0}^{N-1} F_r G_{-r} \tag{10.6.4}$$

Changing r to k on the right-hand side of (10.6.4) gives (10.6.1). Equation (10.6.2) follows from (10.6.1) since $\text{DFT}\{\bar{g}_n\} = \bar{G}_{-k}$. Equation 10.6.3 is obtained by letting $f_n = g_n$ in (10.6.2).

Q.E.D.

Equation 10.6.3 shows that the energy in the sequence $f_0, \ldots, f_{N-1}$ can be calculated either in the time or frequency domains.

10.7 THE FAST FOURIER TRANSFORM, AN EFFICIENT ALGORITHM FOR COMPUTING THE DFT

Direct computation of the DFT of an N-point complex sequence using (10.2.1) requires approximately N^2 operations if an operation is defined to be a complex multiplication and addition. In 1965, Cooley and Tukey [26] published a paper showing how to compute the DFT using approximately $N \log_2 N$ operations when N is a power of two. They presented similar results for the case where N is the product of arbitrary factors. The basic idea used by Cooley and Tukey had been discovered earlier by others but had not become generally known [24]. The Cooley-Tukey paper generated a great deal of excitement because $N \log_2 N$ becomes increasingly smaller relative to N^2 as N becomes large. For example, $N^2/(N \log_2 N) = 102.4$ for $N = 2^{10} = 1024$. This computational savings greatly reduced the time required for spectral analysis and made a frequency domain approach to digital filtering using the DFT competitive with the time domain difference equation approach in many instances. Numerous papers presenting variations, explanations, and applications of the Cooley-Tukey algorithm appeared. A selection of these papers can be found in Reference 111. Methods for computing DFTs requiring $N \log N$ operations became known as *fast Fourier transforms* (FFTs).

Many different FFT algorithms have been derived, each with its own advantages and disadvantages in terms of programming and hardware implementation. We discuss only two algorithms for the case where N is a power of two in detail to show how the computational efficiency arises. To simplify the notation somewhat, we let

$$W_N = e^{-j2\pi/N} \tag{10.7.1}$$

and write the DFT formula (10.2.1) as

$$F_k = \sum_{n=0}^{N-1} f_n W_N^{nk} \quad \text{for} \quad k = 0, \ldots, N-1 \tag{10.7.2}$$

DECIMATION IN TIME

One commonly used FFT algorithm is known as the *decimation in time* algorithm. Decimation is a term used in mathematics for the process of reordering a sequence according to some prescribed rule. We will now assume that N is a power of two.

The first step in deriving the decimation in time algorithm is to split the sum in (10.7.2) into one for n even and one for n odd and write F_k as

$$F_k = \sum_{n=0}^{(N/2)-1} f_{2n} W_N^{2nk} + \sum_{n=0}^{(N/2)-1} f_{2n+1} W_N^{(2n+1)k}$$

and, since $W_{N/2} = W_N^2$,

$$F_k = \sum_{n=0}^{(N/2)-1} f_{2n} W_{N/2}^{nk} + W_N^k \sum_{n=0}^{(N/2)-1} f_{2n+1} W_{N/2}^{nk} \qquad (10.7.3)$$

Notice that

$$G_k = \sum_{n=0}^{(N/2)-1} f_{2n} W_{N/2}^{nk} \qquad (10.7.4)$$

is the DFT of the $N/2$-point sequence $g_n = f_{2n}$ for $n = 0, \ldots, (N/2)-1$ consisting of the even-numbered elements of f_n, and

$$H_k = \sum_{n=0}^{(N/2)-1} f_{2n+1} W_{N/2}^{nk} \qquad (10.7.5)$$

is the DFT of the $N/2$-point sequence $h_n = f_{2n+1}$ for $n = 0, \ldots, (N/2)-1$ consisting of the odd-numbered elements of f_n. Thus the total N-point DFT can be computed by combining the two $N/2$-point DFTs according to the rule

$$F_k = G_k + W_N^k H_k \qquad \text{for} \qquad k = 0, \ldots, N-1 \qquad (10.7.6)$$

Since $G_{k+(N/2)} = G_k$, $H_{k+(N/2)} = H_k$, and $W_N^{k+(N/2)} = -W_N^k$, (10.7.6) is equivalent to the pair of equations

$$F_k = G_k + W_N^k H_k \qquad \text{for} \qquad k = 0, \ldots, \frac{N}{2}-1 \qquad (10.7.7)$$

and

$$F_{k+(N/2)} = G_k - W_N^k H_k \qquad \text{for} \qquad k = 0, \ldots, \frac{N}{2}-1 \qquad (10.7.8)$$

Assuming that G_k and H_k are already known for $k = 0, \ldots, (N/2)-1$, computation of the N-point DFT using (10.7.7) and (10.7.8) requires N complex additions. The term $W_N^k H_k$ can be calculated once for each k and used in both (10.7.7) and (10.7.8). Therefore, only $N/2$ complex multiplications are required if the special cases where $W_N^k = 1$, j, etc., are counted as full complex multiplications.

The second step is to compute the $N/2$-point DFTs G_k and H_k by combining $N/4$-point DFTs. For example, G_k can be computed by combining the DFTs of the $N/4$-point sequences $a_n = g_{2n} = f_{4n}$ and $b_n = g_{2n+1} = f_{4n+2}$ for $n = 0, \ldots, (N/4) - 1$. From (10.7.7) and (10.7.8) with N replaced by $N/2$, it follows that

$$G_k = A_k + W_N^{2k} B_k \qquad \text{for} \qquad k = 0, \ldots, \frac{N}{4} - 1 \qquad (10.7.9)$$

and

$$G_{k+(N/4)} = A_k - W_N^{2k} B_k \qquad \text{for} \qquad k = 0, \ldots, \frac{N}{4} - 1 \qquad (10.7.10)$$

H_k can be computed similarly. Using the same reasoning as before, all values of G_k can be computed by (10.7.9) and (10.7.10) with $N/2$ complex additions and $N/4$ complex multiplications if the A_k's and B_k's are already known. Therefore, computation of both G_k and H_k for $k = 0, \ldots, N/2$ requires N complex additions and $N/2$ complex multiplications just as in the first step.

The process of computing N'-point DFTs by combining $N'/2$-point DFTs can be continued until 1-point DFTs are reached. The DFT of a single point is just the point itself. At the rth step, 2^{r-1} pairs of $N/2^r$-point DFTs are combined to form $N/2^{r-1}$-point DFTs. If $N = 2^m$, 1-point DFTs are reached in $m = \log_2 N$ steps. Going from step to step, the number of DFTs to be computed doubles but the number of points per DFT decreases by a factor of two. Therefore, N complex additions and $N/2$ complex multiplications are required at each step to compute all combinations. Thus $N \log_2 N$ complex additions and $(N/2) \log_2 N$ complex multiplications are required to compute the entire DFT.

A flow graph can be used to display the operations required for computing the FFT. A basic flow graph element is shown in Fig. 10.7.1. It is similar to a block diagram except that adders are indicated by dots called nodes and the branch transfer functions are written next to the branches. A branch with no value next to it is understood to have unity transmission. A flow graph of (10.7.7) and (10.7.8) for an arbitrary value of k is shown in Fig. 10.7.2. This structure is known as a *butterfly*. The sequence of steps used to generate a complete flow graph for an $N = 8$-point decimation in time FFT is shown in Figs. 10.7.3,

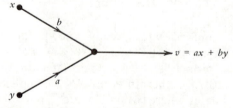

FIGURE 10.7.1.   A basic flow graph element.

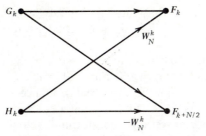

FIGURE 10.7.2. A decimation in time FFT butterfly.

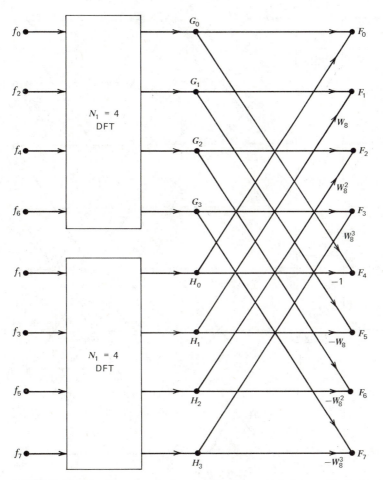

FIGURE 10.7.3. First step in forming the flow graph for an $N = 8$-point decimation in time FFT.

10.7.4, and 10.7.5. Notice that the 2-point DFTs require no multiplications. Since $W_4 = -j$, the 4-point DFTs do not actually require complex multiplications either.

In Fig. 10.7.5, F_k appears in its natural order going down the set of output nodes. This will always be the case for any allowable N because of the method of constructing the flow graph. The original sequence f_n appears in a permuted order at the input nodes. This order is known as the *bit reversed* order since it can be obtained by representing the decimal values of n in binary form,

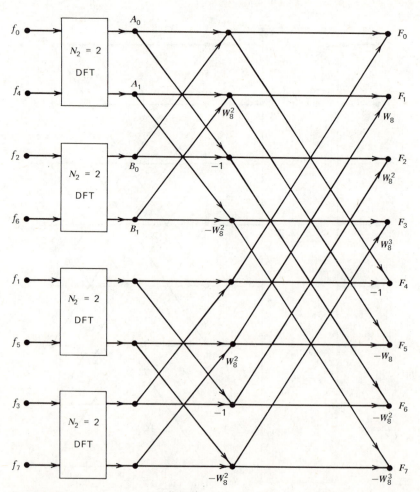

FIGURE 10.7.4. Second step in forming the flow graph for an $N = 8$-point decimation in time FFT.

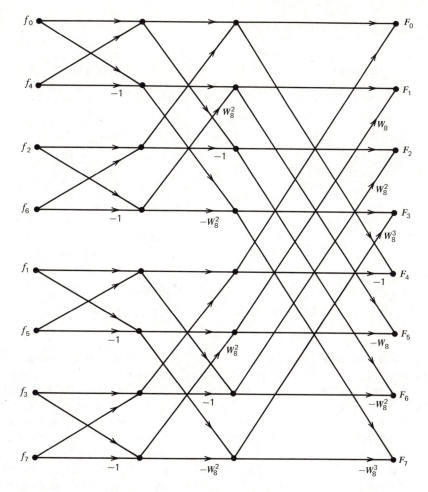

FIGURE 10.7.5. Complete flow graph for $N = 8$-point decimation in time FFT.

reversing the bits in each binary representation, and then recalculating the equivalent decimal numbers. The bit reversal procedure is shown in Table 10.7.1.

The bit reversal procedure gives the proper input ordering for any N that is a power of two. Suppose that a vertical array with locations labeled sequentially from 0 to $N-1$ starting at the top is used to store the input sequence and that the inputs are initially stored in their natural order, that is, f_n is stored in location n. If $N = 2^m$, the integers $0, \ldots, N-1$ can be represented as binary

TABLE 10.7.1   Bit Reversal

n	Binary Representation	Bit Reversed Binary Representation	Decimal Equivalent of Bit Reversed Binary Representation
0	000	000	0
1	001	100	4
2	010	010	2
3	011	110	6
4	100	001	1
5	101	101	5
6	110	011	3
7	111	111	7

$m$tuples. As usual, we will assign the decimal number

$$a = \sum_{r=0}^{m-1} a_r 2^r$$

to the binary $m$tuple $(a_{m-1}, \ldots, a_1, a_0)$. Suppose that n has the binary representation $(n_{m-1}, \ldots, n_1, n_0)$. In the first step of forming the flow graph, all the even-numbered points are moved to the top half of the array and all the odd-numbered points are moved to the bottom half. The points are even-numbered if $n_0 = 0$ and odd-numbered if $n_0 = 1$. Specifically, f_n is moved to location $(n_0, n_{m-1}, \ldots, n_1)$. In the second step, points in the top half of the array are moved to the top half of this half if $n_1 = 0$ and to the bottom half of this half if $n_1 = 1$. Points in the bottom half of the array are treated similarly. This permutation can be compactly described by saying that f_n is moved to location $(n_0, n_1, n_{m-1}, \ldots, n_2)$ at the second step. Continuing this line of reasoning, we find that f_n ends up in the bit reversed location $(n_0, n_1, \ldots, n_{m-1})$ after $m - 1$ steps.

In efficient FFT programs, sequences are usually permuted into bit reversed order by working down the array, successively interchanging f_n and f_i where i is the bit reversed value of n. The interchange of points that have already been interchanged is prevented by comparing n and i. If i is less than n, then f_n and f_i have already been interchanged. This procedure, known as *in-place* bit reversal, only requires one storage location in addition to the N locations required for the input sequence.

Computation of the FFT can also be performed in place. The basic computation in the FFT has the butterfly structure shown in Fig. 10.7.2. In computing one vertical array in the flow graph from the previous vertical array, the pair of inputs to a particular butterfly connecting the arrays is not used in any of the other butterflies. Therefore, the pair of butterfly outputs can be written over the inputs. An FFT program is typically organized to work down an array computing butterflies and writing outputs over inputs. Only one storage location in addition to the N-point array is required. It should be noted that one complex location actually requires two machine locations. In-place computation is particularly convenient for conserving memory space when N is large. When N is extremely large, other techniques can be used [121].

The flow graph for $N = 8$ in Fig. 10.7.5 or for other N can be changed so that the input is in natural order and the output is in bit reversed order by rearranging each vertical set of N nodes into bit reversed order. The new structure also allows in-place computation. It corresponds to the original Cooley-Tukey algorithm.

DECIMATION IN FREQUENCY

Another FFT algorithm known as the *decimation in frequency* algorithm can be derived by first splitting the DFT formula into a sum over the first half and a sum over the second half of the sequence to give

$$F_k = \sum_{n=0}^{(N/2)-1} f_n W_N^{nk} + \sum_{n=0}^{(N/2)-1} f_{n+(N/2)} W_N^{[n+(N/2)]k}$$

$$= \sum_{n=0}^{(N/2)-1} [f_n + (-1)^k f_{n+(N/2)}] W_N^{nk} \qquad \text{for} \qquad k = 0, \dots, N-1 \quad (10.7.11)$$

The even-numbered DFT points can be written as

$$F_{2k} = \sum_{n=0}^{(N/2)-1} [f_n + f_{n+(N/2)}] W_{N/2}^{nk} \qquad \text{for} \qquad k = 0, \dots, \frac{N}{2}-1 \quad (10.7.12)$$

and the odd-numbered points can be written as

$$F_{2k+1} = \sum_{n=0}^{(N/2)-1} [(f_n - f_{n+(N/2)}) W_N^n] W_{N/2}^{nk} \qquad \text{for} \qquad k = 0, \dots, \frac{N}{2}-1 \quad (10.7.13)$$

Thus F_{2k} is the DFT of the $N/2$-point sequence

$$g_n = f_n + f_{n+(N/2)} \qquad \text{for} \qquad n = 0, \dots, \frac{N}{2}-1 \qquad (10.7.14)$$

and F_{2k+1} is the DFT of the $N/2$-point sequence

$$h_n = (f_n - f_{n+(N/2)}) W_N^n \qquad \text{for} \qquad n = 0, \dots, \frac{N}{2}-1 \qquad (10.7.15)$$

The computation required to find g_n and h_n from f_n and $f_{n+(N/2)}$ has a butterfly structure. N complex additions and $N/2$ complex multiplications are required to compute all values of g_n and h_n if the special cases where $W_N^n = 1$, j, etc., are counted as full complex multiplications. The problem has now been reduced to the computation of two $N/2$-point DFTs.

The procedure just described can be repeated until 1-point DFTs are reached. This requires $\log_2 N$ steps. Thus, the decimation in frequency FFT also requires $N \log_2 N$ complex additions and $(N/2) \log_2 N$ complex multiplications. The complete flow graph for $N = 8$ is shown in Fig. 10.7.6. In this form of the algorithm, the inputs are required in natural order and the outputs appear in

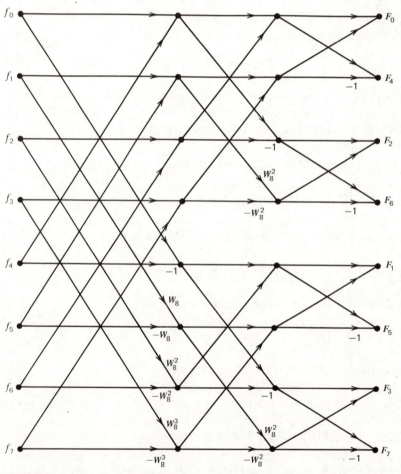

FIGURE 10.7.6. Flow graph for an $N = 8$-point decimation in frequency FFT.

bit reversed order. The flow graph can be changed so that the inputs are required in bit reversed order and the outputs appear in natural order by rearranging each vertical array of nodes into bit reversed order. In either case, the computation can be performed in place. It is interesting to observe that the flow graphs in Figs. 10.7.5 and 10.7.6 are transpose pairs as defined in Section 6.6.

ADDITIONAL COMMENTS

The two FFT algorithms presented in this section were derived by successively expressing N'-point DFTs in terms of $N'/2$-point DFTs. These algorithms and other variations can also be derived by expressing n and k in binary form and writing the DFT formula as a $\log_2 N$-fold summation over the binary digits of n [13,26]. The resulting set of FFT algorithms are called *radix* or *base* 2 algorithms. Either approach can be easily generalized to find radix p algorithms when N is a power of an arbitrary positive integer p. It has been found that radix 4, 8, 16, etc., algorithms are more efficient than radix 2 algorithms if the symmetries as well as the periodicity of the trigonometric functions are exploited [6,36]. The greatest gains are achieved in going from radix 2 to radix 4 or 8. This suggests using a mixed radix algorithm even when N is a power of two [122]. For example, as many radix 4 iterations as possible can be performed followed by a radix 2 iteration if necessary.

Computer programs for almost every conceivable FFT variation have been written and published. A FORTRAN program for a radix 2 decimation in time FFT with in-place computation and bit reversal can be written using only 32 statements [25]. Most computer centers have FFT library subroutines.

The roundoff noise introduced in an actual FFT computation can be analyzed using the techniques discussed in Chapter 9. In general, each complex product is equivalent to four real products. Each real product can be replaced by the appropriate noisy product and the total noise power at a given output node can be calculated.

10.8 EFFICIENT USE OF AN FFT PROGRAM WHEN THE INPUT SEQUENCE IS REAL

FFT programs are frequently written to operate on complex input sequences so that the same program can be used for both the DFT and IDFT. When the input sequence is real, some efficiency is lost because complex arithmetic is still always used. We will now see how to efficiently utilize an FFT program by combining a pair of real input sequences into a complex sequence.

Suppose that g_n and h_n are two real N-point sequences. If we form the complex sequence $a_n = g_n + jh_n$, then by linearity

$$\text{DFT}\{a_n\} = A_k = G_k + jH_k \tag{10.8.1}$$

According to (10.4.8) and linearity

$$\text{DFT}\{\bar{a}_n\} = \bar{A}_{N-k} = G_k - jH_k \tag{10.8.2}$$

Adding (10.8.1) and (10.8.2) yields

$$G_k = \frac{A_k + \bar{A}_{N-k}}{2} \tag{10.8.3}$$

Subtracting (10.8.2) from (10.8.1) gives

$$H_k = \frac{A_k - \bar{A}_{N-k}}{2j} \tag{10.8.4}$$

Thus the DFTs of the two real sequences can be easily determined from the DFT of the complex sequence.

Computation of G_k and H_k for $k = 0, \ldots, N-1$ by separate radix 2 FFTs requires approximately $2N \log_2 N$ complex additions and $N \log_2 N$ complex multiplications. Simultaneous computation by taking the FFT of a_n requires $(N/2) \log_2 N$ complex multiplications and $N \log_2 N$ complex additions for the FFT, and $2N$ complex additions to perform the separation using (10.8.3) and (10.8.4). Division by 2 requires only a simple shift. Thus the number of complex multiplications has been reduced by a factor of two. Comparing $2N + N \log_2 N$ with $2N \log_2 N$, it follows that the number of complex additions is also less when N is greater than 4.

The DFT of a $2N$-point real sequence f_n can be efficiently computed using an N-point FFT. Let $g_n = f_{2n}$, $h_n = f_{2n+1}$, and $a_n = g_n + jh_n$ for $n = 0, \ldots, N-1$. Then G_k and H_k can be computed from the FFT of the N-point complex sequence a_n. Replacing N by $2N$ in (10.7.7) and (10.7.8), we see that the total DFT can be computed by combining G_k and H_k according to the equations

$$F_k = G_k + W_{2N}^k H_k \qquad \text{for} \qquad k = 0, \ldots, N-1 \tag{10.8.5}$$

and

$$F_{k+N} = G_k - W_{2N}^k H_k \qquad \text{for} \qquad k = 0, \ldots, N-1 \tag{10.8.6}$$

Computation of the $2N$-point DFT directly with a radix 2 FFT requires approximately $2N \log_2 (2N) = 2N + 2N \log_2 N$ complex additions and $N \log_2 (2N) = N + N \log_2 N$ complex multiplications. Computation by the method described in the previous paragraph requires $N \log_2 N$ complex additions and $(N/2) \log_2 N$ complex multiplications for the N-point FFT, $2N$

complex additions to perform the separation using (10.8.3) and (10.8.4), and 2N complex additions and N complex multiplications to combine G_k and H_k using (10.8.5) and (10.8.6) for a total of $4N + N \log_2 N$ complex additions and $N + (N/2) \log_2 N$ complex multiplications. The ratio of the number of additions in the indirect and direct methods is

$$c_1 = (4 + \log_2 N)/(2 + 2 \log_2 N) \tag{10.8.7}$$

The ratio of the number of multiplications in the indirect and direct methods is

$$c_2 = (2 + \log_2 N)/(2 + 2 \log_2 N) \tag{10.8.8}$$

If $N = 2^{10}$, for example, the ratios are $c_1 = 7/11$ and $c_2 = 6/11$. For large N, we see that the indirect method is nearly twice as efficient as the direct method.

10.9 ORDINARY CONVOLUTION AND CORRELATION OF FINITE DURATION SEQUENCES USING THE FFT

Previously, we have defined the discrete-time convolution of two signals $f(n)$ and $g(n)$ as the signal

$$h(n) = \sum_{r=-\infty}^{\infty} f(r)g(n-r) \tag{10.9.1}$$

Discrete-time convolution is also known as ordinary, noncyclic, or aperiodic convolution. Suppose that $f(n)$ and $g(n)$ are nonzero for $0 \le n \le N_1 - 1$ and $0 \le n \le N_2 - 1$, respectively, and are zero elsewhere. Without loss of generality, we will assume that $N_2 \le N_1$. Taking into account the finite durations of $f(n)$ and $g(n)$, we find that

$$h(n) = 0 \quad\quad \text{for} \quad\quad n < 0 \text{ or } n > N_1 + N_2 - 2 \tag{10.9.2}$$

$$h(n) = \sum_{r=0}^{n} f(r)g(n-r) \quad\quad \text{for} \quad\quad 0 \le n \le N_2 - 2 \tag{10.9.3}$$

$$h(n) = \sum_{r=n-N_2+1}^{n} f(r)g(n-r) \quad\quad \text{for} \quad\quad N_2 - 1 \le n \le N_1 - 1 \tag{10.9.4}$$

and

$$h(n) = \sum_{r=n-N_2+1}^{N_1-1} f(r)g(n-r) \quad\quad \text{for} \quad\quad N_1 \le n \le N_1 + N_2 - 2 \tag{10.9.5}$$

Direct evaluation of $h(n)$ for $n = 0, 1, \ldots, N_1 + N_2 - 2$ using (10.9.3), (10.9.4), and (10.9.5) requires a total of $N_1 N_2$ multiplications and $(N_1 - 1)(N_2 - 1)$ additions.

The values of $h(n)$ for $n = 0, \ldots, N_1 + N_2 - 2$ can be evaluated indirectly using the DFT. The first step is to form the N-point sequences

$$f_n = f(n) \qquad \text{for} \qquad n = 0, \ldots, N-1 \qquad (10.9.6)$$

and

$$g_n = g(n) \qquad \text{for} \qquad n = 0, \ldots, N-1 \qquad (10.9.7)$$

with

$$N \geq N_1 + N_2 - 1 \qquad (10.9.8)$$

N can be chosen as the smallest power of 2 satisfying (10.9.8) so that a radix 2 FFT can be used. Notice that, since $f(n)$ has duration N_1 and $g(n)$ has duration N_2, $f_n = 0$ for $n = N_1, \ldots, N-1$ and $g_n = 0$ for $n = N_2, \ldots, N-1$. As usual, we will periodically extend f_n and g_n by letting $f_{n+N} = f_n$ and $g_{n+N} = g_n$ for all n. The next step is to compute the N-point DFTs $F_k = \text{DFT}\{f_n\}$ and $G_k = \text{DFT}\{g_n\}$ and form the sequence $H_k = F_k G_k$ for $k = 0, \ldots, N-1$. The last step is to compute $h_n = \text{IDFT}\{H_k\}$. It will be shown in the next paragraph that $h_n = h(n)$ for $n = 0, \ldots, N-1$. The first $N_1 + N_2 - 1$ points are the desired points. The remaining $N - (N_1 + N_2 - 1)$ points are all zero according to (10.9.2).

According to Theorem 10.5.1, the Cyclic Convolution Theorem,

$$h_n = \sum_{r=0}^{N-1} f_r g_{n-r} \qquad (10.9.9)$$

The sequences f_r and g_{-r} for $r = 0, \ldots, N-1$ are illustrated in Fig. 10.9.1. Rearranging (10.9.8), we see that $N - N_2 + 1 \geq N_1$. Therefore, the $N_2 - 1$ non-zero points of g_{-r} that wrap around the right-hand side overlap the zero portion of f_r. The sequence g_{n-r} can be obtained by cyclically shifting g_{-r} to the right n positions. From Fig. 10.9.1, we can see that the contributions to the cyclic convolution sum resulting from the portion of g_{n-r} that wraps around the right-hand side are zero and that h_n is given by (10.9.2) through (10.9.5) for $n = 0, \ldots, N-1$.

Assuming that $f(n)$ and $g(n)$ are complex, efficient computation of $h(n)$ by the indirect method requires three FFTs for finding $\text{DFT}\{f_n\}$, $\text{DFT}\{g_n\}$, and $\text{IDFT}\{H_k\}$, and N complex multiplications for calculating $H_k = F_k G_k$ for $k = 0, \ldots, N-1$. If a radix 2 FFT algorithm is used, the total time required to compute the convolution by this method can be approximated by an equation of the form

$$T_1 = 3 k_f N \log_2 N + k_m N \qquad (10.9.10)$$

The first term accounts for the three FFTs and the second term for the multiplications to compute $H_k = F_k G_k$. The constants k_f and k_m depend on the hardware and software used to implement the procedure. Using (10.9.8), we find that

$$T_1 \geq 3 k_f (N_1 + N_2 - 1) \log_2 (N_1 + N_2 - 1) + k_m (N_1 + N_2 - 1) \qquad (10.9.11)$$

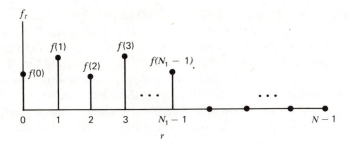

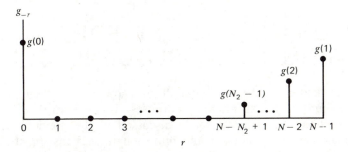

FIGURE 10.9.1. Illustration of f_r and g_{-r} for $r = 0, \ldots, N-1$.

If $f(n)$ and $g(n)$ are real, time can be saved by combining them into a complex sequence and using one FFT to find both F_k and G_k as described in Section 10.8.

The correlation of $f(n)$ and $g(n)$ is defined to be the signal

$$c(n) = \sum_{r=-\infty}^{\infty} f(r+n)\bar{g}(r) \tag{10.9.12}$$

Correlation is quite similar to convolution. From Theorem 10.5.2, it follows that the procedure for indirectly computing the convolution of finite duration sequences can be used to compute their correlation by simply replacing H_k by $H'_k = F_k \bar{G}_k$. The time required to compute the correlation is essentially the same as that required to compute the convolution.

The time required to compute a convolution or correlation by the indirect method just described can be larger or smaller than the time required by the direct method. For example, suppose that the correlation of two signals with $N_1 = N_2$ is desired. As N_1 becomes large, the predominant term on the right-hand side of (10.9.11) becomes proportional to $N_1 \log_2 N_1$. The computation time required by the direct method is essentially proportional to $N_1 N_2 = N_1^2$. Therefore, in this case, the indirect method becomes more efficient than the direct method fairly quickly as N_1 increases.

As another example, suppose that $g(n)$ is the pulse response of a nonrecursive filter with N_2 taps and that $f(n)$ is the signal to be filtered. If N_2 is held fixed and N_1 is allowed to increase, the continuing improvement in the efficiency of the indirect over the direct method observed in the example in the previous paragraph does not occur. In fact, the opposite occurs since $N_1 \log_2 N_1$ approaches and eventually exceeds $N_1 N_2$. In the next section, we see how to remedy this situation by subdividing the data sequence into shorter segments, convolving $g(n)$ with the shorter segments using the indirect method, and then piecing the results together.

10.10 USING THE FFT FOR FILTERING LONG SEQUENCES

In many real- and nonreal-time digital filtering applications, the input sequence has infinite duration for all practical purposes. To perform the filtering using the FFT frequency domain approach discussed in the previous section, the input sequence must be blocked into finite length sections. In both the real- and nonreal-time cases, the section length is clearly limited by the computer memory capacity. In real-time applications, the length is also constrained by the allowable delay. The delay is equal to the time required to receive a complete section plus the time required to process the section. In addition, we observed in Section 10.9 that when the filter pulse response has a fixed finite duration, the indirect frequency domain method of filtering, that is, performing the convolution, decreases in efficiency relative to the direct method as the section length becomes large.

In this section, we examine two methods for blocking the input sequence into sections, filtering the individual sections, and then piecing the results together so that there are no transient effects at the boundaries. We only consider nonrecursive filtering. The FFT approach can be modified to realize recursive filters also [15,38,135]. However, direct difference equation implementation of recursive filters has been found to be more efficient than indirect FFT implementation when the filter order is less than about 30. Most desired recursive filters have orders less than 30. The frequency domain approach does not eliminate the filter design problem. The desired nonrecursive filter must still be designed using the techniques discussed in Chapter 8.

THE OVERLAP-ADD METHOD

The overlap-add method of sectioning is based on the technique for convolving finite duration sequences discussed in Section 10.9. Suppose that $g(n)$ is the pulse response of a desired filter and can be nonzero only for $0 \le n \le N_2 - 1$. We will designate the filter input by $f(n)$ and output by $y(n)$ and will assume that the duration of $f(n)$ is large. The input can be decomposed into nonoverlapping sections of length N_1 and written as

$$f(n) = \sum_{i=-\infty}^{\infty} f_i(n) \tag{10.10.1}$$

where

$$f_i(n) = \begin{cases} f(n) & \text{for} \quad iN_1 \le n \le (i+1)N_1 - 1 \\ 0 & \text{elsewhere} \end{cases} \tag{10.10.2}$$

We will assume that $N_1 \ge N_2$. The desired output is

$$y(n) = f(n) * g(n) = \sum_{i=-\infty}^{\infty} f_i(n) * g(n) = \sum_{i=-\infty}^{\infty} y_i(n) \tag{10.10.3}$$

where

$$y_i(n) = f_i(n) * g(n) = \sum_{r=-\infty}^{\infty} f_i(r) g(n-r) \tag{10.10.4}$$

Using the same reasoning as in the beginning of Section 10.9, it follows that the output section $y_i(n)$ has duration $N_1 + N_2 - 1$ in general and can be nonzero only for $iN_1 \le n \le (i+1)N_1 + N_2 - 2$. The $N_2 - 1$ points $y_i(n)$ for $n = iN_1, \ldots, iN_1 + N_2 - 2$ are generated as $g(n-r)$ slides onto the left-hand side of the nonzero portion of $f_i(n)$ and the $N_2 - 1$ points $y_i(n)$ for $n = (i+1)N_1, \ldots, (i+1)N_1 + N_2 - 2$ are generated as $g(n-r)$ slides off the right-hand side. Since the nonzero portion of $y_{i+1}(n)$ starts at $n = (i+1)N_1$, the last $N_2 - 1$ points of the nonzero portion of $y_i(n)$ overlap the first $N_2 - 1$ points of the nonzero portion of $y_{i+1}(n)$. Thus, in computing the total output using (10.10.3), only the $N_2 - 1$ overlapping points of adjacent output sections actually need to be added. These concepts are illustrated in the following simple example.

Example 10.10.1

Suppose that

$$f(n) = \begin{cases} 1 & \text{for} \quad n \ge 0 \\ 0 & \text{elsewhere} \end{cases}$$

and

$$g(n) = \begin{cases} 1 & \text{for} \quad 0 \le n \le 2 \\ 0 & \text{elsewhere} \end{cases}$$

In this example, $N_2 = 3$. If we let $N_1 = 6$, then $N_1 + N_2 - 1 = 8$ and $N = 8$-point FFTs can be used. The signals $f_0(n)$, $f_1(n)$, $y_0(n)$, $y_1(n)$, and $y(n)$ are shown in

Fig. 10.10.1. Notice that the output sections $y_0(n)$ and $y_1(n)$ overlap by $N_2 - 1 = 2$ points. ◄

The indirect frequency domain approach presented in Section 10.9 can be used to compute the output sections. Assuming that N_2 is fixed and N is a convenient value for an FFT algorithm, usually a power of two, N_1 can be chosen so that

$$N = N_1 + N_2 - 1 \qquad (10.10.5)$$

In filtering applications, the N-point sequence $g_n = g(n)$ for $n = 0, \ldots, N-1$ is known a priori so that $G_k = \text{DFT}\{g_n\}$ can be precomputed and stored. N-point input sequences can be formed according to the rule

$$f_{i,n} = \begin{cases} f(n + iN_1) & \text{for} \quad 0 \le n \le N_1 - 1 \\ 0 & \text{for} \quad N_1 \le n \le N - 1 \end{cases} \qquad (10.10.6)$$

which is equivalent to

$$f_{i,n} = f_i(n + iN_1) \qquad \text{for} \quad 0 \le n \le N - 1 \qquad (10.10.7)$$

The next steps are to compute $F_{i,k} = \text{DFT}\{f_{i,n}\}$, $Y_{i,k} = F_{i,k}G_k$ for $k = 0, \ldots, N-1$, and $y_{i,n} = \text{IDFT}\{Y_{i,k}\}$. It follows that

$$y_i(n + iN_1) = y_{i,n} \qquad \text{for} \quad 0 \le n \le N - 1 \qquad (10.10.8)$$

The sequences $y_{i-1,n}$ and $y_{i,n}$ must be overlapped by $N_2 - 1$ points and added.

Assuming that the filter response G_k has been precomputed and a radix 2 FFT algorithm is employed, the time required to process a section can be approximated by the equation

$$T_2 = 2k_f N \log_2 N + k_a N \qquad (10.10.9)$$

The first term is the time required to perform the two FFTs to compute $F_{i,k} = \text{DFT}\{f_{i,n}\}$ and $y_{i,n} = \text{IDFT}\{Y_{i,k}\}$. The second term is predominantly the time to compute $Y_{i,k} = F_{i,k}G_k$ for $k = 0, \ldots, N-1$ but also includes the time to structure the input sequence and overlap and add the output sequence. N_1 complete filtered output points are generated for each processed section. Therefore, the processing time per filtered output point is

$$\tau_2 = T_2/N_1 \text{ s/output point} \qquad (10.10.10)$$

The processing rate can be maximized by choosing $N = 2^m$ and corresponding section length $N_1 = N - N_2 + 1$ to minimize τ_2. Later in this section, we will see that the processing time can be reduced by a factor of two when the input sequence is real. With this additional factor of two improvement for real input sequences, numerous people have found experimentally that the indirect FFT method of filtering becomes more efficient than the direct method when N_2,

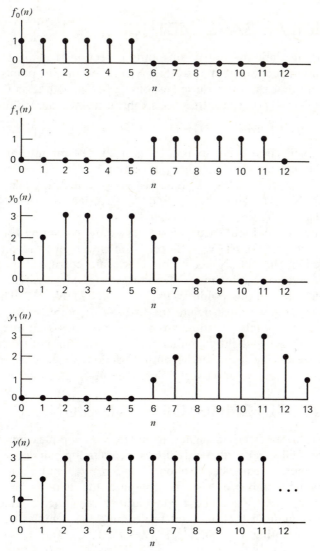

FIGURE 10.10.1. Signals generated in the overlap-add
method for Example 10.10.1.

the duration of the filter pulse response, exceeds a number in the order of 32. The efficiency of the indirect method relative to the direct method increases as N_2 increases.

THE OVERLAP-SAVE METHOD

The need for overlapping and adding output sections can be eliminated by overlapping input sections. Suppose that N_1 filtered output points per section are desired and that the filter pulse response $g(n)$ has duration N_2 as before. The $N = N_1 + N_2 - 1$ point input sections can be formed according to the rule

$$f_{i,n} = f(n + iN_1 - N_2 + 1) \quad \text{for} \quad n = 0, \ldots, N-1 \quad (10.10.11)$$

As usual, we will choose N_1 so that N is a convenient integer for an FFT algorithm. The first point in the ith input section if $f_{i,0} = f(iN_1 - N_2 + 1)$ and the last point is $f_{i,N-1} = f[(i+1)N_1 - 1]$. The last point in the previous input section is $f_{i-1,N-1} = f(iN_1 - 1)$. Thus, $f_{i-1,n}$ and $f_{i,n}$ overlap by the $N_2 - 1$ points $f(iN_1 - N_2 + 1), \ldots, f(iN_1 - 1)$.

Suppose that the ith input section is processed by computing $F_{i,k} = \text{DFT}\{f_{i,n}\}$, $Y_{i,k} = F_{i,k}G_k$, and $y_{i,n} = \text{IDFT}\{Y_{i,k}\}$. Since no zeros were appended to the end of the input section, the first $N_2 - 1$ points in $y_{i,n}$ are corrupted by wraparound products. However, the remaining N_1 points $y_{i,n}$ for $n = N_2 - 1, \ldots, N-1$ are the desired filtered output points $y(iN_1), \ldots, y[(i+1)N_1 - 1]$. Thus, to complete the processing, we must delete the first $N_2 - 1$ points in $y_{i,n}$ and append the remaining N_1 points to the end of the previous output section. The overlap-save method is slightly simpler than the overlap-add method since no additions are necessary to piece the output sections together. The processing speeds of the overlap-save and overlap-add methods are essentially the same.

EFFICIENT FILTERING OF REAL SIGNALS

FFT programs are normally designed to operate on complex input sequences. Therefore, the FFT can be efficiently used by combining pairs of real sequences into single complex sequences. Suppose that a_n, b_n, and c_n are three real N-point sequences and we wish to compute the two cyclic convolutions $x_n = a_n \circledast c_n$ and $y_n = b_n \circledast c_n$. These convolutions can be computed simultaneously using the indirect FFT method by forming the complex sequence $d_n = a_n + jb_n$ and computing $C_k = \text{DFT}\{c_n\}$, $D_k = \text{DFT}\{d_n\}$, and $e_n = \text{IDFT}\{C_kD_k\}$. But

$$\text{IDFT}\{C_kD_k\} = c_n \circledast d_n = a_n \circledast c_n + jb_n \circledast c_n \quad (10.10.12)$$

Therefore

$$x_n = \text{Re } e_n \quad \text{and} \quad y_n = \text{Im } e_n \quad (10.10.13)$$

This result can be used in three ways. First, one real signal can be filtered at twice the rate indicated previously by letting c_n be the pulse response of the desired filter and a_n and b_n be adjacent input sections. Second, two different real signals can be filtered simultaneously by the pulse response c_n by letting a_n be a section of one signal and b_n a section of the other. Third, one real signal can be filtered by two different filters by letting a_n and b_n be the two desired pulse responses and c_n be the input signal.

The technique of performing convolution by the indirect FFT method is frequently called fast or high-speed convolution. However, it should always be kept in mind that filtering by nonrecursive filters with less than about 30 taps or by most desired recursive filters can be performed faster by the direct difference equation approach. The techniques presented in this section can also be modified for efficiently computing correlation functions for long sequences.

11

ESTIMATION OF POWER SPECTRAL DENSITIES

11.1 INTRODUCTION

Digital signal processing techniques have been widely applied to the estimation of power spectral densities. This chapter is only intended to be an introduction to the subject. Additional discussions can be found in a number of other books, including References [4,10,41,42,58, and 78].

Not surprisingly, we will find that the estimation of power spectral densities is closely related to the estimation of autocorrelation functions. We will see that good estimates of the autocorrelation function at lags that are small relative to the data record length can be obtained by forming the obvious time average of lagged products.

A seemingly obvious estimate of the power spectral density is the magnitude squared of the Z-transform of the data record evaluated on the unit circle. This function normalized by the length of the data record is called the periodogram. Surprisingly, the periodogram gives poor estimates of the power spectral density no matter how long the data record becomes and, in fact, actually fluctuates more wildly as the record length increases! Two methods for overcoming this problem are presented. The first method involves blocking the data record into a number of shorter length sections and then taking the arithmetic average of the periodograms of the individual sections at each frequency. This method has become popular since the discovery of the FFT. The second method is based on smoothing the periodogram of the entire data record by convolving it with a suitable function called the spectral window. We will see that the inverse Z-transform of this convolution is the product of an estimate of the autocorrelation function and a sequence called the lag window, which is the inverse transform of the spectral window. This second method was most popular before the discovery of the FFT because the lag window could be chosen to truncate the estimated autocorrelation function at a lag much less than the record length. The required autocorrelation function samples were estimated by directly averaging lagged products, since the indirect frequency domain approach was considered impractical. The FFT can now be used to compute a required set of mean lagged products efficiently. The power spectral density estimates are then usually computed by taking the transform of the product of the lag window and the estimated·autocorrelation function samples rather than by performing a frequency domain convolution.

Estimates of a power spectral density obtained by either averaging or smoothing periodograms have similar properties. In particular, a trade-off between frequency resolution and the variance of the estimates must be made for a fixed record length. High resolution implies larger variance and vice versa.

Finally, a third method for estimating power spectral densities based on modeling the observed data by a white Gaussian noise sequence passed through a digital filter of the form $1/D(z)$ where $D(z)$ is a finite degree polynomial in z^{-1} is discussed. Parameter estimation equations are derived by the maximum likelihood method. This model is known as an autoregressive model. The method is somewhat more complicated than the first two methods and has not been used extensively.

11.2 ESTIMATION OF AUTOCORRELATION FUNCTIONS

Power spectral densities and autocorrelation functions are related by the Fourier transform for continuous-time random processes and by the Z-transform for discrete-time random processes. Therefore, estimation of power spectral densities is intimately related to estimation of autocorrelation functions. In some cases, even though estimation of a power spectral density is the primary goal, an estimate of the autocorrelation function is desired as an intermediate step.

Suppose that a discrete-time, wide-sense stationary, random process $f(n)$ has the unknown autocorrelation function

$$R(m) = E\{f(n+m)f(n)\} \qquad (11.2.1)$$

and that an estimate of $R(m)$ using an observation of the N random variables $f(0), f(1), \ldots, f(N-1)$ is desired. Throughout this chapter, we assume that the sampling period is $T = 1$ to simplify the notation. One estimator for $R(m)$ is

$$R_N(m) = \frac{1}{N} \sum_{n=0}^{N-1-|m|} f(n+|m|)f(n) \qquad \text{for} \qquad |m| \le N - 1 \qquad (11.2.2)$$

This estimator is often called the *sample autocorrelation function.*
   The question of the goodness of the estimator $R_N(m)$ immediately arises. A good estimator should have an expected value that is close to the true value of the quantity being estimated. The difference between the true value of the quantity and the expected value of its estimator is called the *bias* of the estimator. The estimator is said to be *unbiased* if the bias is zero.
   The expected value of the sample autocorrelation function is

$$E\{R_N(m)\} = \frac{1}{N} \sum_{n=0}^{N-1-|m|} E\{f(n+|m|)f(n)\} = \frac{N-|m|}{N} R(m) \qquad \text{for} \qquad |m| \le N - 1$$

$$(11.2.3)$$

Therefore, $R_N(m)$ is a biased estimator. For fixed m, the bias converges to zero as N becomes infinite. An estimator with this property is said to be *asymptotically unbiased.* An unbiased estimator could have been obtained by using $N - |m|$ as the divisor in (11.2.2) instead of N. However, we will find that $R_N(m)$ is a more convenient estimator to use in power spectral density estimation.
   A good estimator should have a small variance in addition to having a small bias. The variance of the sample autocorrelation function can be computed as

$$\text{var}\{R_N(m)\} = E\{R_N^2(m)\} - E^2\{R_N(m)\} \qquad (11.2.4)$$

Multiplying the right-hand side of (11.2.2) by the same expression except with n replaced by k and interchanging expectation and summation, it follows that

$$E\{R_N^2(m)\} = \frac{1}{N^2} \sum_{n=0}^{N-1-|m|} \sum_{k=0}^{N-1-|m|} E\{f(n+|m|)f(n)f(k+|m|)f(k)\} \quad (11.2.5)$$

To proceed further, additional assumptions about the statistics of $f(n)$ must be made. Let us assume that $f(n)$ is a zero-mean Gaussian process. Equation 11.2.5 can be evaluated by using the fact that if X_1, X_2, X_3 and X_4 are zero-mean, jointly Gaussian, random variables, then

$$E\{X_1 X_2 X_3 X_4\} = E\{X_1 X_2\}E\{X_3 X_4\} + E\{X_1 X_3\}E\{X_2 X_4\}$$
$$+ E\{X_1 X_4\}E\{X_2 X_3\} \quad (11.2.6)$$

Combining results, it follows that

$$\text{var}\{R_N(m)\} = \frac{1}{N^2} \sum_{n=0}^{N-1-|m|} \sum_{k=0}^{N-1-|m|} R^2(n-k) + R(n-k+m)R(n-k-m)$$

$$(11.2.7)$$

Letting $n - k = i$ (11.2.7) can be written as

$$\text{var}\{R_N(m)\} = \frac{1}{N} \sum_{i=-(N-1-|m|)}^{N-1-|m|} \left[1 - \frac{|m|+|i|}{N} \right] [R^2(i) + R(i+m)R(i-m)]$$

(11.2.8)

In most cases, $R(m)$ is square summable so that we find from (11.2.8) that var$\{R_N(m)\}$ approaches zero essentially as $1/N$ for large N. For non-Gaussian processes, additional terms appear in (11.2.8) that are usually negligible.

An estimator is said to be *consistent* if it converges in probability to the true value of the quantity being estimated as N becomes infinite. Because of the Chebyshev inequality, a sufficient condition for an estimator to be consistent is that its bias and variance both converge to zero as N becomes infinite. Therefore, $R_N(m)$ is a consistent estimator of $R(m)$ for fixed finite m in practical situations.

11.3 USING THE FFT TO COMPUTE CORRELATION FUNCTIONS

In Section 10.9, we saw that the FFT can be used to efficiently compute the complete convolution or correlation of two sequences of duration N_1 samples by appending at least $N_1 - 1$ zeros to the ends of the original sequences to eliminate wraparound products. Obviously, the only change required to compute the sample autocorrelation function as defined by (11.2.2) is to divide the results obtained with the method of Section 10.9 by the factor N_1.

Frequently, values of the sample autocorrelation function are desired only for $|m| \le M$ where M is much smaller than the length of the observed data record. The reason is that $R_N(m)$ is a reliable estimate of $R(m)$ only when m is moderately small relative to the record length N as we have seen in the previous section. When N is large, it is not efficient and often not practical to use the method described in Section 10.9 to compute the entire sample autocorrelation function only to use the values for m small. We will now examine an efficient method for computing $R_N(m)$ for a small number of lags relative to N by using a sectioning technique similar to the overlap-save method presented in Section 10.10 for filtering long data sequences.

Suppose that the observed data sequence is $f(n)$ for $n = 0, \ldots, N-1$ with $N = KL$. Using the convention that $f(n) = 0$ for $n < 0$ and $n > N-1$, the sample autocorrelation function can be written as

$$R_N(m) = \frac{1}{N} \sum_{n=0}^{N-1} f(n+m)f(n)$$

(11.3.1)

This sum can be decomposed and written as

$$R_N(m) = \frac{1}{N} \sum_{i=0}^{K-1} \sum_{n=0}^{L-1} f(n+iL+m)f(n+iL) \qquad (11.3.2)$$

or as

$$R_N(m) = \frac{1}{N} \sum_{i=0}^{K-1} v_i(m) \qquad (11.3.3)$$

where

$$v_i(m) = \sum_{n=0}^{L-1} f(n+iL+m)f(n+iL) \qquad (11.3.4)$$

Let us assume that $R_N(m)$ is required only for $|m| \leq M$. Then, it is only necessary to compute $R_N(m)$ for $0 \leq m \leq M$ since $R_N(-m) = R_N(m)$. From (11.3.4) we can see that the partial correlation function $v_i(m)$ for $0 \leq m \leq M$ can be computed from the data $f(iL), \ldots, f[(i+1)L+M-1]$.

The required values of $R_N(m)$ can be efficiently computed using the FFT. The first step is to form the sequences

$$a_{i,n} = f(n+iL) \qquad \text{for} \qquad 0 \leq n \leq P-1 \qquad (11.3.5)$$

and

$$b_{i,n} = \begin{cases} f(n+iL) & \text{for} & 0 \leq n \leq L-1 \\ 0 & \text{for} & L \leq n \leq P-1 \end{cases} \qquad (11.3.6)$$

where P is chosen as an integer convenient for an FFT algorithm, usually a power of two, such that

$$P \geq L+M \qquad (11.3.7)$$

As usual when working with the FFT, we will use the convention that $a_{i,n+P} = a_{i,n}$ and $b_{i,n+P} = b_{i,n}$ for all n. Notice that when P is chosen to satisfy (11.3.7), there are at least M zeros at the end of the sequence $b_{i,n}$. Therefore

$$v_{i,m} = \sum_{n=0}^{P-1} a_{i,n+m}b_{i,n} = v_i(m) \qquad \text{for} \qquad 0 \leq m \leq M \qquad (11.3.8)$$

so that, according to (11.3.3)

$$R_N(m) = \frac{1}{N} \sum_{i=0}^{K-1} v_{i,m} \qquad \text{for} \qquad 0 \leq m \leq M \qquad (11.3.9)$$

The values obtained from the right-hand side of (11.3.9) for $M+1 \leq m \leq P-1$ contain wraparound products and should be discarded.

The right-hand side of (11.3.9) can be efficiently computed using the FFT by observing that

$$\frac{1}{N} \sum_{i=0}^{K-1} v_{i,n} = \frac{1}{N} \text{IDFT} \left\{ \sum_{i=0}^{K-1} V_{i,k} \right\} \qquad (11.3.10)$$

where $V_{i,k} = \text{DFT}\{v_{i,n}\}$. Letting $A_{i,k} = \text{DFT}\{a_{i,n}\}$ and $B_{i,k} = \text{DFT}\{b_{i,n}\}$, we see from (10.5.4) that

$$V_{i,k} = A_{i,k}\bar{B}_{i,k} \qquad \text{for} \qquad i = 0, \ldots, K-1 \qquad (11.3.11)$$

For real data sequences (11.3.10) can be evaluated using $K+1$ P-point FFTs. Using the pairing technique described in Section 10.8, the $A_{i,k}$'s and $B_{i,k}$'s can be computed using K FFTs. The IDFT in (11.3.10) accounts for the additional FFT.

Rader [113] discovered that it is particularly convenient to choose M, L, and P so that $M = L = P/2$. With this choice, we see from (11.3.6) that the first half of the sequence $b_{i,n}$ consists of M data points and the second half consists of M zeros. Also

$$a_{i,n} = b_{i,n} + b_{i+1,n-M} \qquad (11.3.12)$$

According to (10.4.6), the cyclic time shift property

$$\text{DFT}\{b_{i+1,n-M}\} = e^{-j2\pi Mk/(2M)} B_{i+1,k} = (-1)^k B_{i+1,k} \qquad (11.3.13)$$

Therefore,

$$A_{i,k} = B_{i,k} + (-1)^k B_{i+1,k} \qquad (11.3.14)$$

and

$$V_{i,k} = [B_{i,k} + (-1)^k B_{i+1,k}]\bar{B}_{i,k} \qquad (11.3.15)$$

When $f(n)$ is real, $B_{i,k}$ and, consequently, $V_{i,k}$ for $i = 0, \ldots, K-1$ can be computed using only $K/2$ FFTs when K is even or $(K-1)/2 + 1$ FFTs when K is odd by using the pairing technique described in Section 10.8. This allows nearly a 50 percent reduction in computation over the case where M is not chosen equal to L and K is large.

When N is large and $M \ll N$, direct computation of $R_N(m)$ for $0 \le m \le M$ requires approximately $N(M+1)$ multiplications: The computation required by Rader's method when K is large is predominantly the calculation of essentially $K/2$ $2M$-point FFTs. Since $N = KL = KM$, the computation time for Rader's method is roughly proportional to

$$(K/2)2M \log_2(2M) = N(1 + \log_2 M)$$

Thus, indirect computation using the FFT quickly becomes more efficient than direct computation as M increases.

Cross-correlation functions can also be efficiently computed using the methods described in this section.

11.4 THE PERIODOGRAM

Suppose that $f(n)$ is a wide-sense stationary, discrete-time, random process with the autocorrelation function $R(m)$ and sampled power spectral density

$$S^*(\omega) = \sum_{m=-\infty}^{\infty} R(m)e^{-j\omega m}$$

We will now begin to examine the problem of estimating $S^*(\omega)$ from an observation of $f(0), \ldots, f(N-1)$. To include an additional degree of flexibility, suppose that the observed sequence is modified to form the signal

$$f_N(n) = a(n)f(n) \tag{11.4.1}$$

where $a(n) = 0$ for $n < 0$ and $n \geq N$. The sequence $a(n)$ is frequently called the *data window*. Additional desirable properties of $a(n)$ will be discussed shortly. We will denote the Z-transform of $f_N(n)$ by

$$F_N(z) = \sum_{n=-\infty}^{\infty} f_N(n)z^{-n} = \sum_{n=0}^{N-1} a(n)f(n)z^{-n} \tag{11.4.2}$$

and let

$$F_N^*(\omega) = F_N(e^{j\omega}) \tag{11.4.3}$$

The sample autocorrelation function for the modified observed sequence can be written as

$$\tilde{R}_N(n) = \frac{1}{N} \sum_{k=-\infty}^{\infty} f_N(k+n)f_N(k) = \frac{1}{N} f_N(n) * f_N(-n) \tag{11.4.4}$$

Notice that, as a result of the truncation property of the data window, $\tilde{R}_N(n) = 0$ for $|n| \geq N$. Denoting the Z-transform of $\tilde{R}_N(n)$ by $I_N(z)$, it follows from the Convolution Theorem and Time Reversal Theorem that

$$I_N(z) = \sum_{n=-\infty}^{\infty} \tilde{R}_N(n)z^{-n} = \frac{1}{N} F_N(z)F_N(z^{-1}) \tag{11.4.5}$$

and so

$$I_N^*(\omega) = \frac{1}{N} F_N^*(\omega)F_N^*(-\omega) = \frac{1}{N}|F_N^*(\omega)|^2 \tag{11.4.6}$$

The function $I_N^*(\omega)$ is known as the *periodogram* for $f_N(n)$.

For the moment, let us choose as the data window the rectangular function

$$a_r(n) = \begin{cases} 1 & \text{for} \quad 0 \leq n \leq N-1 \\ 0 & \text{elsewhere} \end{cases} \tag{11.4.7}$$

Then $\tilde{R}_N(n) = R_N(n)$. An obvious choice for an estimator of $R(0)$, the expected value of the power in $f(n)$, is

$$R_N(0) = \frac{1}{N} \sum_{n=0}^{N-1} f^2(n) = \frac{1}{2\pi} \int_{-\pi}^{\pi} I_N^*(\omega) \, d\omega \qquad (11.4.8)$$

In Section 11.2, we saw that $R_N(0)$ is usually an unbiased and consistent estimator of $R(0)$. In light of (11.4.6) and (11.4.8), it is tempting to conclude that the periodogram becomes a good estimator of $S^*(\omega)$ as N becomes large. We will see shortly that this conclusion is false, that, in fact, the periodogram is a poor estimator for $S^*(\omega)$! We will find that, although the periodogram is an asymptotically unbiased estimator, its variance remains large.

11.5 EXPECTED VALUE OF THE PERIODOGRAM

From (11.4.5) we find that the expected value of the periodogram is

$$E\{I_N^*(\omega)\} = \sum_{n=-\infty}^{\infty} E\{\tilde{R}_N(n)\} e^{-j\omega n} \qquad (11.5.1)$$

and from (11.4.1) and (11.4.4), we see that

$$E\{\tilde{R}_N(n)\} = \frac{1}{N} \sum_{k=-\infty}^{\infty} a(k+n)a(k)E\{f(k+n)f(k)\} = R(n)q(n) \qquad (11.5.2)$$

where

$$q(n) = \frac{1}{N} \sum_{k=-\infty}^{\infty} a(k+n)a(k) \qquad (11.5.3)$$

The Z-transform of $q(n)$ evaluated on the unit circle is

$$Q^*(\omega) = \frac{1}{N} |A^*(\omega)|^2 \qquad (11.5.4)$$

The right-hand side of (11.5.1) is just the Z-transform of $R(n)q(n)$ evaluated on the unit circle, so according to (4.7.18)

$$E\{I_N^*(\omega)\} = \frac{1}{2\pi} \int_{-\pi}^{\pi} Q^*(\lambda) S^*(\omega - \lambda) \, d\lambda \qquad (11.5.5)$$

Therefore, the expected value of the periodogram is equal to the true power spectral density observed through the *spectral window* $Q^*(\omega)$. In general, the periodogram is a biased estimator.

For the rectangular data window $a_r(n)$ defined by (11.4.7),

$$q_r(n) = \begin{cases} 1 - \dfrac{|n|}{N} & \text{for} \quad |n| \le N-1 \\ 0 & \text{elsewhere} \end{cases} \tag{11.5.6}$$

$$A_r^*(\omega) = \sum_{n=0}^{N-1} e^{-j\omega n} = e^{-j\omega(N-1)/2} \frac{\sin \omega N/2}{\sin \omega/2} \tag{11.5.7}$$

and

$$Q_r^*(\omega) = \frac{1}{N} \left[\frac{\sin \omega N/2}{\sin \omega/2} \right]^2 \tag{11.5.8}$$

Functions similar to $A_r^*(\omega)$ arose in the studies of Gibbs' phenomenon and frequency sampling filters in Chapter 8. The general nature of $Q_r^*(\omega)$ can be determined by observing that

$$Q_r^*(k2\pi/N) = \begin{cases} N & \text{for} \quad k \text{ a multiple of } N \\ 0 & \text{elsewhere} \end{cases} \tag{11.5.9}$$

and

$$\frac{1}{2\pi} \int_{-\pi}^{\pi} Q_r^*(\omega)\, d\omega = q_r(0) = 1 \tag{11.5.10}$$

With a little additional work, it can be shown that

$$\lim_{N\to\infty} Q_r^*(\omega) = 2\pi \sum_{k=-\infty}^{\infty} \delta(\omega - 2\pi k) \tag{11.5.11}$$

Therefore, we find from (11.5.5) that

$$\lim_{N\to\infty} E\{I_N^*(\omega)\} = S^*(\omega) \tag{11.5.12}$$

Consequently, the periodogram is an asymptotically unbiased estimator of $S^*(\omega)$ when a rectangular data window is used.

In terms of minimizing bias, the considerations involved in choosing a data window are essentially the same as those involved in choosing a window function for the design of digital filters by the Fourier series method presented in Section 8.8. The data window should be chosen so that the spectral window has a narrow central lobe and small amplitude sidelobes. The spectral window should converge to the train of impulse functions given by the right-hand side of (11.5.11) as N becomes large so that the estimator is asymptotically unbiased. For each finite N, the window should be normalized so that

$$\frac{1}{2\pi} \int_{-\pi}^{\pi} Q^*(\omega)\, d\omega = q(0) = \frac{1}{N} \sum_{n=0}^{N-1} a^2(n) = 1 \tag{11.5.13}$$

From (11.5.5), we can see that this choice results in unbiased estimates when $S^*(\omega)$ is flat. Since $a(n)$ has duration N, the sum in (11.5.13) can also be computed as

$$\frac{1}{N}\sum_{n=0}^{N-1}a^2(n)=\frac{1}{N^2}\sum_{k=0}^{N-1}|A_k|^2 \tag{11.5.14}$$

where $A_k = \mathrm{DFT}\{a(n)\}$. In addition, the data window is usually chosen to have even symmetry, that is, with $a(N-1-n)=a(n)$ for $n=0,\ldots,N-1$, and to monotonically decrease from a peak at the center of the sequence.

We observed in Section 8.8 that a compromise must be made between the width of the central lobe and the amplitude of the sidelobes of the spectral window. The width of the central lobe of the spectral window corresponding to the rectangular data window is $4\pi/N=2\omega_s/N$. Although this width is small, the amplitudes of the sidelobes are relatively large and decay slowly. The amplitude of the first sidelobe of $|A_r^*(\omega)|$ is approximately 20 percent of the amplitude of the central lobe. This can result in significant bias, especially in the vicinity of jumps or large peaks in $S^*(\omega)$ resulting from leakage through the sidelobes.

A variety of better windows have been proposed. One popular data window is the Hanning window

$$a_{Hn}(n)=c_{Hn}0.5\left\{1+\cos\left[\left(n-\frac{N-1}{2}\right)\frac{2\pi}{N}\right]\right\}a_r(n) \tag{11.5.15}$$

with

$$A_{Hn}^*(\omega)=c_{Hn}e^{-j\omega(N-1)/2}[0.5P_N^*(\omega)+0.25P_N^*(\omega-2\pi/N)+0.25P_N^*(\omega+2\pi/N)] \tag{11.5.16}$$

where

$$P_N^*(\omega)=\frac{\sin\omega N/2}{\sin\omega/2} \tag{11.5.17}$$

and $c_{Hn}=(3/8)^{-1/2}$ is the normalization constant required to satisfy (11.5.13). A very similar window is the Hamming window

$$a_H(n)=c_H\left\{0.54+0.46\cos\left[\left(n-\frac{N-1}{2}\right)\frac{2\pi}{N}\right]\right\}a_r(n) \tag{11.5.18}$$

with

$$A_H^*(\omega)=c_He^{-j\omega(N-1)/2}[0.54P_N^*(\omega)+0.23P_N^*(\omega-2\pi/N)+0.23P_N^*(\omega+2\pi/N)] \tag{11.5.19}$$

and $1/c_H^2=(0.54)^2+(0.46)^2/2=0.3974$. The width of the central lobe of either the Hanning or Hamming spectral window is $8\pi/N=4\omega_s/N$. The amplitudes of the sidelobes of $|A_{Hn}^*(\omega)|$ are no more than 3 percent of the amplitude of the central lobe, and the amplitudes of the sidelobes of $|A_H^*(\omega)|$ are no more than 1 percent of the amplitude of the central lobe. On the other hand, the amplitudes of the sidelobes of the Hanning window fall off more rapidly than those of the Hamming window. The choice between the two is a matter of preference.

Another popular data window is the triangular or Bartlett window. One form of the window is

$$a_B(n) = \begin{cases} c_B(n+1) & \text{for} & 0 \le n \le (N-1)/2 \\ c_B(N-n) & \text{for} & (N-1)/2 \le n \le N-1 \\ 0 & \text{elsewhere} \end{cases} \qquad (11.5.20)$$

Assuming that N is even, it follows from the identity

$$\sum_{k=1}^{L} k^2 = L(L+1)(2L+1)/6 \qquad (11.5.21)$$

that the value of c_B that satisfies (11.5.13) is $c_B = [(N+1)(N+2)/12]^{-1/2}$. The spectral window can be determined by observing that, for N even, $a_B(n)$ is the convolution of the rectangular sequences

$$h_1(n) = \begin{cases} c_B & \text{for} & 0 \le n \le N/2 - 1 \\ 0 & \text{elsewhere} \end{cases}$$

and

$$h_2(n) = \begin{cases} 1 & \text{for} & 0 \le n \le N/2 \\ 0 & \text{elsewhere} \end{cases}$$

Thus, for N even

$$A_B^*(\omega) = H_1^*(\omega)H_2^*(\omega) = c_B e^{-j\omega(N-1)/2} \frac{\sin(\omega N/4) \sin[\omega(N+2)/4]}{\sin^2(\omega/2)} \qquad (11.5.22)$$

For N moderately large, $N \cong N+2$, and

$$A_B^*(\omega) \cong c_B e^{-j\omega(N-1)/2} \frac{\sin^2(\omega N/4)}{\sin^2(\omega/2)} \qquad (11.5.23)$$

Since the zeros of $\sin(\omega N/4)$ occur when ω is a multiple of $4\pi/N$, we find that the width of the central lobe of $A_B^*(\omega)$ is also $8\pi/N = 4\omega_s/N$. The peak sidelobe amplitude for $A_B^*(\omega)$ is approximately 4 percent of the peak amplitude of the central lobe.

The more complicated Blackman and Kaiser windows presented in Section 8.8 can be modified in a similar manner and used as data windows.

We can see from (11.5.5) that, even with a good spectral window, large biases will occur in the vicinity of sharp peaks in $S^*(\omega)$. This problem can be helped by preprocessing the data to eliminate any known discrete spectral components. A high-pass filter can be used to eliminate dc components and slow trends.

The resolution of a spectrum estimator can be loosely defined as the minimum frequency separation between two discrete spectral components, that is, impulses, in $S^*(\omega)$ that still gives rise to two distinct peaks in the estimate. From (11.5.5), we can see that the resolution of the periodogram of the modified data sequence is approximately half the width of the central lobe of the spectral window $Q^*(\omega)$. This width is inversely proportional to the record length N.

11.6 VARIANCE OF THE PERIODOGRAM

The covariance of the periodogram evaluated at the frequencies ω_1 and ω_2 can be computed using the formula

$$\text{cov}\{I_N^*(\omega_1),\, I_N^*(\omega_2)\} = E\{I_N^*(\omega_1) I_N^*(\omega_2)\} - E\{I_N^*(\omega_1)\} E\{I_N^*(\omega_2)\} \quad (11.6.1)$$

As a first step, we will derive an alternate formula for $E\{I_N^*(\omega)\}$. According to (11.4.2) and (11.4.6)

$$E\{I_N^*(\omega)\} = E\left\{ \frac{1}{N} \sum_{n=-\infty}^{\infty} a(n) f(n) e^{-j\omega n} \sum_{k=-\infty}^{\infty} a(k) f(k) e^{j\omega k} \right\}$$

$$= \frac{1}{N} \sum_{n=-\infty}^{\infty} \sum_{k=-\infty}^{\infty} a(n) a(k) R(n-k) e^{-j\omega(n-k)} \quad (11.6.2)$$

Again using (11.4.2) and (11.4.6), we find that

$$E\{I_N^*(\omega_1) I_N^*(\omega_2)\} = \frac{1}{N^2} \sum_n \sum_k \sum_p \sum_q a(n) a(k) a(p) a(q) E\{f(n) f(k) f(p) f(q)\}$$

$$\times e^{-j\omega_1(n-k)} e^{-j\omega_2(p-q)} \quad (11.6.3)$$

where for convenience, it will be understood that all sums extend from $-\infty$ to ∞. To proceed further, the statistics of $f(n)$ must be specified. Let us assume that $f(n)$ is a zero-mean Gaussian process. Using (11.2.6) to evaluate the expectation on the right-hand side of (11.6.3), rearranging terms, and using (11.6.2) yields

$$E\{I_N^*(\omega_1) I_N^*(\omega_2)\} = E\{I_N^*(\omega_1)\} E\{I_N^*(\omega_2)\}$$

$$+ \left| \frac{1}{N} \sum_n \sum_p a(n) a(p) R(n-p) e^{-j(\omega_1 n + \omega_2 p)} \right|^2$$

$$+ \left| \frac{1}{N} \sum_n \sum_q a(n) a(q) R(n-q) e^{-j(\omega_1 n - \omega_2 q)} \right|^2 \quad (11.6.4)$$

A more informative version of (11.6.4) can be obtained by observing that

$$\sum_n \sum_p a(n)a(p)R(n-p)e^{-j(\omega_1 n+\omega_2 p)} = \sum_p a(p)\left[\sum_n a(n)R(n-p)e^{-j\omega_1 n}\right]e^{-j\omega_2 p}$$

$$= \sum_p a(p)\frac{1}{2\pi}\int_{-\pi}^{\pi} S^*(\lambda)A^*(\omega_1-\lambda)e^{-j\lambda p}\,d\lambda e^{-j\omega_2 p}$$

$$= \frac{1}{2\pi}\int_{-\pi}^{\pi} S^*(\lambda)\bar{A}^*(\lambda-\omega_1)\sum_p a(p)e^{-j(\lambda+\omega_2)p}\,d\lambda$$

$$= \frac{1}{2\pi}\int_{-\pi}^{\pi} S^*(\lambda)\bar{A}^*(\lambda-\omega_1)A^*(\lambda+\omega_2)\,d\lambda$$

$$(11.6.5)$$

A similar expression for the double sum on the far right-hand side of (11.6.4) can be obtained by simply replacing ω_2 by $-\omega_2$ in (11.6.5). Therefore

$$\text{cov}\{I_N^*(\omega_1), I_N^*(\omega_2)\} = \left|\frac{1}{2\pi}\int_{-\pi}^{\pi} S^*(\lambda)\frac{1}{N}\bar{A}^*(\lambda-\omega_1)A^*(\lambda+\omega_2)\,d\lambda\right|^2$$

$$+ \left|\frac{1}{2\pi}\int_{-\pi}^{\pi} S^*(\lambda)\frac{1}{N}\bar{A}^*(\lambda-\omega_1)A^*(\lambda-\omega_2)\,d\lambda\right|^2 \quad (11.6.6)$$

Letting $\omega_1 = \omega_2$ gives

$$\text{var}\{I_N^*(\omega_1)\} = \left|\frac{1}{2\pi}\int_{-\pi}^{\pi} S^*(\lambda)\frac{1}{N}\bar{A}^*(\lambda-\omega_1)A^*(\lambda+\omega_1)\,d\lambda\right|^2 + [E\{I_N^*(\omega_1)\}]^2$$

$$(11.6.7)$$

Thus the variance of the periodogram is greater than or equal to the square of its mean no matter how large N becomes.

If a good data window is chosen, then we can assume that $A^*(\omega) \cong 0$ for $B = \alpha 2\pi/N < |\omega| < \pi$ where $2B$ is essentially the width of the central lobe of the spectral window. When $B < \omega_1 < \pi - B$, $\bar{A}^*(\lambda-\omega_1)A^*(\lambda+\omega_1) \cong 0$ in (11.6.7). Thus

$$\text{var}\{I_N^*(\omega_1)\} \cong [E\{I_N^*(\omega_1)\}]^2 \quad \text{for} \quad B < \omega_1 < \pi - B \quad (11.6.8)$$

If both ω_1 and ω_2 are restricted to the interval $(0, \pi - B)$, then it follows from (11.6.6) that

$$\text{cov}\{I_N^*(\omega_1), I_N^*(\omega_2)\} \cong 0 \quad \text{for} \quad |\omega_2-\omega_1| > 2B \quad (11.6.9)$$

Thus values of the periodogram at frequencies separated by at least $2B$ are uncorrelated for all practical purposes. As N becomes large, B becomes small. Therefore, we would expect the periodogram to fluctuate more rapidly as N increases since the frequency separation between uncorrelated values decreases.

Example 11.6.1

Suppose that $f(n)$ is a white noise sequence with the sampled power spectral density $S^*(\omega) = \sigma^2$ and that a rectangular data window is used. Then (11.6.6) reduces to

$$\text{cov}\{I_N^*(\omega_1), I_N^*(\omega_2)\} = \sigma^4 \left| \frac{1}{2\pi} \int_{-\pi}^{\pi} \frac{1}{N} \bar{A}_r^*(\lambda - \omega_1) A_r^*(\lambda + \omega_2) \, d\lambda \right|^2$$

$$+ \sigma^4 \left| \frac{1}{2\pi} \int_{-\pi}^{\pi} \frac{1}{N} \bar{A}_r^*(\lambda - \omega_1) A_r^*(\lambda - \omega_2) \, d\lambda \right|^2$$

It can be shown using Parseval's Theorem that

$$\frac{1}{2\pi} \int_{-\pi}^{\pi} \frac{1}{N} \bar{A}_r^*(\lambda - \omega_1) A_r^*(\lambda + \omega_2) \, d\lambda = \frac{1}{N} \sum_n |a_r(n)|^2 e^{-j(\omega_1 + \omega_2)n}$$

$$= \frac{1}{N} \sum_{n=0}^{N-1} e^{-j(\omega_1 + \omega_2)n} = e^{-j(\omega_1 + \omega_2)(N-1)/2} \frac{\sin[(\omega_1 + \omega_2)N/2]}{N \sin[(\omega_1 + \omega_2)/2]}$$

Therefore

$$\text{cov}\{I_N^*(\omega_1), I_N^*(\omega_2)\} = \sigma^4 \left\{ \frac{\sin^2[(\omega_1 + \omega_2)N/2]}{N^2 \sin^2[(\omega_1 + \omega_2)/2]} + \frac{\sin^2[(\omega_1 - \omega_2)N/2]}{N^2 \sin^2[(\omega_1 - \omega_2)/2]} \right\}$$

$$\tag{11.6.10}$$

Letting $\omega_1 = \omega_2$ yields

$$\text{var}\{I_N^*(\omega_1)\} = \sigma^4 \left\{ 1 + \frac{\sin^2 \omega_1 N}{N^2 \sin^2 \omega_1} \right\} \tag{11.6.11}$$

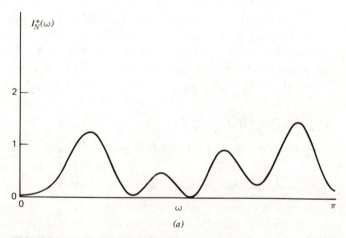

FIGURE 11.6.1.   Periodograms for a Gaussian white noise sequence with a rectangular data window and $N = 16$, 32, and 64. (a) $N = 16$; (b) $N = 32$; (c) $N = 64$.

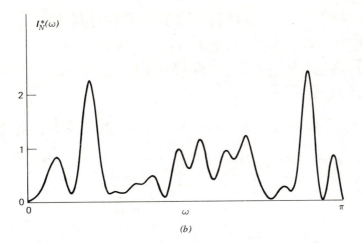

(b)

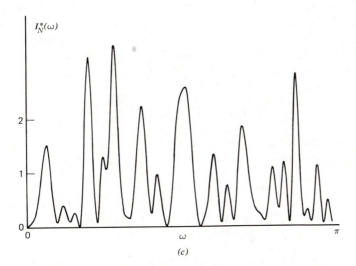

(c)

FIGURE 11.6.1.   (cont'd.)

If $\omega_1 = k2\pi/N$, $\omega_2 = p2\pi/N$, and k and p are integers such that $k+p$ and $k-p$ are not multiples of N, then we find from (11.6.10) that cov $\{I_N^*(\omega_1),\ I_N^*(\omega_2)\} = 0$.

The increasing fluctuations in the periodogram with N are illustrated in Figs. 11.6.1a, b, and c. These periodograms were computed from $N = 16$, 32, and 64 points taken from a computer-generated white Gaussian noise sequence with unity variance. The rectangular data window was used. ◄

11.7 ESTIMATION OF POWER SPECTRAL DENSITIES BY AVERAGING SEPARATE PERIODOGRAMS

If $X_1, X_2, \ldots, X_L$ are uncorrelated random variables, each with expected value μ and variance σ^2, then their arithmetic mean $(X_1 + X_2 + \cdots + X_L)/L$ has expected value μ and variance σ^2/L. This fact suggests that a spectrum estimator with its variance reduced by essentially a factor of L over the periodogram can be obtained by splitting up the N-point observed data sequence into L nonoverlapping M-point sections and averaging the periodograms of the individual sections. If the autocorrelation function $R(m)$ becomes negligible for m small relative to M, then we would expect that the periodograms of separate sections are nearly uncorrelated. This approach with a rectangular data window is often credited to Bartlett [4]. Welch [137] has suggested that additional improvement can be obtained by using a better data window and allowing overlapping sections.

To be more specific, suppose that the observed data is $f(n)$ for $n = 0, \ldots, N-1$. Let us form the L M-point sections

$$f_{M,p}(n) = a(n)f(n+pK) \qquad \text{for} \qquad \begin{array}{l} p = 0, \ldots, L-1 \\ n = 0, \ldots, M-1 \end{array} \qquad (11.7.1)$$

where $a(n)$ is the data window and K is a positive integer. To include only observed data, we must require that

$$(L-1)K + M \leq N \qquad (11.7.2)$$

The sections overlap if $K < M$ and do not if $K \geq M$. It will be convenient to let n take on any integral value and consider $a(n)$ to be identically zero for $n < 0$ and $n > M-1$. The corresponding periodograms are

$$I_{M,p}^*(\omega) = \frac{1}{M}|F_{M,p}^*(\omega)|^2 \qquad \text{for} \qquad p = 0, \ldots, L-1 \qquad (11.7.3)$$

where

$$F_{M,p}^*(\omega) = \sum_{n=0}^{M-1} a(n)f(n+pK)e^{-j\omega n} \qquad (11.7.4)$$

Assuming that $a(n)$ has been normalized to satisfy (11.5.13) with N replaced by M, the desired spectrum estimator is

$$\hat{S}^*(\omega) = \frac{1}{L}\sum_{p=0}^{L-1} I_{M,p}^*(\omega) \qquad (11.7.5)$$

We will call this the *averaged periodogram* estimator.

If $f(n)$ is a stationary random process, the expected values of the individual periodograms are all identical. Therefore, $E\{\hat{S}^*(\omega)\}$ is given by (11.5.5) with N replaced by M. The bias of $\hat{S}^*(\omega)$ is greater than that of $I_N^*(\omega)$ since the width of the spectral window is greater by the factor N/M.

To compute the variance of $\hat{S}^*(\omega)$, first let us observe that

$$\hat{S}^*(\omega) - E\{\hat{S}^*(\omega)\} = \frac{1}{L} \sum_{n=0}^{L-1} [I_{M,n}^*(\omega) - E\{I_{M,n}^*(\omega)\}] \tag{11.7.6}$$

Multiplying the right-hand side of (11.7.6) by the same sum but with n replaced by k and then taking expected values gives

$$\text{var}\{\hat{S}^*(\omega)\} = \frac{1}{L^2} \sum_{n=0}^{L-1} \sum_{k=0}^{L-1} \text{cov}\{I_{M,n}^*(\omega), I_{M,k}^*(\omega)\} \tag{11.7.7}$$

If $f(n)$ is a stationary random process, the covariances in (11.7.7) depend only on $n - k$. Letting

$$\Gamma_r(\omega) = \text{cov}\{I_{M,n}^*(\omega), I_{M,n+r}^*(\omega)\} \tag{11.7.8}$$

(11.7.7) can be converted into the single sum

$$\text{var}\{\hat{S}^*(\omega)\} = \frac{1}{L} \text{var}\{I_M^*(\omega)\} \sum_{r=-(L-1)}^{L-1} \left[1 - \frac{|r|}{L}\right] \frac{\Gamma_r(\omega)}{\Gamma_0(\omega)} \tag{11.7.9}$$

The quantity $\Gamma_r(\omega)/\Gamma_0(\omega)$ is the correlation coefficient for $I_{M,n}^*(\omega)$ and $I_{M,n+r}^*(\omega)$. If the correlation between periodograms is small, then we see from (11.7.9) that the variance of the arithmetic average of L periodograms of different M-point data sections is essentially reduced by a factor of L over that of the periodogram of a single M-point section. If M is large enough so that $S^*(\omega)$ is smooth relative to the spectral window, then it follows from (11.5.5) and (11.6.7) that $\text{var}\{I_M^*(\omega)\} \cong \text{var}\{I_N^*(\omega)\}$ for Gaussian processes. Results for other types of random processes are similar. Therefore, the variance of $\hat{S}^*(\omega)$ is also less than that of $I_N^*(\omega)$ by nearly a factor of L when the separate periodograms are uncorrelated.

The price paid for the decrease in variance is a decrease in resolution by a factor of N/M and an increase in bias. This trade-off between bias and variance for N fixed is an inherent property of spectrum estimators.

With the same methods used in deriving (11.6.4), it can be shown that when $f(n)$ is a zero-mean Gaussian process with autocorrelation function $R(m)$

$$\Gamma_r(\omega) = \left| \frac{1}{M} \sum_{n=0}^{M-1} \sum_{p=0}^{M-1} a(n)a(p)R(n-p-rK)e^{-j\omega(n+p)} \right|^2$$

$$+ \left| \frac{1}{M} \sum_{n=0}^{M-1} \sum_{p=0}^{M-1} a(n)a(p)R(n-p-rK)e^{-j\omega(n-p)} \right|^2 \tag{11.7.10}$$

This result can be converted into a form similar to (11.6.6).

In the special case of white noise with $R(m) = \sigma^2 \delta_{m,0}$ (11.7.10) reduces to

$$\Gamma_r(\omega) = \sigma^4 \left| \frac{1}{M} \sum_{n=0}^{M-1} a(n)a(n-rK)e^{-j2\omega n} \right|^2 + \sigma^4 \left| \frac{1}{M} \sum_{n=0}^{M-1} a(n)a(n-rK) \right|^2$$

(11.7.11)

If $K \geq M$, the data sections do not overlap and we see that $\Gamma_r(\omega) = 0$ for $r \neq 0$ and $\text{var}\{\hat{S}^*(\omega)\} = \text{var}\{I_M^*(\omega)\}/L$ as expected. The correlation between periodograms increases as overlap is allowed. On the other hand, the number of M-point sections that can be formed also increases. These two effects tend to counteract each other as far as the variance of $\hat{S}^*(\omega)$ is concerned. With a good data window, the increased number of sections has the stronger effect until the overlap gets quite large. Using the white noise case, Welch has suggested that a 50 percent overlap is a reasonable choice for reducing the variance when N and M are fixed.

Example 11.7.1

To illustrate the type of results that can be expected using the averaged periodogram estimator, a zero-mean Gaussian random process with the sampled power spectral density

$$S(z) = \frac{1}{(1 - az^{-1})(1 - az)} \qquad a = 1 - \sqrt{0.1}$$

was generated by passing a zero-mean white Gaussian noise sequence with

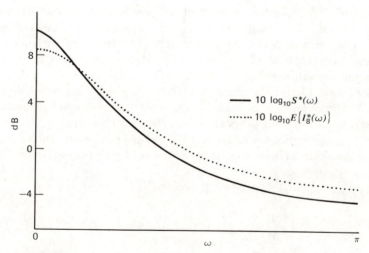

FIGURE 11.7.1. $S^*(\omega)$ and $E\{I_8^*(\omega)\}$ for Example 11.7.1.

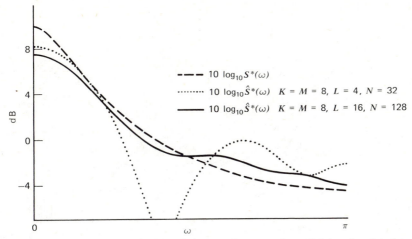

FIGURE 11.7.2. Averaged periodogram estimator using adjoining 8-point data sections for Example 11.7.1.

unity variance through the filter $1/(1 - az^{-1})$. For simplicity, the rectangular data window was chosen for this example. $E\{I_8^*(\omega)\}$ and $S^*(\omega)$ are shown in Fig. 11.7.1 to illustrate the bias that can be expected for $M = 8$. The function $E\{I_8^*(\omega)\}$ was computed using (11.5.1) and (11.5.2) rather than by the convolution integral in (11.5.5).

The averages of 4 and 16 periodograms of directly adjacent 8-point data sections are shown in Fig. 11.7.2. The effect of the larger variance for $L = 4$ is apparent. Comparing the $L = 16$ curve with Fig. 11.7.1, it appears that a large part of the error in this estimate may be attributable to bias rather than variance.

The averages of 2 and 8 periodograms of directly adjacent 16-point data sections are shown in Fig. 11.7.3. Again the effect of L is apparent. However, both the $L = 2$ and $L = 8$ curves fluctuate significantly and it is tempting to conjecture that the errors are mainly due to variance.

Comparing Figs. 11.7.2 and 11.7.3, it appears that the best choice in this problem is to sacrifice bias for reduced variance and use the $L = 16$, $M = 8$ estimator rather than the $L = 8$, $M = 16$ estimator. These are the types of judgments that must be made when dealing with actual data. Of course, the true power spectral density is unknown in practice. However, the spectral window is known and some a priori information about $S^*(\omega)$ is also usually known. By varying M and L and using a priori knowledge, a good choice can be usually made. ◀

The averaged periodogram estimator is particularly well-suited for use with the FFT. The periodograms can be evaluated at any number of equally spaced

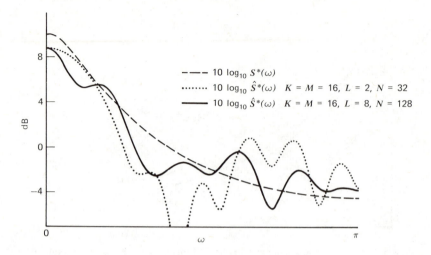

FIGURE 11.7.3. Averaged periodogram estimator using adjoining 16-point data sections for Example 11.7.1.

frequencies by augmenting the data sections with the appropriate number of zeros and then taking FFTs. Sectioning of long data records fits in naturally. The sum of pairs of periodograms for real data sections can be computed efficiently by a simple modification of the method described in Section 10.8. Let us denote two real P-point augmented data sections by g_n and h_n and form the complex sequence $d_n = g_n + jh_n$. Then, according to (10.8.3) and (10.8.4)

$$ G_k = \frac{D_k + \bar{D}_{P-k}}{2} \quad \text{and} \quad H_k = \frac{D_k - \bar{D}_{P-k}}{2j} $$

With a little algebra, we find that

$$ |G_k|^2 + |H_k|^2 = \frac{|D_k|^2 + |D_{P-k}|^2}{2} \tag{11.7.12} $$

Thus, the sum of the periodograms of the two real sequences can be computed simply from the FFT of the single complex sequence. Another technique that can be used to reduce computation time is to select an FFT algorithm that accepts the input data in natural order and generates the output data in bit-reversed order. The L periodograms can be accumulated in bit-reversed order and then a single bit reversal can be performed rather than L separate bit reversals.

11.8 ESTIMATION OF A POWER SPECTRAL DENSITY BY SMOOTHING A SINGLE PERIODOGRAM

Estimation of power spectral densities by averaging separate periodograms did not become popular until after the FFT was discovered and widely known. Before then, the most common approach was to use an estimator of the form

$$\tilde{S}^*(\omega) = \sum_{n=-M}^{M} v(n) R_N(n) e^{-j\omega n} \tag{11.8.1}$$

with M much less than the record length N. The required values of $R_N(n)$ were computed directly using (11.2.2), since it was not considered practical to compute them by the indirect frequency domain method. Of course, they can now be efficiently computed using the techniques described in Section 11.3. From one point of view, the philosophy was to obtain a more stable spectrum estimator by including only the most reliable estimates of $R(n)$, that is, $R_N(n)$ for $|n| \ll N$, in the Fourier series. This was done by truncating the sample autocorrelation function by the *lag window* $v(n)$. To maintain the symmetry of $R_N(n)$ and obtain real spectrum estimates, $v(n)$ must be even. It should also rise to a peak at $n=0$ to weight the most reliable estimates of $R(n)$ most heavily.

According to the frequency domain convolution formula, an alternate form for (11.8.1) is

$$\tilde{S}^*(\omega) = \frac{1}{2\pi} \int_{-\pi}^{\pi} V^*(\lambda) I_N^*(\omega - \lambda) \, d\lambda \tag{11.8.2}$$

where

$$V^*(\omega) = \sum_{n=-M}^{M} v(n) e^{-j\omega n} \tag{11.8.3}$$

$V^*(\omega)$ is called the *spectral window* corresponding to $v(n)$. Interchanging time and frequency, we see that $\tilde{S}^*(\omega)$ can be thought of as being obtained by passing the periodogram $I_N^*(\omega)$ through a filter with the "impulse response" $V^*(\omega)$ and "frequency response" $v(n)$. Since $v(n)$ has low-pass characteristics, $\tilde{S}^*(\omega)$ is a smoothed version of the periodogram. Therefore, we will call $\tilde{S}^*(\omega)$ the *smoothed periodogram estimator*.

The expected value of the smoothed periodogram estimator is

$$E\{\tilde{S}^*(\omega)\} = \frac{1}{2\pi} \int_{-\pi}^{\pi} V^*(\lambda) E\{I_N^*(\omega - \lambda)\} \, d\lambda \tag{11.8.4}$$

The periodogram is asymptotically unbiased. Therefore, for large N we can use the approximation

$$E\{\tilde{S}^*(\omega)\} \cong \frac{1}{2\pi} \int_{-\pi}^{\pi} V^*(\lambda) S^*(\omega - \lambda) \, d\lambda \qquad (11.8.5)$$

Thus the biases of the averaged and smoothed periodogram estimators are approximately the same when the spectral windows are identical. When $S^*(\omega)$ is essentially constant over a region that is somewhat wider than the central lobe of $V^*(\omega)$ (11.8.5) can be approximated by

$$S^*(\omega) \frac{1}{2\pi} \int_{-\pi}^{\pi} V^*(\lambda) \, d\lambda = S^*(\omega) v(0) \qquad (11.8.6)$$

Therefore, we should require that

$$\frac{1}{2\pi} \int_{-\pi}^{\pi} V^*(\lambda) \, d\lambda = v(0) = 1 \qquad (11.8.7)$$

to obtain properly scaled estimates.

The data windows discussed in Section 11.7 can be simply modified to form lag windows. The Hanning lag window is

$$v_{Hn}(n) = 0.5[1 + \cos{(2\pi n/L)}]v_r(n) \qquad (11.8.8)$$

where

$$L = 2M + 1 \qquad (11.8.9)$$

and

$$v_r(n) = \begin{cases} 1 & \text{for} \quad |n| \le M \\ 0 & \text{elsewhere} \end{cases} \qquad (11.8.10)$$

The corresponding spectral window is

$$V_{Hn}^*(\omega) = 0.5 V_r^*(\omega) + 0.25 V_r^*(\omega - 2\pi/L) + 0.25 V_r^*(\omega + 2\pi/L) \qquad (11.8.11)$$

where

$$V_r^*(\omega) = \frac{\sin{\omega L/2}}{\sin{\omega/2}} \qquad (11.8.12)$$

The Hamming windows are

$$v_H(n) = [0.54 + 0.46 \cos{(2\pi n/L)}]v_r(n) \qquad (11.8.13)$$

and

$$V_H^*(\omega) = 0.54 V_r^*(\omega) + 0.23 V_r^*(\omega - 2\pi/L) + 0.23 V_r^*(\omega + 2\pi/L) \qquad (11.8.14)$$

The Bartlett windows are

$$v_B(n) = \left(1 - \frac{|n|}{M+1}\right)v_r(n) \qquad (11.8.15)$$

and

$$V_B^*(\omega) = \frac{\sin^2 [\omega(M+1)/2]}{(M+1) \sin^2 (\omega/2)} \tag{11.8.16}$$

The Hanning and Hamming spectral windows have some negative portions. This can result in negative spectrum estimates in the vicinity of large peaks in $S^*(\omega)$. The Bartlett spectral window is nonnegative and always results in positive estimates. On the other hand, the sidelobes of the Bartlett window are larger than those of the Hanning and Hamming windows. Thus, the Bartlett window can allow positive biases that are larger in magnitude than the negative biases with the Hanning and Hamming windows in the vicinity of a spectral peak. If the data is preprocessed to eliminate large spectral peaks, then the choice of windows is not really critical and is mainly a matter of preference.

It follows from (11.8.2) and (11.8.4) that

$$\text{cov} \{\tilde{S}^*(\omega_1), \tilde{S}^*(\omega_2)\} = E\{[\tilde{S}^*(\omega_1) - E\{\tilde{S}^*(\omega_1)\}][\tilde{S}^*(\omega_2) - E\{\tilde{S}^*(\omega_2)\}]\}$$

$$= \frac{1}{(4\pi)^2} \int_{-\pi}^{\pi} \int_{-\pi}^{\pi} V^*(\lambda) V^*(\mu) \, \text{cov} \{I_N^*(\omega_1 - \lambda), I_N^*(\omega_2 - \mu)\} \, d\lambda \, d\mu \tag{11.8.17}$$

To gain some insight into the significance of this formula, let us assume that the observed data comes from a white Gaussian noise sequence with the power spectral density $S^*(\omega) = \sigma^2$. Then according to (11.6.10)

$$\text{cov} \{I_N^*(\omega_1 - \lambda), I_N^*(\omega_2 - \mu)\} = \frac{\sigma^4}{N} [Q_r^*(\omega_1 + \omega_2 - \lambda - \mu) + Q_r^*(\omega_1 - \omega_2 - \lambda + \mu)] \tag{11.8.18}$$

where

$$Q_r^*(\omega) = \frac{\sin^2 \omega N/2}{N \sin^2 \omega/2} \tag{11.8.19}$$

Let us assume that $0 < \omega_1, \omega_2 < \pi$ and N is large enough so that (11.5.11) is a good approximation to $Q_r^*(\omega)$. Then integrating over λ in (11.8.17) yields

$$\text{cov} \{\tilde{S}^*(\omega_1), \tilde{S}^*(\omega_2)\} \cong \frac{\sigma^4}{2\pi N} \int_{-\pi}^{\pi} V^*(\mu)[V^*(\omega_1 + \omega_2 - \mu) + V^*(\omega_1 - \omega_2 + \mu)] \, d\mu \tag{11.8.20}$$

As a further approximation, let us assume that the width of the central lobe of $V^*(\omega)$ is small enough to neglect $V^*(\mu) V^*(\omega_1 + \omega_2 - \mu)$. Then

$$\text{cov} \{\tilde{S}^*(\omega_1), \tilde{S}^*(\omega_2)\} \cong \frac{\sigma^4}{2\pi N} \int_{-\pi}^{\pi} V^*(\mu) V^*(\omega_1 - \omega_2 + \mu) \, d\mu \tag{11.8.21}$$

For a good spectral window, we see that the estimates at ω_1 and ω_2 essentially are uncorrelated when $|\omega_1 - \omega_2|$ is greater than the width of the central lobe of the window. This width is approximately proportional to $1/M$ for the windows discussed here. Thus the estimates at adjacent frequencies become more correlated as $R_N(n)$ is truncated at smaller values of M.

Letting $\omega_1 = \omega_2$ in (11.8.21) gives

$$\text{var}\{\tilde{S}^*(\omega_1)\} \cong \frac{\sigma^4}{2\pi N} \int_{-\pi}^{\pi} [V^*(\mu)]^2 \, d\mu \qquad (11.8.22)$$

Since $v(n)$ is even, we see from Parseval's Theorem that

$$\Lambda = \frac{1}{2\pi N} \int_{-\pi}^{\pi} [V^*(\mu)]^2 \, d\mu = \frac{1}{N} \sum_{n=-M}^{M} v^2(n) \qquad (11.8.23)$$

For large N and $0 < \omega_1 < \pi$, we find from (11.6.11) that the variance of the periodogram is approximately σ^4. Therefore, Λ is the ratio of the variances of the smoothed and unsmoothed periodograms. For example, with the Hanning window

$$\Lambda_{Hn} = \frac{3}{8} \frac{2M+1}{N} \qquad (11.8.24)$$

As expected, the variance of the smoothed periodogram estimator decreases as M decreases relative to N.

If $S^*(\omega)$ is smooth relative to the spectral window and M is small relative to N, then $\text{cov}\{\tilde{S}^*(\omega_1), \tilde{S}^*(\omega_2)\}$ and $\text{var}\{\tilde{S}^*(\omega_1)\}$ are approximately given by (11.8.21) and (11.8.22) with σ^4 replaced by $S^*(\omega_1)S^*(\omega_2)$.

Once again, we see that there is a trade-off between variance and resolution for fixed N. A small variance requires a small M while high resolution requires a large M. The averaged and smoothed periodogram estimators are comparable in terms of bias and variance when the spectral windows are similar.

Finally, $\tilde{S}^*(\omega)$ can be computed at the frequencies $k2\pi/K$ for $k = 0, \ldots, K-1$ with $K \geq 2M+1$ by first forming the K-point sequence

$$s_n = \begin{cases} v(n)R_N(n) & \text{for} & 0 \leq n \leq M \\ 0 & \text{for} & M < n < K-M \\ v(n-K)R_N(n-K) & \text{for} & K-M \leq n \leq K-1 \end{cases} \qquad (11.8.25)$$

Then

$$\text{DFT}\{s_n\} = \tilde{S}^*(k2\pi/K) \qquad \text{for} \qquad k = 0, \ldots, K-1 \qquad (11.8.26)$$

11.9 ESTIMATION OF POWER SPECTRAL DENSITIES IN TERMS OF AUTOREGRESSIVE MODELS

The averaged and smoothed periodogram estimators approximate the true power spectral density by a function of the form $A(z)A(z^{-1})$ where $A(z)$ is a finite degree polynomial in z^{-1} with real coefficients. The degree of $A(z)$ must be large to give good estimates when the true power spectral density has sharp peaks. Intuitively, we would expect that better estimates using fewer parameters could be obtained in this case by including poles in the spectral density model. In this section, we will examine a method for estimating the parameters of a model that only has poles except for zeros at $z = 0$ and ∞.

A discrete-time random process $x(n)$ with a sampled power spectral density of the form

$$S_{xx}(z) = \frac{\beta^2}{D(z)D(z^{-1})} \tag{11.9.1}$$

where

$$D(z) = \sum_{i=0}^{N} d_i z^{-i}, \qquad d_0 = 1 \tag{11.9.2}$$

and $D(z)$ has all its zeros inside the unit circle is called an *autoregressive* process. This process can be generated by passing a white noise sequence $v(n)$ with the power spectral density $S_{vv}(z) = \beta^2$ through the filter $1/D(z)$. Thus $x(n)$ and $v(n)$ are related by the difference equation

$$\sum_{k=0}^{N} d_k x(n-k) = v(n) \tag{11.9.3}$$

If $v(n)$ is also a Gaussian process, the joint probability density function for $v(0), \ldots, v(M-1)$ is

$$f(v_0, v_1, \ldots, v_{M-1}) = (2\pi\beta^2)^{-M/2} \exp\left[-\frac{1}{2\beta^2} \sum_{n=0}^{M-1} v_n^2\right] \tag{11.9.4}$$

The Jacobian of the transformation between $v(0), \ldots, v(M-1)$ and $x(0), \ldots, x(M-1)$ given $x(-N), \ldots, x(-1)$ is unity. Therefore

$$f[x(0), \ldots, x(M-1)/x(-N), \ldots, x(-1)]$$

$$= (2\pi\beta^2)^{-M/2} \exp\left\{-\frac{1}{2\beta^2} \sum_{n=0}^{M-1} \left[\sum_{k=0}^{N} d_k x(n-k)\right]^2\right\} \tag{11.9.5}$$

The joint density for $x(-N), \ldots, x(M-1)$ can be obtained by multiplying (11.9.5) by the joint density for $x(-N), \ldots, x(-1)$. However, in practical applications, $M \gg N$ and the change is negligible. Therefore, the maximum likelihood estimates of the parameters $\beta^2, d_1, \ldots, d_N$ given an observation of $x(-N), \ldots, x(M-1)$ are essentially specified by the equations

$$\frac{\partial}{\partial d_m} \ln f[x(0), \ldots, x(M-1)/x(-N), \ldots, x(-1)] = 0 \qquad \text{for} \qquad m = 1, \ldots, N$$

$$\tag{11.9.6}$$

and

$$\frac{\partial}{\partial \beta^2} \ln f[x(0), \ldots, x(M-1)/x(-N), \ldots, x(-1)] = 0 \tag{11.9.7}$$

Carrying out the indicated operations yields

$$\sum_{k=0}^{N} \hat{d}_k c_{k,m} = 0 \qquad \text{for} \qquad m = 1, \ldots, N \tag{11.9.8}$$

and

$$\hat{\beta}^2 = \frac{1}{M} \sum_{n=0}^{M-1} \left[\sum_{k=0}^{N} \hat{d}_k x(n-k) \right]^2 \tag{11.9.9}$$

where

$$\hat{d}_0 = 1 \qquad \text{and} \qquad c_{k,m} = \sum_{n=0}^{M-1} x(n-k)x(n-m) \tag{11.9.10}$$

Writing the squared term in (11.9.9) as a double sum, interchanging orders of summation, and using (11.9.8) gives

$$\hat{\beta}^2 = \frac{1}{M} \sum_{k=0}^{N} \hat{d}_k c_{k,0} \tag{11.9.11}$$

The symbol, ^, is used to distinguish between the true and estimated parameters.

Before solving the estimation equations, we will make one additional approximation and replace $c_{k,m}$ by

$$MR_M(k-m) = \sum_{n=0}^{M-1-|k-m|} x(n)x(n+|k-m|) \tag{11.9.12}$$

Then the solution to (11.9.8) becomes

$$\hat{\mathbf{d}} = -\mathbf{R}^{-1}\mathbf{r} \tag{11.9.13}$$

where

$$\hat{\mathbf{d}} = \begin{bmatrix} d_1 \\ d_2 \\ \cdot \\ \cdot \\ \cdot \\ d_N \end{bmatrix} \qquad \mathbf{r} = \begin{bmatrix} R_M(1) \\ R_M(2) \\ \cdot \\ \cdot \\ \cdot \\ R_M(N) \end{bmatrix} \tag{11.9.14}$$

and

$$\mathbf{R} = \begin{bmatrix} R_M(0) & R_M(1) & \cdots & R_M(N-1) \\ R_M(1) & R_M(0) & \cdots & R_M(N-2) \\ \cdot & \cdot & & \cdot \\ \cdot & \cdot & & \cdot \\ \cdot & \cdot & & \cdot \\ R_M(N-1) & \cdots & & R_M(0) \end{bmatrix} \tag{11.9.15}$$

Also

$$\hat{\beta}^2 = \sum_{k=0}^{N} \hat{d}_k R_M(k) \qquad (11.9.16)$$

It is interesting to observe that the estimation equations derived here have exactly the same form as those derived in Section 7.6 for the optimum linear predictor using the last N samples. This is not surprising since $v(n)$ corresponds to the prediction error and β^2 corresponds to the variance of the prediction error. The estimation equations can be efficiently solved by using the recursive method derived in Section 7.6. It was also shown in Section 7.6 that $\hat{D}(z) = 1 + \hat{d}_1 z^{-1} + \cdots + \hat{d}_N z^{-N}$ is guaranteed to have all its zeros inside the unit circle.

The autoregressive model can be used to estimate power spectral densities that have both poles and zeros since a numerator factor of the form $1 - az^{-1}$ can be approximated by a large enough number of poles. In practice, we are faced with the problem of choosing an appropriate value for N. One approach is to start with a small value of N and plot $\hat{\beta}^2$ as N increases. When $\hat{\beta}^2$ appears to level off at a minimum value, then an appropriate value of N has been reached. The partial correlation coefficient defined in Section 7.6 is usually a more sensitive indicator and can be used in the same manner. Both of these parameters appear naturally in the recursive solution of the estimation equations.

The maximum likelihood approach can be applied to estimating the parameters of more general models that include both poles and zeros. However, the estimation equations turn out to be nonlinear and must be solved by numerical methods [133].

Example 11.9.1

Suppose that $x(n)$ has the spectral density

$$S(z) = \frac{\beta^2}{(1 + d_1 z^{-1})(1 + d_1 z)} \qquad (11.9.17)$$

Then the true autocorrelation function is

$$R(n) = \frac{\beta^2}{1 - d_1^2} (-d_1)^{|n|} \qquad (11.9.18)$$

Notice that

$$d_1 = -R(1)/R(0) \qquad \text{and} \qquad \beta^2 = R(0)(1 - d_1^2) \qquad (11.9.19)$$

From (11.9.13) and (11.9.16) with $N = 1$, we find that

$$\hat{d}_1 = -R_M(1)/R_M(0) \qquad \text{and} \qquad \hat{\beta}^2 = R_M(0)(1 - \hat{d}_1^2) \qquad (11.9.20)$$

These estimates are similar in form to the true parameters as given by (11.9.19).   ◀

12

FUNDAMENTALS OF HILBERT SPACES

12.1 INTRODUCTION

Hilbert spaces are abstractions of ordinary three-dimensional Euclidean vector spaces. In general, Hilbert spaces are infinite-dimensional. The "vectors" in a Hilbert space might be random variables, sequences, or ordinary functions, among other things. The generalized measure of distance in a Hilbert space is called a norm and the generalization of the dot product is called an inner product. Almost all the properties of Hilbert spaces can be visualized in terms of the geometric properties of three-dimensional Euclidean spaces.

The goal of this chapter is to present the fundamental definitions and results in the theory of Hilbert spaces and, in particular, to prove the Projection Theorem. (More detailed treatments of the theory of Hilbert spaces can be found in References 31, 77, 88, and 115.) The Projection Theorem is a generalization of the fact in Euclidean geometry that the shortest distance between a point and a plane is obtained by dropping a perpendicular from the point to the plane.

The Projection Theorem provides a unified approach to solving the linear estimation problems discussed in Chapters 13 and 14. These problems all involve finding linear estimates that minimize a quadratic cost function. These problems can be solved by other standard techniques, such as by setting derivatives of the cost function with respect to the unknown parameters to zero. However, the Projection Theorem more easily gives necessary and sufficient conditions that the optimum estimates must satisfy. In addition, the

Projection Theorem provides an appealing geometric insight into the optimization problems. This insight is particularly helpful in solving the recursive estimation problem discussed in Chapter 14.

12.2 VECTOR SPACES AND SUBSPACES

The familiar properties of vectors in a three-dimensional Euclidean space can be abstracted and applied to many situations. In an abstract sense, a *vector space* consists of:

1. a field F of scalars
2. a set V of objects called vectors
3. an operation called vector addition that assigns to each pair of vectors $\mathbf{x}$ and $\mathbf{y}$ in V a vector $\mathbf{x} + \mathbf{y}$ in V such that
    a. $\mathbf{x} + \mathbf{y} = \mathbf{y} + \mathbf{x}$ (commutative law)
    b. $\mathbf{w} + (\mathbf{x} + \mathbf{y}) = (\mathbf{w} + \mathbf{x}) + \mathbf{y}$ (associative law)
    c. there is a unique vector $\boldsymbol{\theta}$ called the zero vector such that $\mathbf{x} + \boldsymbol{\theta} = \mathbf{x}$ for all $\mathbf{x}$ in V
    d. each vector $\mathbf{x}$ in V has a unique inverse vector $-\mathbf{x}$ in V such that $\mathbf{x} + (-\mathbf{x}) = \boldsymbol{\theta}$.
4. an operation called scalar multiplication that assigns to each scalar c in F and vector $\mathbf{x}$ in V a vector $c\mathbf{x}$ in V such that
    a. $1\mathbf{x} = \mathbf{x}$
    b. $(c_1 c_2)\mathbf{x} = c_1(c_2\mathbf{x})$ (associative law)
    c. $c(\mathbf{x} + \mathbf{y}) = c\mathbf{x} + c\mathbf{y}$ $\left.\right\}$(distributive laws)
    d. $(c_1 + c_2)\mathbf{x} = c_1\mathbf{x} + c_2\mathbf{x}$

The entire set of items specified in the vector space definition will be referred to simply as the vector space V.

A number of simple facts follow almost immediately from the definition of a vector space. For example, according to 3(c) and 4(c)

$$c\boldsymbol{\theta} = c(\boldsymbol{\theta} + \boldsymbol{\theta}) = c\boldsymbol{\theta} + c\boldsymbol{\theta}$$

Adding $-(c\boldsymbol{\theta})$ to both sides and using 3(b), (c), and (d) gives

$$c\boldsymbol{\theta} = \boldsymbol{\theta} \qquad (12.2.1)$$

Similarly, it can be shown that

$$0\mathbf{x} = \boldsymbol{\theta} \qquad (12.2.2)$$

Also

$$\boldsymbol{\theta} = 0\mathbf{x} = (1-1)\mathbf{x} = \mathbf{x} + (-1)\mathbf{x}$$

so that

$$-\mathbf{x} = (-1)\mathbf{x} \qquad (12.2.3)$$

Vector addition has been defined as an operation on pairs of vectors. However, the sum of any number of vectors can be written without parentheses because the associative law for vector addition requires that the same result must be obtained regardless of the order in which vectors are combined.

Numerous examples of vector spaces exist. Three common examples are presented below.

Example 12.2.1 *n-Dimensional Euclidean Space*

In an n-dimensional Euclidean space, the field F of scalars is the field of real numbers. Vectors are $n$tuples of the form

$$\mathbf{x} = (x_1, \ldots, x_i, \ldots, x_n)$$

where x_i is a real number and is called the ith component of $\mathbf{x}$. Vector addition is defined as

$$\mathbf{x} + \mathbf{y} = (x_1 + y_1, \ldots, x_i + y_i, \ldots, x_n + y_n)$$

where addition of components is ordinary addition. Scalar multiplication is defined as

$$c\mathbf{x} = (cx_1, \ldots, cx_i, \ldots, cx_n)$$

where ordinary multiplication is used on the right. The zero vector is $\mathbf{\theta} = (0, \ldots, 0)$ and the inverse of any vector $\mathbf{x}$ is $-\mathbf{x} = (-x_1, \ldots, -x_n)$.

These spaces are encountered frequently and are often denoted by the symbol R^n. If the scalars and components are allowed to be complex numbers, the resulting spaces are often denoted by C^n. ◀

Example 12.2.2 *Polynomials of Degree n or Less*

A polynomial of degree n over a field F has the form

$$c_0 + c_1 t + \cdots + c_n t^n$$

where the coefficients c_0, $c_1, \ldots, c_n$ are in F and $c_n \neq 0$. The set of polynomials of degree n or less over F with ordinary polynomial addition and scalar multiplication is a vector space. ◀

Example 12.2.3 *Finite Mean and Variance Random Variables*

Let F be the field of real numbers and V be the set of finite mean and variance random variables. If $c_1\mathbf{x} + c_2\mathbf{y}$ is defined in the ordinary sense, then a vector space is obtained. ◀

A nonempty subset M of a vector space V is called a *subspace* of V if every vector of the form $c_1\mathbf{x} + c_2\mathbf{y}$ is in M for all $\mathbf{x}$ and $\mathbf{y}$ in M and c_1 and c_2 in F. It is easy to verify that a subspace of a vector space V is itself a vector space with the same rules for vector addition and scalar multiplication as in V.

Example 12.2.4

The set of vectors from R^n with $x_i = 0$ for $i = k, \ldots, n$ where k is a fixed positive integer is a subspace of R^n. ◄

Any nonempty subset $G = \{\mathbf{g}_k\}$ of V can be turned into a subspace of V by adding to G all possible linear combinations of vectors in G, that is, all vectors of the form $\sum_k c_k \mathbf{g}_k$. We will denote the resulting subspace by $[G]$ and will call it the subspace or *linear manifold* generated by G. It is also called the subspace *spanned* by G.

If a particular vector $\mathbf{x}_0$ in a vector space V is added to every vector in a subspace M of V, the resulting set of vectors is called a *linear variety*. We will denote the linear variety by $\mathbf{x}_0 + M$. In geometric terms, a linear variety is a translation of a subspace. We will see that linear varieties arise naturally in some constrained optimization problems.

12.3 LINEAR INDEPENDENCE AND DIMENSION

The vectors $\mathbf{x}_1, \ldots, \mathbf{x}_n$ are said to be *linearly independent* if and only if

$$c_1\mathbf{x}_1 + c_2\mathbf{x}_2 + \cdots + c_n\mathbf{x}_n = \mathbf{\theta}$$

implies that $c_1 = c_2 = \cdots = c_n = 0$. Vectors that are not linearly independent are called *linearly dependent*.

A finite set G of linearly independent vectors is called a *basis* for a space V if G generates V. Any vector space with a finite basis is said to be *finite-dimensional*. All other vector spaces are called *infinite-dimensional*. It can be shown that any two bases for a finite-dimensional vector space contain the same numbers of elements [47].

The *dimension* of any finite-dimensional vector space V is defined to be the number of vectors in any basis for V.

Example 12.3.1

Consider the unit vectors

$$\mathbf{u}_1 = (1, 0, 0, \ldots, 0)$$
$$\mathbf{u}_2 = (0, 1, 0, \ldots, 0)$$
.
.
.
$$\mathbf{u}_n = (0, 0, \ldots, 0, 1)$$

in R^n. Now

$$c_1\mathbf{u}_1 + c_2\mathbf{u}_2 + \cdots + c_n\mathbf{u}_n = (c_1, c_2, \ldots, c_n) = \mathbf{0}$$

implies that $c_1 = c_2 = \cdots = c_n = 0$ so that the unit vectors are linearly independent. Also, any $n$tuple can be generated by a linear combination of the n unit vectors so that they form a basis for R^n. Thus R^n has dimension n. ◀

12.4 NORMED VECTOR SPACES AND CONVERGENCE

A *norm* for a vector space V in which the field F of scalars is the field of real or complex numbers is any function that assigns to each $\mathbf{x}$ in V a real number $\|\mathbf{x}\|$, called the norm of $\mathbf{x}$, and satisfies the following axioms:

1. $\|\mathbf{x}\| \geq 0$ for all $\mathbf{x}$ in V, and $\|\mathbf{x}\| = 0$ if and only if $\mathbf{x} = \mathbf{0}$.
2. $\|\mathbf{x} + \mathbf{y}\| \leq \|\mathbf{x}\| + \|\mathbf{y}\|$ for each $\mathbf{x}$ and $\mathbf{y}$ in V (triangle inequality).
3. $\|c\mathbf{x}\| = |c| \cdot \|\mathbf{x}\|$ for all c in F and $\mathbf{x}$ in V.

The norm is an abstraction of the concept of length in ordinary three-dimensional Euclidean space. Axiom 2 is called the triangle inequality since it is a generalization of the fact that the sum of the lengths of two sides of a triangle is always greater than or equal to the length of the third side.

Example 12.4.1

Consider the space C^n of complex $n$tuples. One norm for this space is

$$\|\mathbf{x}\|_1 = \sum_{i=1}^{n} |x_i|$$

Axioms 1 and 3 are clearly satisfied. Axiom 2 is satisfied since $|x_i + y_i| \leq |x_i| + |y_i|$.

Another norm is

$$\|\mathbf{x}\|_2 = \left(\sum_{i=1}^{n} |x_i|^2 \right)^{1/2}$$

In R^3, this norm is just the length of a vector. Again, we can easily see that Axioms 1 and 3 are satisfied. The fact that Axiom 2 is satisfied is shown in Section 12.5. ◀

An infinite sequence of vectors $\{\mathbf{x}_n\}$ in a normed vector space is said to *converge* to a limit $\mathbf{x}$ if the sequence of real numbers $\{\|\mathbf{x} - \mathbf{x}_n\|\}$ converges to zero. More formally, we say that $\mathbf{x}_n$ converges to the limit $\mathbf{x}$ if for every $\epsilon > 0$ there exists an N such that $\|\mathbf{x}_n - \mathbf{x}\| < \epsilon$ for $n > N$. It is common to use the notation $\mathbf{x}_n \to \mathbf{x}$ to indicate that $\mathbf{x}_n$ converges to $\mathbf{x}$.

THEOREM 12.4.1. Continuity of the Norm
In a normed vector space if $\mathbf{x}_n \to \mathbf{x}$, then $\|\mathbf{x}_n\| \to \|\mathbf{x}\|$.

Proof

$$\|\mathbf{x}_n\| = \|\mathbf{x}_n - \mathbf{x} + \mathbf{x}\| \le \|\mathbf{x}_n - \mathbf{x}\| + \|\mathbf{x}\|$$

or

$$\|\mathbf{x}_n\| - \|\mathbf{x}\| \le \|\mathbf{x}_n - \mathbf{x}\|$$

Interchanging $\mathbf{x}_n$ and $\mathbf{x}$ yields

$$\|\mathbf{x}\| - \|\mathbf{x}_n\| \le \|\mathbf{x}_n - \mathbf{x}\|$$

so that

$$|\,\|\mathbf{x}_n\| - \|\mathbf{x}\|\,| \le \|\mathbf{x}_n - \mathbf{x}\| \tag{12.4.1}$$

Q.E.D.

A class of sequences known as *Cauchy sequences* is very important in the theory of normed spaces. Any infinite sequence $\{\mathbf{x}_n\}$ in a normed vector space is called a Cauchy sequence if $\|\mathbf{x}_n - \mathbf{x}_m\| \to 0$ as n and $m \to \infty$.

Any sequence that converges is a Cauchy sequence since, when $\mathbf{x}_n \to \mathbf{x}$

$$\|\mathbf{x}_n - \mathbf{x}_m\| = \|\mathbf{x}_n - \mathbf{x} + \mathbf{x} - \mathbf{x}_m\| \le \|\mathbf{x}_n - \mathbf{x}\| + \|\mathbf{x}_m - \mathbf{x}\| \to 0$$

However, not all Cauchy sequences converge.

A normed vector space in which every Cauchy sequence converges to a limit in the space is called a *complete* space. Complete, normed, vector spaces are known as *Banach* spaces.

12.5 INNER PRODUCTS AND HILBERT SPACES

In addition to the norm which is a generalized measure of distance, we need a generalization of the dot product to complete the geometric analogy between ordinary three-dimensional Euclidean space and an abstract vector space V

in which the field F of scalars is the real or complex numbers. This generalization is known as an *inner product*. An inner product is a function that assigns a scalar $(\mathbf{x}, \mathbf{y})$ to each pair of vectors $\mathbf{x}$ and $\mathbf{y}$ in V and satisfies the following axioms:

1. $(\mathbf{x}, \mathbf{y}) = \overline{(\mathbf{y}, \mathbf{x})}$
2. $(c\mathbf{x}, \mathbf{y}) = c(\mathbf{x}, \mathbf{y})$
3. $(\mathbf{x} + \mathbf{y}, \mathbf{w}) = (\mathbf{x}, \mathbf{w}) + (\mathbf{y}, \mathbf{w})$
4. $(\mathbf{x}, \mathbf{x}) \geq 0$ and $(\mathbf{x}, \mathbf{x}) = 0$ if and only if $\mathbf{x} = \mathbf{0}$

A vector space with an inner product is called an *inner product space*. In an inner product space, it is customary to use the notation

$$(\mathbf{x}, \mathbf{x})^{1/2} = \|\mathbf{x}\| \qquad (12.5.1)$$

In the following sequence of theorems, we will see that $(\mathbf{x}, \mathbf{x})^{1/2}$ is indeed a norm.

THEOREM 12.5.1.   The Cauchy-Schwarz Inequality
For all x and y in an inner product space V

$$|(\mathbf{x}, \mathbf{y})| \leq \|\mathbf{x}\| \, \|\mathbf{y}\| \qquad (12.5.2)$$

Equality holds if and only if $\mathbf{y} = \mathbf{0}$ or $\mathbf{x} = c\mathbf{y}$.

Proof If $\mathbf{y} = \mathbf{0}$, equality holds since it follows from Axiom 2 that $(\mathbf{x}, \mathbf{0}) = 0$ for all $\mathbf{x}$ in V. Now let us assume that $\mathbf{y} \neq \mathbf{0}$. Let us visualize $\mathbf{x}$ and $\mathbf{y}$ as vectors in a two-dimensional Euclidean space as shown in Fig. 12.5.1 and choose the scalar c so that

$$(\mathbf{x} - c\mathbf{y}, \mathbf{y}) = 0 \qquad (12.5.3)$$

If the inner product were a dot product, we would say that the vectors $\mathbf{y}$ and $\mathbf{x} - c\mathbf{y}$ were perpendicular or orthogonal as shown in Fig. 12.5.1. Using Axioms 2 and 3, we find that the required value of c is

$$c = (\mathbf{x}, \mathbf{y})/(\mathbf{y}, \mathbf{y}) = (\mathbf{x}, \mathbf{y})/\|\mathbf{y}\|^2 \qquad (12.5.4)$$

Then

$$\begin{aligned}
\|\mathbf{x}\|^2 = (\mathbf{x}, \mathbf{x}) &= (\mathbf{x} - c\mathbf{y} + c\mathbf{y}, \mathbf{x} - c\mathbf{y} + c\mathbf{y}) \\
&= (\mathbf{x} - c\mathbf{y}, \mathbf{x} - c\mathbf{y}) + (\mathbf{x} - c\mathbf{y}, c\mathbf{y}) + (c\mathbf{y}, \mathbf{x} - c\mathbf{y}) + (c\mathbf{y}, c\mathbf{y}) \\
&= \|\mathbf{x} - c\mathbf{y}\|^2 + \bar{c}(\mathbf{x} - c\mathbf{y}, \mathbf{y}) + c\overline{(\mathbf{x} - c\mathbf{y}, \mathbf{y})} + |c|^2 \|\mathbf{y}\|^2 \\
&= \|\mathbf{x} - c\mathbf{y}\|^2 + |c|^2 \|\mathbf{y}\|^2 \\
&\geq |c|^2 \|\mathbf{y}\|^2 = |(\mathbf{x}, \mathbf{y})|^2/\|\mathbf{y}\|^2
\end{aligned}$$

Multiplying through by $\|\mathbf{y}\|^2$ and taking square roots yields (12.5.2). The

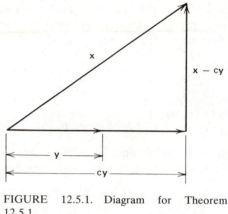

FIGURE 12.5.1. Diagram for Theorem 12.5.1.

inequality arose in replacing $\|\mathbf{x} - c\mathbf{y}\|^2$ by zero. Equality occurs when $\|\mathbf{x} - c\mathbf{y}\|^2 = 0$ which, according to Axiom 4 for inner products, happens if and only if $\mathbf{x} = c\mathbf{y}$.

Q.E.D.

The Cauchy-Schwarz inequality is a generalization of the fact that the magnitude of the dot product of two vectors is less than or equal to the product of their lengths in a Euclidean space. Equality is achieved when the vectors are colinear.

THEOREM 12.5.2. The Triangle Inequality
For all $\mathbf{x}$ and $\mathbf{y}$ in an inner product space

$$\|\mathbf{x} + \mathbf{y}\| \leq \|\mathbf{x}\| + \|\mathbf{y}\| \tag{12.5.5}$$

Proof
$$\|\mathbf{x} + \mathbf{y}\|^2 = (\mathbf{x} + \mathbf{y}, \mathbf{x} + \mathbf{y}) = (\mathbf{x}, \mathbf{x}) + (\mathbf{x}, \mathbf{y}) + (\mathbf{y}, \mathbf{x}) + (\mathbf{y}, \mathbf{y})$$

Taking the magnitude of the right-hand side and using the Cauchy-Schwarz inequality gives

$$\|\mathbf{x} + \mathbf{y}\|^2 \leq \|\mathbf{x}\|^2 + 2\,|(\mathbf{x}, \mathbf{y})| + \|\mathbf{y}\|^2$$
$$\leq \|\mathbf{x}\|^2 + 2\,\|\mathbf{x}\|\,\|\mathbf{y}\| + \|\mathbf{y}\|^2$$
$$= (\|\mathbf{x}\| + \|\mathbf{y}\|)^2$$

Q.E.D.

THEOREM 12.5.3
In an inner product space $(\mathbf{x}, \mathbf{x})^{1/2}$ is a norm.

Proof Axiom 1 for norms is satisfied as a consequence of Axiom 4 for inner products. Axiom 2 for norms is satisfied according to Theorem 12.5.2. Axiom

3 for norms is satisfied since $(c\mathbf{x}, c\mathbf{x}) = |c|^2(\mathbf{x}, \mathbf{x})$ as a result of Axioms 1 and 2 for inner products.

Q.E.D.

THEOREM 12.5.4. The Parallelogram Law
In an inner product space

$$\|\mathbf{x} + \mathbf{y}\|^2 + \|\mathbf{x} - \mathbf{y}\|^2 = 2\|\mathbf{x}\|^2 + 2\|\mathbf{y}\|^2 \tag{12.5.6}$$

Proof The theorem can be proved by simply expanding the left-hand side of (12.5.6) using the axioms for inner products.

Q.E.D.

The two-dimensional geometric interpretation of the parallelogram law is illustrated in Fig. 12.5.2. The sum of the squares of the lengths of the diagonals of a parallelogram is equal to twice the sum of the squares of the lengths of two adjacent sides.

A *Hilbert space* is defined as a complete inner product space. The norm in a Hilbert space is always chosen as $(\mathbf{x}, \mathbf{x})^{1/2}$. In general, it is not easy to prove that a given inner product space is complete. We will ignore this problem and assume that the spaces we deal with in the remainder of this book are complete. It can be shown that these spaces actually are complete.

Three examples of Hilbert spaces are presented below. In each case, it is left to the reader to prove that the inner products defined satisfy the four axioms.

Example 12.5.1. The Space l^2

Let F be the field of complex numbers and V be the set of infinite sequences $\mathbf{x} = \{x_1, x_2, \ldots\}$ of complex numbers such that

$$\sum_{i=1}^{\infty} |x_i|^2 < \infty$$

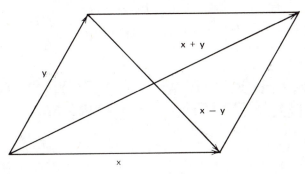

FIGURE 12.5.2. The Parallelogram Law

The inner product will be defined as

$$(\mathbf{x}, \mathbf{y}) = \sum_{i=1}^{\infty} x_i \bar{y}_i$$

This inner product is analogous to the dot product in three-dimensional Euclidean space. The vectors in l^2 are said to be square summable. It follows from the triangle inequality that the sum of two square summable sequences is also square summable. The square of the norm of the difference of two vectors is the sum of the squares of the errors in their components, that is

$$\|\mathbf{x} - \mathbf{y}\|^2 = \sum_{i=1}^{\infty} |x_i - y_i|^2 \qquad \blacktriangleleft$$

Example 12.5.2.   The Space $L^2[a, b]$

Let F be the complex numbers and V be the set of complex functions $x(t)$ such that

$$\int_a^b |x(t)|^2 \, dt < \infty$$

where the integral is taken in the Lebesgue sense. These functions are called square integrable on $[a, b]$. The inner product will be defined as

$$(\mathbf{x}, \mathbf{y}) = \int_a^b x(t) \bar{y}(t) \, dt$$

In $L^2[a, b]$, the square of the norm of the difference of two vectors is their integral square error over $[a, b]$. The fact that $L^2[a, b]$ is complete is known as the Riesz-Fischer Theorem [117]. $\qquad \blacktriangleleft$

Example 12.5.3

Let F be the real numbers and V be the space of zero-mean, finite-variance random variables. The inner product will be defined as

$$(\mathbf{x}, \mathbf{y}) = E\{\mathbf{xy}\}$$

In this space, $\|\mathbf{x}\|^2 = E\{\mathbf{x}^2\} = \text{var } \mathbf{x}$. Also, $\|\mathbf{x} - \mathbf{y}\|^2 = E\{(\mathbf{x} - \mathbf{y})^2\}$ is the mean square error between $\mathbf{x}$ and $\mathbf{y}$. $\qquad \blacktriangleleft$

12.6 ORTHOGONAL VECTORS AND THE PROJECTION THEOREM

Two vectors $\mathbf{x}$ and $\mathbf{y}$ in a Hilbert space are said to be *orthogonal* if $(\mathbf{x}, \mathbf{y}) = 0$. We will use the geometric notation $\mathbf{x} \perp \mathbf{y}$ to indicate that $\mathbf{x}$ and $\mathbf{y}$ are orthogonal. If $\mathbf{x}$ is orthogonal to every vector in a set A, $\mathbf{x}$ is said to be orthogonal to A. This will be abbreviated as $\mathbf{x} \perp A$.

As we might expect, we will now see that two orthogonal vectors in a Hilbert space satisfy a generalization of the Pythagorean Theorem.

THEOREM 12.6.1. The Pythagorean Theorem
Let $\mathbf{x}$ and $\mathbf{y}$ be any two orthogonal vectors in a Hilbert space. Then

$$\|\mathbf{x}+\mathbf{y}\|^2 = \|\mathbf{x}\|^2 + \|\mathbf{y}\|^2 \qquad (12.6.1)$$

Proof

$$\|\mathbf{x}+\mathbf{y}\|^2 = \|\mathbf{x}\|^2 + (\mathbf{x},\mathbf{y}) + (\mathbf{y},\mathbf{x}) + \|\mathbf{y}\|^2$$

But $(\mathbf{x},\mathbf{y})=(\mathbf{y},\mathbf{x})=0$ by hypothesis.

Q.E.D.

We are all familiar with the fact in Euclidean geometry that the shortest distance between a point and a plane is obtained by dropping a perpendicular to the plane. This result can be generalized to Hilbert spaces and is known as the Projection Theorem. We see later that the Projection Theorem provides a unified approach to finding linear estimates that are best in the minimum norm sense.

THEOREM 12.6.2. The Projection Theorem
Let M be a complete subspace of a Hilbert space H. Then, given any vector $\mathbf{x}$ in H, there is a unique vector $\mathbf{m}_0$ in M such that $\|\mathbf{x}-\mathbf{m}_0\|\leq\|\mathbf{x}-\mathbf{m}\|$ for all $\mathbf{m}$ in M. A necessary and sufficient condition that $\mathbf{m}_0$ be the unique minimizing vector is that the error vector $\mathbf{x}-\mathbf{m}_0$ be orthogonal to M.

Proof First let us assume that a vector $\mathbf{m}_0$ in M exists such that $\mathbf{x}-\mathbf{m}_0 \perp M$. Let $\mathbf{m}$ be any vector in M. Then

$$\|\mathbf{x}-\mathbf{m}\|^2 = \|\mathbf{x}-\mathbf{m}_0+\mathbf{m}_0-\mathbf{m}\|^2$$

The vector $\mathbf{m}_0-\mathbf{m}$ is in M so that $(\mathbf{x}-\mathbf{m}_0)\perp(\mathbf{m}_0-\mathbf{m})$. Thus, according to the Pythagorean Theorem

$$\|\mathbf{x}-\mathbf{m}\|^2 = \|\mathbf{x}-\mathbf{m}_0\|^2 + \|\mathbf{m}_0-\mathbf{m}\|^2$$

and

$$\|\mathbf{x}-\mathbf{m}\| > \|\mathbf{x}-\mathbf{m}_0\|$$

for $\mathbf{m}\neq\mathbf{m}_0$. Consequently, $\mathbf{m}_0$ is the unique minimizing vector.

Now let us assume that a unique minimizing vector $\mathbf{m}_0$ in M exists but that there is a vector $\mathbf{m}$ in M that is not orthogonal to $\mathbf{x}-\mathbf{m}_0$. Let us normalize $\mathbf{m}$ so that $\|\mathbf{m}\|=1$ and let $(\mathbf{x}-\mathbf{m}_0,\mathbf{m})=c\neq0$. If we define the vector $\mathbf{m}_1$ in M as $\mathbf{m}_1=\mathbf{m}_0+c\mathbf{m}$, then

$$\|\mathbf{x}-\mathbf{m}_1\|^2 = \|\mathbf{x}-\mathbf{m}_0-c\mathbf{m}\|^2$$
$$= \|\mathbf{x}-\mathbf{m}_0\|^2 - (\mathbf{x}-\mathbf{m}_0,c\mathbf{m}) - (c\mathbf{m},\mathbf{x}-\mathbf{m}_0) + |c|^2$$
$$= \|\mathbf{x}-\mathbf{m}_0\|^2 - |c|^2 < \|\mathbf{x}-\mathbf{m}_0\|^2$$

Thus $\mathbf{m}_0$ is not the minimizing vector which is a contradiction and the necessity of the condition $\mathbf{x} - \mathbf{m}_0 \perp M$ is established.

Now the existence of $\mathbf{m}_0$ will be demonstrated. If $\mathbf{x}$ is in M, then $\mathbf{m}_0 = \mathbf{x}$, obviously. Let us assume that $\mathbf{x}$ is not in M and let

$$\delta = \inf_{\mathbf{m} \text{ in } M} \|\mathbf{x} - \mathbf{m}\|$$

Let $\{\mathbf{m}_i\}$ be a sequence of vectors in M such that $\|\mathbf{x} - \mathbf{m}_i\| \to \delta$. According to the parallelogram law

$$\|(\mathbf{m}_j - \mathbf{x}) + (\mathbf{x} - \mathbf{m}_i)\|^2 + \|(\mathbf{m}_j - \mathbf{x}) - (\mathbf{x} - \mathbf{m}_i)\|^2 = 2\|\mathbf{m}_j - \mathbf{x}\|^2 + 2\|\mathbf{x} - \mathbf{m}_i\|^2 \quad (12.6.2)$$

Taking the second term on the left-hand side of (12.6.2) to the right-hand side gives

$$\|\mathbf{m}_i - \mathbf{m}_j\|^2 = 2\|\mathbf{m}_j - \mathbf{x}\|^2 + 2\|\mathbf{x} - \mathbf{m}_i\|^2 - 4\|\mathbf{x} - (\mathbf{m}_i + \mathbf{m}_j)/2\|^2 \quad (12.6.3)$$

Since $(\mathbf{m}_i + \mathbf{m}_j)/2$ is in M for all i and j, $\|\mathbf{x} - (\mathbf{m}_i + \mathbf{m}_j)/2\| \ge \delta$. Substituting this result in (12.6.3) yields

$$\|\mathbf{m}_i - \mathbf{m}_j\|^2 \le 2\|\mathbf{m}_j - \mathbf{x}\|^2 + 2\|\mathbf{x} - \mathbf{m}_i\|^2 - 4\delta^2$$

By hypothesis, $\|\mathbf{m}_i - \mathbf{x}\| \to \delta$ so that

$$\|\mathbf{m}_i - \mathbf{m}_j\|^2 \to 0 \qquad \text{as } i, j \to \infty$$

Consequently, $\{\mathbf{m}_i\}$ is a Cauchy sequence and converges to some limit $\mathbf{m}_0$ in M. By continuity of the norm, $\|\mathbf{x} - \mathbf{m}_0\| = \delta$ and we have shown that a minimizing vector $\mathbf{m}_0$ in M exists.

Q.E.D.

The Projection Theorem is illustrated in Fig. 12.6.1. Using the Pythagorean Theorem, we see that

$$\|\mathbf{x}\|^2 = \|\mathbf{m}_0\|^2 + \|\mathbf{x} - \mathbf{m}_0\|^2$$

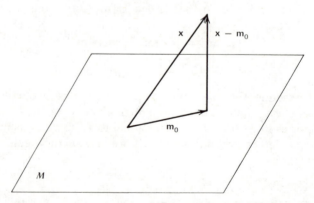

FIGURE 12.6.1.   Illustration of the Projection Theorem.

Thus, the squared norm of the error is

$$\|\mathbf{x} - \mathbf{m}_0\|^2 = \|\mathbf{x}\|^2 - \|\mathbf{m}_0\|^2 \qquad (12.6.4)$$

Also

$$\|\mathbf{m}_0\| \le \|\mathbf{x}\| \qquad (12.6.5)$$

The minimizing vector $\mathbf{m}_0$ is called the projection of $\mathbf{x}$ onto M and will be sometimes denoted as $\mathscr{E}\{\mathbf{x}/M\}$. We will call $\mathscr{E}\{\cdot/M\}$ a projection operator. The following theorem shows that a projection operator is a linear operator.

THEOREM 12.6.3.   Linearity of Projection Operators
Let M be a subspace of a Hilbert space H, $\mathbf{x}_1$ and $\mathbf{x}_2$ be vectors in H, and c_1 and c_2 be scalars. Then

$$\mathscr{E}\{c_1\mathbf{x}_1 + c_2\mathbf{x}_2/M\} = c_1\mathscr{E}\{\mathbf{x}_1/M\} + c_2\mathscr{E}\{\mathbf{x}_2/M\} \qquad (12.6.6)$$

Proof   Let $\hat{\mathbf{x}}_1 = \mathscr{E}\{\mathbf{x}_1/M\}$, $\hat{\mathbf{x}}_2 = \mathscr{E}\{\mathbf{x}_2/M\}$, and $\mathbf{m}$ be any vector in M. Then

$$(c_1\mathbf{x}_1 + c_2\mathbf{x}_2 - c_1\hat{\mathbf{x}}_1 - c_2\hat{\mathbf{x}}_2, \mathbf{m}) = c_1(\mathbf{x}_1 - \hat{\mathbf{x}}_1, \mathbf{m}) + c_2(\mathbf{x}_2 - \hat{\mathbf{x}}_2, \mathbf{m}) = 0$$

Q.E.D.

Some constrained minimization problems can be solved by finding the minimum norm vector in a linear variety. The following modification of the Projection Theorem solves this problem.

THEOREM 12.6.4.   The Dual Projection Theorem
Let M be a complete subspace of a Hilbert space H. Let $\mathbf{x}$ be a fixed vector in H and let B be the linear variety $\mathbf{x} + M$. Then there is a unique vector $\mathbf{b}_0$ in B

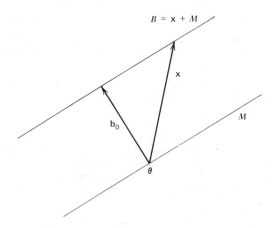

FIGURE 12.6.2.   Illustration of the Dual Projection Theorem.

of minimum norm. A necessary and sufficient condition that $\mathbf{b}_0$ be the minimum norm vector in B is that $\mathbf{b}_0 \perp M$.

Proof Any vector $\mathbf{b}$ in B has the form $\mathbf{b} = \mathbf{x} - \mathbf{m}$ where $\mathbf{m}$ is in M. Therefore, $\|\mathbf{b}\| = \|\mathbf{x} - \mathbf{m}\|$. According to the Projection Theorem, $\|\mathbf{x} - \mathbf{m}\|$ is minimized when $\mathbf{m}$ is $\mathbf{m}_0$, the projection of $\mathbf{x}$ onto M, and then $\mathbf{b}_0 = \mathbf{x} - \mathbf{m}_0$ is orthogonal to M.

Q.E.D.

The Dual Projection Theorem is illustrated in Fig. 12.6.2.

12.7 DECOMPOSITION OF A HILBERT SPACE INTO ORTHOGONAL SUBSPACES

Let M be a complete subspace of a Hilbert space H. The set of all vectors in H orthogonal to M is called the *orthogonal complement* of M and will be denoted by $M^\perp$. The sum of any two vectors orthogonal to M is also orthogonal to M. Therefore, $M^\perp$ is a subspace of H. It can be shown that $M^\perp$ is a complete subspace and that $M^{\perp\perp} = M$.

A vector space C is said to be the *direct sum* of the subspaces A and B if every vector in C can be uniquely represented in the form $\mathbf{a} + \mathbf{b}$ with $\mathbf{a}$ in A and $\mathbf{b}$ in B. The notation $C = A \oplus B$ will be used to denote that C is the direct sum of A and B. The following theorem shows that any Hilbert space can be represented as the direct sum of a subspace and its orthogonal complement.

THEOREM 12.7.1

Let M be a complete subspace of a Hilbert space H. Then $H = M \oplus M^\perp$. That is, any vector in H can be uniquely represented in the form $\hat{\mathbf{x}} + \tilde{\mathbf{x}}$ with $\hat{\mathbf{x}}$ in M and $\tilde{\mathbf{x}}$ in $M^\perp$.

Proof Let $\mathbf{x}$ be a vector in H, $\hat{\mathbf{x}} = \mathscr{E}\{\mathbf{x}/M\}$, and $\tilde{\mathbf{x}} = \mathbf{x} - \hat{\mathbf{x}}$. Then $\mathbf{x} = \tilde{\mathbf{x}} + \tilde{\mathbf{x}}$ with $\hat{\mathbf{x}}$ in M and $\tilde{\mathbf{x}}$ in $M^\perp$ according to the Projection Theorem.

To prove the uniqueness, suppose that there is another representation of the form $\mathbf{x} = \hat{\mathbf{y}} + \tilde{\mathbf{y}}$ with $\hat{\mathbf{y}}$ in M and $\tilde{\mathbf{y}}$ in $M^\perp$. Then $\mathbf{x} - \mathbf{x} = \boldsymbol{\theta} = (\hat{\mathbf{x}} - \hat{\mathbf{y}}) + (\tilde{\mathbf{x}} - \tilde{\mathbf{y}})$. Notice that $\hat{\mathbf{x}} - \hat{\mathbf{y}}$ is in M and $\tilde{\mathbf{x}} - \tilde{\mathbf{y}}$ is in $M^\perp$. Using the Pythagorean Theorem gives $0 = \|\boldsymbol{\theta}\|^2 = \|\hat{\mathbf{x}} - \hat{\mathbf{y}}\|^2 + \|\tilde{\mathbf{x}} - \tilde{\mathbf{y}}\|^2$ which is satisfied if and only if $\hat{\mathbf{x}} = \hat{\mathbf{y}}$ and $\tilde{\mathbf{x}} = \tilde{\mathbf{y}}$.

Q.E.D.

12.8 PROJECTIONS ONTO THE SUBSPACE GENERATED BY A GIVEN FINITE SET OF VECTORS

Suppose that the set of vectors $\mathbf{y}_1, \ldots, \mathbf{y}_n$ in a Hilbert space H generates the subspace M. Then the linear combination of $\mathbf{y}$'s that best approximates an arbitrary vector $\mathbf{x}$ in H in the minimum norm sense is the projection of $\mathbf{x}$ onto M. Let us denote this projection by

$$\hat{\mathbf{x}} = \mathscr{E}\{\mathbf{x}/M\} = \sum_{i=1}^{n} c_i \mathbf{y}_i \qquad (12.8.1)$$

According to the Projection Theorem, $(\mathbf{x} - \hat{\mathbf{x}}) \perp M$. This is the case if and only if

$$(\mathbf{x} - \hat{\mathbf{x}}, \mathbf{y}_k) = 0 \qquad \text{for} \qquad k = 1, \ldots, n \qquad (12.8.2)$$

Substituting the right-hand side of (12.8.1) into (12.8.2) gives the set of equations

$$\sum_{i=1}^{n} c_i (\mathbf{y}_i, \mathbf{y}_k) = (\mathbf{x}, \mathbf{y}_k) \qquad \text{for} \qquad k = 1, \ldots, n \qquad (12.8.3)$$

These equations are called the *normal equations* and can be put in the matrix form

$$\mathbf{GC} = \mathbf{B} \qquad (12.8.4)$$

where

$$\mathbf{G} = \begin{bmatrix} (\mathbf{y}_1, \mathbf{y}_1) & (\mathbf{y}_2, \mathbf{y}_1) & \cdots & (\mathbf{y}_n, \mathbf{y}_1) \\ (\mathbf{y}_1, \mathbf{y}_2) & (\mathbf{y}_2, \mathbf{y}_2) & \cdots & (\mathbf{y}_n, \mathbf{y}_2) \\ \cdot & & & \\ \cdot & & & \\ \cdot & & & \\ (\mathbf{y}_1, \mathbf{x}_n) & (\mathbf{y}_2, \mathbf{y}_n) & \cdots & (\mathbf{y}_n, \mathbf{y}_n) \end{bmatrix} \qquad (12.8.5)$$

$$\mathbf{C} = \begin{bmatrix} c_1 \\ c_2 \\ \cdot \\ \cdot \\ \cdot \\ c_n \end{bmatrix} \qquad \mathbf{B} = \begin{bmatrix} (\mathbf{x}, \mathbf{y}_1) \\ (\mathbf{x}, \mathbf{y}_2) \\ \cdot \\ \cdot \\ \cdot \\ (\mathbf{x}, \mathbf{y}_n) \end{bmatrix} \qquad (12.8.6)$$

The matrix $\mathbf{G}$ is called the *Gram matrix* of $\mathbf{y}_1, \ldots, \mathbf{y}_n$. The determinant of $\mathbf{G}$ is called the Gram determinant or *Gramian* of $\mathbf{y}_1, \ldots, \mathbf{y}_n$. If the Gram matrix is nonsingular or, equivalently, the Gramian is nonzero, then (12.8.4) has the unique solution

$$C = G^{-1}B \tag{12.8.7}$$

If G is singular, (12.8.4) has many solutions. The following theorem shows when G is nonsingular.

THEOREM 12.8.1
The matrix G is singular or, equivalently, $\det G = 0$, if and only if $y_1, \ldots, y_n$ are linearly dependent.

Proof If $y_1, \ldots, y_n$ are linearly dependent, then there is a set of scalars $a_1, \ldots, a_n$ which are not all zero such that

$$a_1 y_1 + a_2 y_2 + \cdots + a_n y_n = \boldsymbol{0}$$

Thus

$$\sum_{i=1}^{n} a_i (y_i, y_k) = \left(\sum_{i=1}^{n} a_i y_i, y_k \right) = 0 \qquad \text{for} \qquad k = 1, \ldots, n$$

so that the columns of G are linearly dependent and $\det G = 0$.

On the other hand, if $\det G = 0$, the columns of G must be linearly dependent and there is a set of constants $a_1, \ldots, a_n$ such that

$$\sum_{i=1}^{n} a_i (y_i, y_k) = 0 \qquad \text{for} \qquad k = 1, \ldots, n$$

Therefore

$$\left(\sum_{i=1}^{n} a_i y_i, y_k \right) = 0 \qquad \text{for} \qquad k = 1, \ldots, n$$

and

$$\sum_{k=1}^{n} \bar{a}_k \left(\sum_{i=1}^{n} a_i y_i, y_k \right) = \left\| \sum_{i=1}^{n} a_i y_i \right\|^2 = 0$$

Thus

$$\sum_{i=1}^{n} a_i y_i = \boldsymbol{0}$$

and $y_1, \ldots, y_n$ are linearly dependent.

Q.E.D.

12.9 THE DUAL APPROXIMATION PROBLEM

Suppose that the vectors $y_1, \ldots, y_n$ in a Hilbert space H are linearly independent and generate the subspace M. Consider the set D of vectors d in H such that

$$(\mathbf{d}, \mathbf{y}_k) = a_k \qquad \text{for} \qquad k = 1, \ldots, n \qquad (12.9.1)$$

If $a_k = 0$ for $k = 1, \ldots, n$, then $\mathbf{d} \perp M$ and $D = M^\perp$. In general, $D = \mathbf{d}_1 + M^\perp$ where $\mathbf{d}_1$ is any vector that satisfies (12.9.1). Thus D is a linear variety.

The minimum norm vector $\mathbf{d}_0$ in D can be found by applying Theorem 12.6.4, the Dual Projection Theorem. According to this theorem, $\mathbf{d}_0 \perp M^\perp$ so that $\mathbf{d}_0$ is in $M^{\perp\perp} = M$. Thus $\mathbf{d}_0$ must have the form

$$\mathbf{d}_0 = \sum_{i=1}^{n} c_i \mathbf{y}_i \qquad (12.9.2)$$

Substituting (12.9.2) into (12.9.1), we obtain the set of equations

$$\sum_{i=1}^{n} c_i(\mathbf{y}_i, \mathbf{y}_k) = a_k \qquad \text{for} \qquad k = 1, \ldots, n \qquad (12.9.3)$$

or the matrix equation

$$\mathbf{GC} = \mathbf{A} \qquad (12.9.4)$$

where $\mathbf{G}$ and $\mathbf{C}$ are given by (12.8.5) and (12.8.6), respectively, and

$$\mathbf{A} = \begin{bmatrix} a_1 \\ a_2 \\ \cdot \\ \cdot \\ \cdot \\ a_n \end{bmatrix} \qquad (12.9.5)$$

By hypothesis, $\mathbf{y}_1, \ldots, \mathbf{y}_n$ are linearly independent so that the Gram matrix $\mathbf{G}$ is nonsingular. Thus the solution to (12.9.4) is

$$\mathbf{C} = \mathbf{G}^{-1}\mathbf{A} \qquad (12.9.6)$$

12.10 THE GRAM-SCHMIDT PROCEDURE FOR GENERATING A SET OF ORTHONORMAL VECTORS

A set of vectors $\mathbf{x}_1, \ldots, \mathbf{x}_n$ in a Hilbert space is called an *orthogonal set* if $(\mathbf{x}_i, \mathbf{x}_j) = 0$ for $i \neq j$. If, in addition, $\|\mathbf{x}_i\| = 1$ for all i, the set is called an *orthonormal set*.

An important property of a set of nonzero orthogonal vectors $\mathbf{x}_1, \ldots, \mathbf{x}_n$ is that they are linearly independent. Suppose $c_1, \ldots, c_n$ are scalars such that

$$\sum_{i=1}^{n} c_i \mathbf{x}_i = \mathbf{0}$$

Then

$$0 = (\mathbf{0}, \mathbf{x}_k) = \left(\sum_{i=1}^{n} c_i \mathbf{x}_i, \mathbf{x}_k\right) = \sum_{i=1}^{n} c_i(\mathbf{x}_i, \mathbf{x}_k) = c_k(\mathbf{x}_k, \mathbf{x}_k)$$

so that $c_k = 0$ for $k = 1, \ldots, n$ and the $\mathbf{x}$'s are linearly independent.

Orthonormal sets are convenient for both theoretical and practical applications. As an example, consider the projection problem discussed in Section 12.8. If $\mathbf{y}_1, \ldots, \mathbf{y}_n$ is an orthonormal set, then the Gram matrix is the identity matrix and no matrix inversion is required. In the following theorem, we will see that any finite set of n linearly independent vectors can be converted into a set of n orthonormal vectors that generates the same subspace. This technique is equivalent, in a sense, to inverting the Gram matrix.

THEOREM 12.10.1. The Gram-Schmidt Procedure

Let $\{\mathbf{x}_i\}$ be a finite or countable sequence of linearly independent vectors in a Hilbert space H. Then there exists an orthonormal sequence $\{\mathbf{u}_i\}$ such that for each n the space generated by $\mathbf{x}_1, \ldots, \mathbf{x}_n$ is the same as that generated by $\mathbf{u}_1, \ldots, \mathbf{u}_n$.

Proof The orthonormal sequence can be generated recursively. First, let

$$\mathbf{u}_1 = \mathbf{x}_1/\|\mathbf{x}_1\|$$

Clearly, $\mathbf{u}_1$ has unity norm and generates the same space as $\mathbf{x}_1$. Let us denote the space generated by $\mathbf{u}_i$ as U_i. Next, let

$$\mathbf{v}_2 = \mathbf{x}_2 - \mathscr{E}\{\mathbf{x}_2/U_1\} = \mathbf{x}_2 - (\mathbf{x}_2, \mathbf{u}_1)\mathbf{u}_1$$

and

$$\mathbf{u}_2 = \mathbf{v}_2/\|\mathbf{v}_2\|$$

According to the Projection Theorem, $\mathbf{u}_2 \perp \mathbf{u}_1$. The vector $\mathbf{v}_2$ cannot be zero since $\mathbf{x}_2$ and $\mathbf{u}_1$ are linearly independent. Also, $\mathbf{u}_1$ and $\mathbf{u}_2$ generate the same space as $\mathbf{x}_1$ and $\mathbf{x}_2$ since $\mathbf{x}_2$ can be expressed as a linear combination of $\mathbf{u}_1$ and $\mathbf{u}_2$. The remaining $\mathbf{u}$'s can be generated recursively according to the rule

$$\mathbf{v}_n = \mathbf{x}_n - \sum_{i=1}^{n-1} \mathscr{E}\{\mathbf{x}_n/U_i\} = \mathbf{x}_n - \sum_{i=1}^{n-1} (\mathbf{x}_n, \mathbf{u}_i)\mathbf{u}_i \qquad (12.10.1)$$

and

$$\mathbf{u}_n = \mathbf{v}_n/\|\mathbf{v}_n\| \qquad (12.10.2)$$

By direct computation, it can be shown that $(\mathbf{v}_n, \mathbf{u}_k) = 0$ for $k < n$. Also, $\mathbf{v}_n$

cannot be zero since it can be expressed as a linear combination of independent vectors. It follows by induction that $\mathbf{u}_1, \ldots, \mathbf{u}_n$ and $\mathbf{x}_1, \ldots, \mathbf{x}_n$ generate the same space.

Q.E.D.

12.11 FOURIER SERIES

Let $\{\mathbf{u}_i\}$ be an infinite sequence of orthonormal vectors in a Hilbert space H, M_n be the subspace generated by $\mathbf{u}_1, \ldots, \mathbf{u}_n$, and $\mathbf{x}$ be any vector in H. According to the Projection Theorem, $\|\mathbf{x} - \mathbf{x}_n\|$ is minimized for $\mathbf{x}_n$ in M_n by the vector

$$\hat{\mathbf{x}}_n = \mathscr{E}\{\mathbf{x}/M_n\}$$

The Gram matrix is the identity matrix in this case. Therefore, we find from (12.8.1) and (12.8.4) that

$$\hat{\mathbf{x}}_n = \sum_{i=1}^{n} (\mathbf{x}, \mathbf{u}_i)\mathbf{u}_i \tag{12.11.1}$$

Notice that $(\mathbf{x}, \mathbf{u}_i)\mathbf{u}_i$ is the projection of $\mathbf{x}$ onto the space U_i generated by $\mathbf{u}_i$. Therefore

$$\hat{\mathbf{x}}_n = \sum_{i=1}^{n} \mathscr{E}\{\mathbf{x}/U_i\}$$

This is simply a generalization of the direct sum decomposition of a Hilbert space discussed in Section 12.7 and we can write that $M_n = U_1 \oplus U_2 \oplus \cdots \oplus U_n$. Using (12.6.4) gives

$$\|\mathbf{x} - \hat{\mathbf{x}}_n\|^2 = \|\mathbf{x}\|^2 - \sum_{i=1}^{n} |(\mathbf{x}, \mathbf{u}_i)|^2 \tag{12.11.2}$$

The numbers $f_i = (\mathbf{x}, \mathbf{u}_i)$ are called the *Fourier coefficients* of $\mathbf{x}$ relative to the set $\{\mathbf{u}_i\}$. The series

$$\sum_{i=1}^{\infty} f_i \mathbf{u}_i$$

is called the *Fourier series* of $\mathbf{x}$ relative to the set $\{\mathbf{u}_i\}$.

It follows from (12.11.2) that

$$\sum_{i=1}^{n} |f_i|^2 \leq \|\mathbf{x}\|^2 \qquad \text{for all } n$$

Therefore

$$\sum_{i=1}^{\infty} |f_i|^2 \leq \|\mathbf{x}\|^2 \tag{12.11.3}$$

This is called *Bessel's inequality*.

The orthonormal sequence $\{\mathbf{u}_i\}$ is said to be *complete* if

$$\sum_{i=1}^{\infty} |f_i|^2 = \|\mathbf{x}\|^2 \qquad (12.11.4)$$

for all $\mathbf{x}$ in H. This is called *Parseval's equality*. It follows from (12.11.2) that an orthonormal sequence is complete if and only if the partial sums $\hat{\mathbf{x}}_n$ of the Fourier series of every vector $\mathbf{x}$ in H converge to $\mathbf{x}$.

Bessel's inequality implies that it is necessary that the series

$$\sum_{i=1}^{\infty} |f_i|^2$$

converge if $\{f_i\}$ are the Fourier coefficients of a vector in H. The following theorem shows that this condition is also sufficient.

THEOREM 12.11.1

Let $\{\mathbf{u}_i\}$ be an orthonormal set of vectors in a Hilbert space H and let $\{c_i\}$ be a sequence of numbers such that

$$\sum_{i=1}^{\infty} |c_i|^2 < \infty \qquad (12.11.5)$$

Then there exists a vector $\hat{\mathbf{x}}$ in H such that

$$c_i = (\hat{\mathbf{x}}, \mathbf{u}_i)$$

and

$$\sum_{i=1}^{\infty} |c_i|^2 = \|\hat{\mathbf{x}}\|^2 \qquad (12.11.6)$$

Proof Let

$$\mathbf{x}_n = \sum_{i=1}^{n} c_i \mathbf{u}_i$$

Then, for $n < m$

$$\|\mathbf{x}_n - \mathbf{x}_m\|^2 = \left\| \sum_{i=n+1}^{m} c_i \mathbf{u}_i \right\|^2 = \sum_{i=n+1}^{m} |c_i|^2 \qquad (12.11.7)$$

The right-hand side of (12.11.7) converges to zero as $n \to \infty$ as a result of (12.11.5). Therefore, $\mathbf{x}_n$ is a Cauchy sequence and converges to an element $\hat{\mathbf{x}}$ in H. Now

$$(\hat{\mathbf{x}}, \mathbf{u}_k) = (\mathbf{x}_n, \mathbf{u}_k) + (\hat{\mathbf{x}} - \mathbf{x}_n, \mathbf{u}_k) \qquad (12.11.8)$$

The first term on the right-hand side of (12.11.8) becomes c_k for $n \geq k$. Since

$$|(\hat{\mathbf{x}} - \mathbf{x}_n, \mathbf{u}_k)| \leq \|\hat{\mathbf{x}} - \mathbf{x}_n\| \|\mathbf{u}_k\|$$

the second term becomes zero as $n \to \infty$. The left-hand side of (12.11.8) is independent of n so that letting $n \to \infty$ on the right-hand side gives

$$(\hat{\mathbf{x}}, \mathbf{u}_k) = c_k$$

Equation 12.11.6 follows from the fact that $\|\hat{\mathbf{x}} - \mathbf{x}_n\| \to 0$ as $n \to \infty$.

Q.E.D.

13

LINEAR PARAMETER
ESTIMATION

13.1 INTRODUCTION

Three common types of parameter estimation problems that can be formulated as minimum norm problems in Hilbert space and, consequently, solved by applications of the Projection Theorem are discussed in this chapter. The first problem discussed is the approximation of a measured N-point sequence by a linear combination of a given set of N-point sequences to minimize the sum of the squares of the errors. This is called the linear least-squares approximation problem. In the second problem, an N-point observed data sequence is assumed to be the sum of a linear transformation of a set of unknown but deterministic parameters and a sequence of random variables representing the observation noise. The problem is to find the linear combinations of the observed data that are the minimum-variance unbiased estimates of the unknown parameters. In the third problem, both the parameters to be estimated and the observed data are considered to be random variables and the problem is to find the linear combinations of the observed data that are the minimum mean square error estimates of the parameters.

In each of the three problems, the estimates are constrained to be linear functions of the observed data and are optimum in the sense of minimizing a quadratic cost function. Explicit formulas for the optimum estimates can easily be found in this framework. Estimates that are not constrained to be linear and that satisfy other optimality criteria can be used. The most common of these are the Bayes', maximum likelihood, and maximum a posteriori probability estimates. Detailed discussions of these estimates can be found in References

30 and 118. In general, these estimates may be difficult to find. However, they reduce to the linear estimates discussed in this chapter when the random variables involved are Gaussian.

The method of least-squares estimation has been traced back to publications by Legendre in 1806 [83] and Gauss in 1809 [35]. The method was proposed and used for calculating planetary orbits.

13.2 LINEAR LEAST-SQUARES APPROXIMATION

The problem of approximating an N-point measured data sequence by a linear combination of K preselected N-point sequences with $K \ll N$ arises in a variety of situations. An intuitively reasonable criterion is to choose the approximation to minimize the sum of the squares of the errors. This problem can be formulated in the space of complex N-dimensional column vectors with the inner product of any two vectors $\mathbf{u}$ and $\mathbf{v}$ given by

$$(\mathbf{u}, \mathbf{v}) = \mathbf{v}^{\dagger}\mathbf{u} = \sum_{n=1}^{N} \bar{v}_n u_n \qquad (13.2.1)$$

where $\dagger$ denotes conjugate transpose. Suppose that the measured data is arranged in the column vector

$$\mathbf{x} = \begin{bmatrix} x_1 \\ x_2 \\ \cdot \\ \cdot \\ \cdot \\ x_N \end{bmatrix}$$

and that the preselected sequences are the N-dimensional column vectors $\mathbf{y}_1, \ldots, \mathbf{y}_K$. Then, the problem is to find the constants $c_1, \ldots, c_K$ such that the vector

$$\hat{\mathbf{x}} = \sum_{k=1}^{K} c_k \mathbf{y}_k = \mathbf{Yc} \qquad (13.2.2)$$

where

$$\mathbf{c} = \begin{bmatrix} c_1 \\ c_2 \\ \cdot \\ \cdot \\ \cdot \\ c_K \end{bmatrix} \qquad \text{and} \qquad \mathbf{Y} = [\mathbf{y}_1 \vdots \mathbf{y}_2 \vdots \cdots \vdots \mathbf{y}_K]$$

minimizes

$$\epsilon^2 = \|\mathbf{x} - \hat{\mathbf{x}}\|^2 = (\mathbf{x} - \hat{\mathbf{x}})^\dagger(\mathbf{x} - \hat{\mathbf{x}}) = \sum_{n=1}^{N} |x_n - \hat{x}_n|^2 \qquad (13.2.3)$$

According to Theorem 12.6.2, the Projection Theorem, ϵ^2 is minimized when $\hat{\mathbf{x}}$ is the projection of $\mathbf{x}$ onto the subspace generated by the columns of $\mathbf{Y}$. The Gram matrix $\mathbf{G}$ defined by (12.8.5) and the matrix $\mathbf{B}$ defined in (12.8.6) can be written as $\mathbf{G} = \mathbf{Y}^\dagger\mathbf{Y}$ and $\mathbf{B} = \mathbf{Y}^\dagger\mathbf{x}$. We will assume that the columns of $\mathbf{Y}$ are linearly independent so that $\mathbf{G}$ is nonsingular. Then according to (12.8.7), the optimum coefficient vector is

$$\mathbf{c} = (\mathbf{Y}^\dagger\mathbf{Y})^{-1}\mathbf{Y}^\dagger\mathbf{x} \qquad (13.2.4)$$

The corresponding value for ϵ^2 can be found using the Pythagorean theorem. An alternate method is to remember that $\mathbf{x} - \hat{\mathbf{x}}$ is orthogonal to $\hat{\mathbf{x}}$ and, consequently

$$\epsilon^2 = (\mathbf{x} - \hat{\mathbf{x}}, \mathbf{x} - \hat{\mathbf{x}}) = (\mathbf{x} - \hat{\mathbf{x}}, \mathbf{x}) = (\mathbf{x}, \mathbf{x}) - (\hat{\mathbf{x}}, \mathbf{x}) \qquad (13.2.5)$$

Substituting (13.2.2) and (13.2.4) into (13.2.5) gives

$$\epsilon^2 = \mathbf{x}^\dagger\mathbf{x} - \mathbf{x}^\dagger\mathbf{Y}(\mathbf{Y}^\dagger\mathbf{Y})^{-1}\mathbf{Y}^\dagger\mathbf{x} \qquad (13.2.6)$$

Example 13.2.1 *Polynomial Fits*

Suppose that a real signal $x(t)$ is sampled at times $t_1, \ldots, t_N$ and we wish to find the polynomial

$$\hat{x}(t) = \sum_{k=1}^{K} c_k t^{k-1}$$

which minimizes

$$\sum_{n=1}^{N} [x(t_n) - \hat{x}(t_n)]^2$$

We can arrange the samples of $x(t)$ in the column vector

$$\mathbf{x} = \begin{bmatrix} x(t_1) \\ x(t_2) \\ \cdot \\ \cdot \\ \cdot \\ x(t_N) \end{bmatrix}$$

The samples of $\hat{x}(t)$ can be expressed as

$$
\begin{bmatrix}
\hat{x}(t_1) \\
\hat{x}(t_2) \\
\cdot \\
\cdot \\
\cdot \\
\hat{x}(t_N)
\end{bmatrix}
=
\begin{bmatrix}
1 & t_1 & t_1^2 & \cdots & t_1^{K-1} \\
1 & t_2 & t_2^2 & \cdots & t_2^{K-1} \\
\cdot & \cdot & \cdot & & \cdot \\
\cdot & \cdot & \cdot & \cdots & \cdot \\
\cdot & \cdot & \cdot & & \cdot \\
1 & t_N & t_N^2 & \cdots & t_N^{K-1}
\end{bmatrix}
\begin{bmatrix}
c_1 \\
c_2 \\
\cdot \\
\cdot \\
\cdot \\
c_K
\end{bmatrix}
$$

or as

$$\hat{\mathbf{x}} = \mathbf{Yc}$$

with the obvious matrix correspondences. The optimum set of coefficients is then given by (13.2.4). ◀

13.3 VECTOR RANDOM VARIABLES

The least-squares parameter estimation problem discussed in the previous section was formulated using a deterministic point of view. In Sections 13.4 and 13.5, we will consider the observed data and, consequently, the parameter estimates to be random variables. It will be convenient to represent a set of N random variables $x_1, x_2, \ldots, x_N$ as a column vector

$$
\mathbf{x} =
\begin{bmatrix}
x_1 \\
x_2 \\
\cdot \\
\cdot \\
\cdot \\
x_N
\end{bmatrix}
$$

We will call $\mathbf{x}$ an *N-dimensional vector random variable*.

The *expected value* or *mean* of $\mathbf{x}$ is defined as the column vector whose components are the expected values of the corresponding components of $\mathbf{x}$, that is

$$
E\{\mathbf{x}\} =
\begin{bmatrix}
E\{x_1\} \\
\cdot \\
\cdot \\
\cdot \\
E\{x_N\}
\end{bmatrix}
\tag{13.3.1}
$$

More generally, the expected value of an $N \times M$ matrix of random variables is defined as the $N \times M$ matrix of corresponding expected values.

The *covariance matrix* for an N-dimensional vector random variable $\mathbf{x}$ with mean $\boldsymbol{\mu}$ is defined as the $N \times N$ matrix

$$\text{cov}\{\mathbf{x}\} = E\{(\mathbf{x}-\boldsymbol{\mu})(\mathbf{x}-\boldsymbol{\mu})^{\dagger}\} = \mathbf{V}_{\mathbf{xx}} \tag{13.3.2}$$

The ijth element of the covariance matrix is $\text{cov}\{x_i, x_j\}$. Thus the diagonal elements are the variances of the components of $\mathbf{x}$. Since the operations E and $\dagger$ commute, it follows that $\mathbf{V}_{\mathbf{xx}}^{\dagger} = \mathbf{V}_{\mathbf{xx}}$. A matrix that is equal to its conjugate transpose is said to be *Hermitian*. If $\mathbf{x}$ is real, then the covariance matrix is simply *symmetric*, that is, $\mathbf{V}_{\mathbf{xx}}^{t} = \mathbf{V}_{\mathbf{xx}}$ where the superscript t denotes transpose.

The covariance matrix $\mathbf{V}_{\mathbf{xx}}$ is also *nonnegative definite*, that is

$$\mathbf{c}^{\dagger}\mathbf{V}_{\mathbf{xx}}\mathbf{c} \geq 0 \tag{13.3.3}$$

for all N-dimensional column vectors $\mathbf{c}$ of complex numbers. This can be shown as follows

$$0 \leq E\{|\mathbf{c}^{\dagger}(\mathbf{x}-\boldsymbol{\mu})|^{2}\} = E\{\mathbf{c}^{\dagger}(\mathbf{x}-\boldsymbol{\mu})(\mathbf{x}-\boldsymbol{\mu})^{\dagger}\mathbf{c}\} = \mathbf{c}^{\dagger}E\{\mathbf{x}-\boldsymbol{\mu})(\mathbf{x}-\boldsymbol{\mu})^{\dagger}\}\mathbf{c} = \mathbf{c}^{\dagger}\mathbf{V}_{\mathbf{xx}}\mathbf{c} \tag{13.3.4}$$

$\mathbf{V}_{\mathbf{xx}}$ is said to be *positive definite* when equality can be achieved in (13.3.3) if and only if $\mathbf{c} = \boldsymbol{0}$. It can be shown that any positive definite matrix is nonsingular. It follows from (13.3.4) that if equality can be achieved for a nonzero $\mathbf{c}$, then $\mathbf{c}^{\dagger}(\mathbf{x}-\boldsymbol{\mu})$ is the zero random variable and the components of $\mathbf{x}$ are linearly related. In the future, we will assume that this situation is not present so that $\mathbf{V}_{\mathbf{xx}}$ is positive definite.

The cross-covariance matrix for an N-dimensional vector random variable $\mathbf{x}$ with mean $\boldsymbol{\mu}$ and an M-dimensional vector random variable $\mathbf{w}$ with mean $\boldsymbol{\lambda}$ is defined as the $N \times M$ matrix

$$\text{cov}\{\mathbf{x}, \mathbf{w}\} = E\{(\mathbf{x}-\boldsymbol{\mu})(\mathbf{w}-\boldsymbol{\lambda})^{\dagger}\} = \mathbf{V}_{\mathbf{xw}} \tag{13.3.5}$$

Since E and $\dagger$ commute, $\mathbf{V}_{\mathbf{xw}}^{\dagger} = \mathbf{V}_{\mathbf{wx}}$.

Linear transformations of vector random variables are often formed. Suppose that $\mathbf{x}$ is an N-dimensional vector random variable with mean $\boldsymbol{\mu}$ and covariance matrix $\mathbf{V}_{\mathbf{xx}}$. A new M-dimensional vector random variable $\mathbf{w}$ is formed by the transformation $\mathbf{w} = \mathbf{A}\mathbf{x}$ where $\mathbf{A}$ is an $M \times N$ constant matrix. Then

$$\boldsymbol{\lambda} = E\{\mathbf{w}\} = E\{\mathbf{A}\mathbf{x}\} = \mathbf{A}E\{\mathbf{x}\} = \mathbf{A}\boldsymbol{\mu} \tag{13.3.6}$$

Therefore

$$\mathbf{w} - \boldsymbol{\lambda} = \mathbf{A}(\mathbf{x}-\boldsymbol{\mu})$$

and

$$\text{cov}\{\mathbf{w}\} = E\{(\mathbf{w}-\boldsymbol{\lambda})(\mathbf{w}-\boldsymbol{\lambda})^{\dagger}\} = \mathbf{A}E\{(\mathbf{x}-\boldsymbol{\mu})(\mathbf{x}-\boldsymbol{\mu})^{\dagger}\}\mathbf{A}^{\dagger} = \mathbf{A}\mathbf{V}_{\mathbf{xx}}\mathbf{A}^{\dagger} \tag{13.3.7}$$

13.4 LINEAR MINIMUM-VARIANCE UNBIASED ESTIMATES

In many situations, it is reasonable to assume that an observed data vector $\mathbf{x}$ is related to an unknown but deterministic parameter vector $\boldsymbol{\beta}$ by an equation of the form

$$\mathbf{x} = \mathbf{Y}\boldsymbol{\beta} + \mathbf{e} \qquad (13.4.1)$$

where $\mathbf{x}$ is an N-dimensional column vector, $\boldsymbol{\beta}$ is a K-dimensional column vector, $\mathbf{Y}$ is a known $N \times K$ matrix, and $\mathbf{e}$ is an N-dimensional vector random variable with zero mean and the positive definite covariance matrix $\mathbf{Q}$. The problem is to estimate $\boldsymbol{\beta}$ by a linear transformation of $\mathbf{x}$.

To add some generality, suppose that we wish to estimate

$$\mathbf{b} = \mathbf{F}\boldsymbol{\beta} \qquad (13.4.2)$$

where $\mathbf{F}$ is a known $L \times K$ matrix using an estimator of the form

$$\hat{\mathbf{b}} = \mathbf{D}\mathbf{x} \qquad (13.4.3)$$

where $\mathbf{D}$ is an $L \times N$ constant matrix. We will require the estimator to be unbiased, that is

$$E\{\hat{\mathbf{b}}\} = \mathbf{b} = \mathbf{F}\boldsymbol{\beta} \qquad (13.4.4)$$

Using (13.4.1) and (13.4.3), we have

$$E\{\hat{\mathbf{b}}\} = E\{\mathbf{D}\mathbf{x}\} = \mathbf{D}E\{\mathbf{Y}\boldsymbol{\beta} + \mathbf{e}\} = \mathbf{D}\mathbf{Y}\boldsymbol{\beta} \qquad (13.4.5)$$

Comparing (13.4.4) and (13.4.5), we see that $\mathbf{D}$ must satisfy the constraint

$$\mathbf{D}\mathbf{Y} = \mathbf{F} \qquad (13.4.6)$$

With the constraint $\mathbf{D}\mathbf{Y} = \mathbf{F}$, we will select $\mathbf{D}$ to minimize the mean square errors in the estimates of the components of $\mathbf{b}$, that is

$$E\{|\hat{b}_i - b_i|^2\} = \text{var } \hat{b}_i \qquad \text{for} \qquad i = 1, \ldots, L$$

The error between $\hat{\mathbf{b}}$ and $\mathbf{b}$ is

$$\hat{\mathbf{b}} - \mathbf{b} = \mathbf{D}\mathbf{x} - \mathbf{F}\boldsymbol{\beta} = \mathbf{D}\mathbf{Y}\boldsymbol{\beta} + \mathbf{D}\mathbf{e} - \mathbf{F}\boldsymbol{\beta} = \mathbf{D}\mathbf{e}$$

Therefore, the error covariance matrix is

$$\boldsymbol{\Gamma}_{\mathbf{b}} = E\{(\hat{\mathbf{b}} - \mathbf{b})(\hat{\mathbf{b}} - \mathbf{b})^\dagger\} = \mathbf{D}\mathbf{Q}\mathbf{D}^\dagger \qquad (13.4.7)$$

The mean square errors are the diagonal terms of $\boldsymbol{\Gamma}_{\mathbf{b}}$. If we denote the ith rows of $\mathbf{D}$ and $\mathbf{F}$ by $\mathbf{d}_i$ and $\mathbf{f}_i$, respectively, then the problem becomes to minimize

$$\text{var } \hat{b}_i = \mathbf{d}_i \mathbf{Q} \mathbf{d}_i^\dagger \qquad (13.4.8)$$

subject to the constraint

$$\mathbf{d}_i \mathbf{Y} = \mathbf{f}_i \qquad (13.4.9)$$

for $i = 1, \ldots, L$.

This problem can be formulated as a set of dual approximation problems in the space of N-dimensional row vectors over the field of complex numbers. Let us define the inner product of any two vectors $\mathbf{u}$ and $\mathbf{v}$ in this space as

$$(\mathbf{u}, \mathbf{v}) = \mathbf{u} \mathbf{Q} \mathbf{v}^\dagger \qquad (13.4.10)$$

Then (13.4.8) can be written as

$$\text{var } \hat{b}_i = (\mathbf{d}_i, \mathbf{d}_i) = \|\mathbf{d}_i\|^2 \qquad (13.4.11)$$

Equation 13.4.9 can be written as

$$\mathbf{d}_i \mathbf{Q}(\mathbf{Q}^{-1}\mathbf{Y}) = \mathbf{f}_i \qquad (13.4.12)$$

Let us denote the kth row of $(\mathbf{Q}^{-1}\mathbf{Y})^\dagger = \mathbf{Y}^\dagger \mathbf{Q}^{-1}$ by $\mathbf{y}_k$. Then (13.4.12) is equivalent to

$$(\mathbf{d}_i, \mathbf{y}_k) = f_{ik} \qquad \text{for} \qquad k = 1, \ldots, K \qquad (13.4.13)$$

where f_{ik} is the ikth element of $\mathbf{F}$. Thus, the parameter estimation problem can be considered to be the L dual approximation problems of finding the minimum norm vectors $\mathbf{d}_i$ subject to the constraints of (13.4.13) for $i = 1, \ldots, L$.

According to the discussion in Section 12.9, the ith row of the optimum $\mathbf{D}$ must have the form

$$\mathbf{d}_i = \sum_{k=1}^{K} h_{ik} \mathbf{y}_k = \mathbf{h}_i \mathbf{Y}^\dagger \mathbf{Q}^{-1} \qquad (13.4.14)$$

where $\mathbf{h}_i = [h_{i1}, \ldots, h_{iK}]$ and $\{h_{ik}\}$ is a set of constants. The transpose of the Gram matrix $\mathbf{G}$ defined by (12.8.5) is

$$\mathbf{G}^t = (\mathbf{Q}^{-1}\mathbf{Y})^\dagger \mathbf{Q}(\mathbf{Q}^{-1}\mathbf{Y}) = \mathbf{Y}^\dagger \mathbf{Q}^{-1} \mathbf{Y} \qquad (13.4.15)$$

The matrices $\mathbf{C}$ and $\mathbf{A}$ defined in Section 12.9 correspond to $\mathbf{h}_i^t$ and $\mathbf{f}_i^t$, respectively. Therefore, according to (12.9.6), the optimum choice for $\mathbf{h}_i$ is

$$\mathbf{h}_i = (\mathbf{G}^{-1}\mathbf{f}_i^t)^t = \mathbf{f}_i(\mathbf{G}^t)^{-1} = \mathbf{f}_i(\mathbf{Y}^\dagger \mathbf{Q}^{-1}\mathbf{Y})^{-1} \qquad (13.4.16)$$

Substituting (13.4.16) into (13.4.14), we find that the optimum value for $\mathbf{d}_i$ is

$$\mathbf{d}_i = \mathbf{f}_i(\mathbf{Y}^\dagger \mathbf{Q}^{-1}\mathbf{Y})^{-1}\mathbf{Y}^\dagger \mathbf{Q}^{-1} \qquad (13.4.17)$$

and, on combining the results for $i = 1, \ldots, L$, that the optimum $\mathbf{D}$ is

$$\mathbf{D} = \mathbf{F}(\mathbf{Y}^\dagger \mathbf{Q}^{-1}\mathbf{Y})^{-1}\mathbf{Y}^\dagger \mathbf{Q}^{-1} \qquad (13.4.18)$$

Substituting (13.4.18) into (13.4.7), it follows that the error covariance matrix for the optimum $\mathbf{D}$ is

$$\mathbf{\Gamma_b} = \mathbf{DQD}^\dagger = \mathbf{F}(\mathbf{Y}^\dagger \mathbf{Q}^{-1} \mathbf{Y})^{-1} \mathbf{F}^\dagger \qquad (13.4.19)$$

Letting $\mathbf{F}$ be the $K \times K$ identity matrix, we find that the optimum estimate of the unknown parameter vector $\boldsymbol{\beta}$ is

$$\hat{\boldsymbol{\beta}} = (\mathbf{Y}^\dagger \mathbf{Q}^{-1} \mathbf{Y})^{-1} \mathbf{Y}^\dagger \mathbf{Q}^{-1} \mathbf{x} \qquad (13.4.20)$$

and that the corresponding error covariance matrix is

$$\mathbf{\Gamma_\beta} = (\mathbf{Y}^\dagger \mathbf{Q}^{-1} \mathbf{Y})^{-1} \qquad (13.4.21)$$

Thus, for an arbitrary $\mathbf{F}$,

$$\hat{\mathbf{b}} = \mathbf{F}\hat{\boldsymbol{\beta}} \qquad \text{and} \qquad \mathbf{\Gamma_b} = \mathbf{F}\mathbf{\Gamma_\beta}\mathbf{F}^\dagger \qquad (13.4.22)$$

In other words, the optimum estimate of a linear transformation of $\boldsymbol{\beta}$ is the linear transformation of the optimum estimate of $\boldsymbol{\beta}$.

Now suppose that the components of the noise vector $\mathbf{e}$ are uncorrelated and all have the same variance σ^2. Then the covariance matrix for $\mathbf{e}$ is $\mathbf{Q} = \sigma^2 \mathbf{I}$ and

$$\hat{\boldsymbol{\beta}} = (\mathbf{Y}^\dagger \mathbf{Y})^{-1} \mathbf{Y}^\dagger \mathbf{x} \qquad (13.4.23)$$

Notice that this estimate has exactly the same form as the linear least-squares estimate given by (13.2.4).

The linear minimum-variance unbiased estimates are also known as Gauss-Markov estimates.

Example 13.4.1

Suppose that an object is subjected to an unknown constant acceleration of magnitude $\ddot{p}(0)$. Then its position $p(t)$ in the direction of the acceleration satisfies the differential equation

$$\ddot{p}(t) = \ddot{p}(0)$$

which has the solution

$$p(t) = p(0) + t\dot{p}(0) + \tfrac{1}{2}t^2 \ddot{p}(0)$$

Suppose that the data

$$x(nT) = p(nT) + e(nT) \qquad \text{for} \qquad n = 0, \ldots, N$$

where $e(nT)$ is a white noise sequence with variance σ^2 is observed. This data can be put in the form

$$\mathbf{x} = \mathbf{Y}\boldsymbol{\beta} + \mathbf{e}$$

where

$$\mathbf{x} = \begin{bmatrix} x(0) \\ x(T) \\ \cdot \\ \cdot \\ \cdot \\ x(NT) \end{bmatrix}, \quad \mathbf{e} = \begin{bmatrix} e(0) \\ e(T) \\ \cdot \\ \cdot \\ \cdot \\ e(NT) \end{bmatrix}, \quad \boldsymbol{\beta} = \begin{bmatrix} p(0) \\ \dot{p}(0) \\ \ddot{p}(0) \end{bmatrix}$$

and

$$\mathbf{Y} = \begin{bmatrix} 1 & 0 & 0 \\ 1 & T & T^2/2 \\ \cdot & \cdot & \cdot \\ \cdot & \cdot & \cdot \\ \cdot & \cdot & \cdot \\ 1 & NT & (NT)^2/2 \end{bmatrix}$$

The covariance matrix for $\mathbf{e}$ is $\mathbf{Q} = \sigma^2 \mathbf{I}$. Therefore, the optimum estimates of the initial position, velocity, and acceleration are given by (13.4.23). The corresponding error covariance matrix is $\boldsymbol{\Gamma_\beta} = \sigma^2 (\mathbf{Y}^\dagger \mathbf{Y})^{-1}$. Now suppose that we wish to estimate $p(t)$ for an arbitrary value of t. The position $p(t)$ can be written in the matrix form

$$p(t) = \mathbf{F}(t)\boldsymbol{\beta}$$

where

$$\mathbf{F}(t) = [1 \quad t \quad t^2/2]$$

Thus, the optimum estimate of $p(t)$ is

$$\hat{p}(t) = \mathbf{F}(t)\hat{\boldsymbol{\beta}} \qquad \blacktriangleleft$$

Example 13.4.2

Suppose that in a particular communication system a signal of the form

$$s(t) = a_k \cos k\omega_d t + b_k \sin k\omega_d t$$

is transmitted for $0 \le t \le (N-1)T$ where a_k and b_k are constants selected to give $s(t)$ a desired amplitude and phase, $\omega_d = \omega_s/N$, $\omega_s = 2\pi/T$, and k is an integer in the interval $(0, N/2)$. The receiver observes the data

$$x(nT) = s(nT) + e(nT) \qquad \text{for} \qquad n = 0, \ldots, N-1$$

where $e(nT)$ is a white noise sequence with variance σ^2. The observed data can be put in the form $\mathbf{x} = \mathbf{Y}\boldsymbol{\beta} + \mathbf{e}$ where

$$\mathbf{x} = \begin{bmatrix} x(0) \\ x(T) \\ \cdot \\ \cdot \\ \cdot \\ x(NT-T) \end{bmatrix} \qquad \mathbf{e} = \begin{bmatrix} e(0) \\ e(T) \\ \cdot \\ \cdot \\ \cdot \\ e(NT-T) \end{bmatrix} \qquad \boldsymbol{\beta} = \begin{bmatrix} a_k \\ b_k \end{bmatrix}$$

and

$$\mathbf{Y} = \begin{bmatrix} 1 & 0 \\ \cos k\omega_d T & \sin k\omega_d T \\ \cdot & \cdot \\ \cdot & \cdot \\ \cdot & \cdot \\ \cos k\omega_d(NT-T) & \sin k\omega_d(NT-T) \end{bmatrix}$$

Using Theorem 10.6.1, Parseval's Theorem for DFTs, it is easy to show that

$$\sum_{n=0}^{N-1} \cos^2 nk\omega_d T = \sum_{n=0}^{N-1} \sin^2 nk\omega_d T = N/2$$

and

$$\sum_{n=0}^{N-1} \cos nk\omega_d T \sin nk\omega_d T = 0$$

Therefore

$$\mathbf{Y}^\dagger\mathbf{Y} = \frac{N}{2}\begin{bmatrix} 1 & 0 \\ 0 & 1 \end{bmatrix}$$

and we find from (13.4.23) that the linear minimum-variance unbiased estimates of a_k and b_k are

$$\hat{a}_k = \frac{2}{N}\sum_{n=0}^{N-1} x(nT)\cos nk\omega_d T \qquad (13.4.24)$$

and

$$\hat{b}_k = \frac{2}{N}\sum_{n=0}^{N-1} x(nT)\sin nk\omega_d T \qquad (13.4.25)$$

According to (13.4.21), the error covariance matrix is

$$\boldsymbol{\Gamma}_\beta = \frac{2\sigma^2}{N}\begin{bmatrix} 1 & 0 \\ 0 & 1 \end{bmatrix} \qquad (13.4.26)$$

This example can be generalized by letting the transmitted signal be

$$s(t) = \sum_{k=k_1}^{k_2} a_k \cos k\omega_d t + b_k \sin k\omega_d t$$

with $0 < k_1 < k_2 < N/2$. A signal of this type is said to be the sum of parallel tones. Digital data is commonly transmitted over an HF radio channel by phase modulating a set of parallel tones. Using the orthogonality properties of the sinusoids, it follows that $\hat{a}_k$ and $\hat{b}_k$ are still given by (13.4.24) and (13.4.25) for $k = k_1, \ldots, k_2$.

It is interesting to observe that

$$\hat{a}_k - j\hat{b}_k = \frac{2}{N} \sum_{n=0}^{N-1} x(nT)e^{-jnk\omega_d T}$$

which is just the kth component of $(N/2)\text{DFT}\{x(nT)\}$. Thus the received signal can be demodulated, that is, a_k and b_k estimated for $k = k_1, \ldots, k_2$ by performing an FFT to compute $\text{DFT}\{x(nT)\}$.   ◀

13.5 LINEAR MINIMUM MEAN SQUARE ERROR ESTIMATES

In the previous section the parameters to be estimated were deterministic while the observed data was a set of random variables. We will now consider the case where both the parameters to be estimated and the observed data are random variables.

Suppose that the observed data vector $\mathbf{x}$ is an N-dimensional vector random variable and the parameter vector $\mathbf{w}$ is a K-dimensional vector random variable. We will assume that $\mathbf{x}$ and $\mathbf{w}$ both have zero mean and that $\mathbf{V}_{xx} = \text{cov}\{\mathbf{x}\}$, $\mathbf{V}_{wx} = \text{cov}\{\mathbf{w}, \mathbf{x}\}$, and $\mathbf{V}_{ww} = \text{cov}\{\mathbf{w}\}$ are known. The problem is to find the constant matrix $\mathbf{A}$ such that the estimate

$$\hat{\mathbf{w}} = \mathbf{A}\mathbf{x} \tag{13.5.1}$$

is optimum in the sense that the mean square errors

$$E\{|w_k - \hat{w}_k|^2\} \qquad \text{for} \qquad k = 1, \ldots, K$$

in the estimates of the individual components of $\mathbf{w}$ are minimized.

This problem can be formulated in the Hilbert space S in which the vectors are finite-variance complex scalar random variables, the set of scalars is the field of complex numbers, and the inner product of two random variables u and v in S is defined as

$$(u, v) = E\{u\bar{v}\} \tag{13.5.2}$$

Then $\|u - v\|^2 = (u - v, u - v) = E\{|u - v|^2\}$ is the mean square error between u and v. Suppose that X is the subspace of S generated by the components of $\mathbf{x}$.

That is, the elements of X are scalar random variables of the form

$$\sum_{n=1}^{N} c_n x_n$$

where $c_1, \ldots, c_N$ are finite complex numbers. Then the optimum estimate of w_k is the projection of w_k onto X. We will denote the K-dimensional column vector whose components are the projections of the corresponding components of $\mathbf{w}$ onto X by

$$\mathscr{E}\{\mathbf{w}/X\} = \begin{bmatrix} \mathscr{E}\{w_1/X\} \\ \cdot \\ \cdot \\ \cdot \\ \mathscr{E}\{w_K/X\} \end{bmatrix} \tag{13.5.3}$$

According to the Projection Theorem, the error $w_k - \hat{w}_k$ for the optimum estimate must be orthogonal to X. If we denote the kth row of $\mathbf{A}$ by $\mathbf{a}_k$, then $\hat{w}_k = \mathbf{a}_k \mathbf{x}$. The orthogonality condition requires the optimum estimates to satisfy the equations

$$E\{(w_k - \mathbf{a}_k \mathbf{x})\bar{x}_n\} = 0 \qquad \text{for} \qquad n = 1, \ldots, N \qquad \text{and} \qquad k = 1, \ldots, K$$

or on expanding the expectation

$$E\{w_k \bar{x}_n\} = \mathbf{a}_k E\{\mathbf{x}\bar{x}_n\} \qquad \text{for} \qquad n = 1, \ldots, N \qquad \text{and} \qquad k = 1, \ldots, K$$

These equations can be arranged in the single matrix equation

$$\mathbf{V}_{\mathbf{wx}} = \mathbf{A}\mathbf{V}_{\mathbf{xx}}.$$

Thus, assuming that $\mathbf{V}_{\mathbf{xx}}$ is nonsingular, the optimum value for $\mathbf{A}$ is

$$\mathbf{A} = \mathbf{V}_{\mathbf{wx}}\mathbf{V}_{\mathbf{xx}}^{-1} \tag{13.5.4}$$

and the optimum estimate of $\mathbf{w}$ is

$$\hat{\mathbf{w}} = \mathbf{V}_{\mathbf{wx}}\mathbf{V}_{\mathbf{xx}}^{-1}\mathbf{x} \tag{13.5.5}$$

The corresponding error covariance matrix is

$$\boldsymbol{\Gamma}_{\mathbf{w}} = E\{(\mathbf{w} - \hat{\mathbf{w}})(\mathbf{w} - \hat{\mathbf{w}})^{\dagger}\} = E\{(\mathbf{w} - \hat{\mathbf{w}})\mathbf{w}^{\dagger}\}$$

$$= E\{\mathbf{w}\mathbf{w}^{\dagger}\} - E\{\hat{\mathbf{w}}\mathbf{w}^{\dagger}\} = \mathbf{V}_{\mathbf{ww}} - \mathbf{V}_{\mathbf{wx}}\mathbf{V}_{\mathbf{xx}}^{-1}\mathbf{V}_{\mathbf{xw}} \tag{13.5.6}$$

The orthogonality condition was used in going from the first to the second expectation in (13.5.6).

Given a constant matrix $\mathbf{F}$, the optimum estimate of $\mathbf{b} = \mathbf{F}\mathbf{w}$ is, by the linearity of projections

$$\hat{\mathbf{b}} = \mathbf{F}\hat{\mathbf{w}} \tag{13.5.7}$$

where $\hat{\mathbf{w}}$ is the optimum estimate of $\mathbf{w}$. The corresponding error covariance matrix is

$$\boldsymbol{\Gamma}_{\mathbf{b}} = \mathbf{F}\boldsymbol{\Gamma}_{\mathbf{w}}\mathbf{F}^{\dagger} \tag{13.5.8}$$

Example 13.5.1

Let $x(n)$ be a discrete-time random process with the autocorrelation function $R(n)$. Suppose that for some integer M we wish to estimate $w = x(n+M)$ from the data $\mathbf{x}^t = [x(n) \quad x(n-1) \quad \cdots \quad x(n-N+1)]$. Then

$$\mathbf{V}_{\mathbf{wx}} = [R(M) \quad R(M+1) \quad \cdots \quad R(M+N-1)]$$

and

$$\mathbf{V}_{\mathbf{xx}} = \begin{bmatrix} R(0) & R(1) & R(2) & \cdots & R(N-1) \\ R(1) & R(0) & R(1) & \cdots & R(N-2) \\ \vdots & & & & \vdots \\ R(N-1) & \cdots & & & R(0) \end{bmatrix}$$

The optimum estimate of $x(n+M)$ is given by (13.5.5) and the corresponding mean square error by (13.5.6). For $M=1$ these equations are equivalent to those derived for one-step prediction in Section 7.6.

As a special case, suppose that $M \geq 0$ and $R(n) = a^{|n|}$ with $-1 < a < 1$. Then

$$\mathbf{V}_{\mathbf{wx}} = a^M[1 \quad a \quad a^2 \quad \cdots \quad a^{N-1}]$$

By direct multiplication, it can be verified that

$$\mathbf{V}_{\mathbf{xx}}^{-1} = \begin{bmatrix} 1 & -a & 0 & \cdots & & & & 0 \\ -a & 1+a^2 & -a & 0 & \cdots & & & 0 \\ 0 & -a & 1+a^2 & -a & 0 & \cdots & & 0 \\ \vdots & & & & & & & \vdots \\ 0 & \cdots & & & 0 & -a & 1+a^2 & -a \\ 0 & \cdots & & & & 0 & -a & 1 \end{bmatrix} \frac{1}{1-a^2}$$

Evaluating (13.5.5) and (13.5.6) gives

$$\hat{w} = \hat{x}(n+M) = a^M x(n) \quad \text{and} \quad \Gamma_w = 1 - a^{2M}$$

Notice that these same results were derived in Example 7.5.3 using the realizable Wiener filter equations ◄

Now let us consider the important special case where

$$\mathbf{x} = \mathbf{Yw} + \mathbf{e} \qquad (13.5.9)$$

where $\mathbf{Y}$ is an $N \times K$ constant matrix and $\mathbf{e}$ is an N-dimensional vector random variable with zero mean and the covariance matrix $\mathbf{Q}$. In addition, we will assume that $E\{\mathbf{we}^{\dagger}\} = \mathbf{0}$ where $\mathbf{0}$ is a $K \times N$ matrix of zeros. In this case

$$\mathbf{V_{xx}} = \mathbf{YV_{ww}Y^{\dagger}} + \mathbf{Q}$$

and

$$\mathbf{V_{wx}} = \mathbf{V_{ww}Y^{\dagger}}$$

According to (13.5.5) and (13.5.6), the optimum estimate of $\mathbf{w}$ is

$$\hat{\mathbf{w}} = \mathbf{V_{ww}Y^{\dagger}}(\mathbf{YV_{ww}Y^{\dagger}} + \mathbf{Q})^{-1}\mathbf{x} \qquad (13.5.10)$$

and the corresponding error covariance matrix is

$$\mathbf{\Gamma_w} = \mathbf{V_{ww}} - \mathbf{V_{ww}Y^{\dagger}}(\mathbf{YV_{ww}Y^{\dagger}} + \mathbf{Q})^{-1}\mathbf{YV_{ww}} \qquad (13.5.11)$$

Equations 13.5.10 and 13.5.11 can be written in an interesting alternate form by using the matrix identity

$$\mathbf{V_{ww}Y^{\dagger}}(\mathbf{YV_{ww}Y^{\dagger}} + \mathbf{Q})^{-1} = (\mathbf{Y^{\dagger}Q^{-1}Y} + \mathbf{V_{ww}^{-1}})^{-1}\mathbf{Y^{\dagger}Q^{-1}} \qquad (13.5.12)$$

This identity can be verified by premultiplying both sides by $\mathbf{Y^{\dagger}Q^{-1}Y} + \mathbf{V_{ww}^{-1}}$ and postmultiplying both sides by $\mathbf{YV_{ww}Y^{\dagger}} + \mathbf{Q}$. Thus (13.5.10) can be written as

$$\hat{\mathbf{w}} = (\mathbf{Y^{\dagger}Q^{-1}Y} + \mathbf{V_{ww}^{-1}})^{-1}\mathbf{Y^{\dagger}Q^{-1}x} \qquad (13.5.13)$$

Using (13.5.12) in (13.5.11) gives

$$\begin{aligned}
\mathbf{\Gamma_w} &= \mathbf{V_{ww}} - (\mathbf{Y^{\dagger}Q^{-1}Y} + \mathbf{V_{ww}^{-1}})^{-1}\mathbf{Y^{\dagger}Q^{-1}YV_{ww}} \\
&= (\mathbf{Y^{\dagger}Q^{-1}Y} + \mathbf{V_{ww}^{-1}})^{-1}[(\mathbf{Y^{\dagger}Q^{-1}Y} + \mathbf{V_{ww}^{-1}})\mathbf{V_{ww}} - \mathbf{Y^{\dagger}Q^{-1}YV_{ww}}] \\
&= (\mathbf{Y^{\dagger}Q^{-1}Y} + \mathbf{V_{ww}^{-1}})^{-1} \qquad (13.5.14)
\end{aligned}$$

Notice that the right-hand sides of (13.5.13) and (13.5.14) are identical to (13.4.20) and (13.4.21) for the minimum-variance unbiased estimates except for the additional term $\mathbf{V_{ww}^{-1}}$. The minimum mean square error estimates in this special case converge to the minimum-variance unbiased estimates as $\mathbf{V_{ww}^{-1}}$ converges to the $K \times K$ zero matrix. This happens when the variances of the components of $\mathbf{w}$ become infinite.

Example 13.5.2

Suppose that the components of the observed data vector have the form

$$x_n = w + e_n \qquad \text{for} \qquad n = 1, \ldots, N$$

where w is a zero-mean random variable with variance γ^2, $e_1, \ldots, e_N$ are mutually uncorrelated zero-mean random variables each with variance σ^2, and $E\{w\bar{e}_n\} = 0$ for $n = 1, \ldots, N$. Then $\mathbf{Q} = \sigma^2 \mathbf{I}$, $\mathbf{V}_{ww} = \gamma^2$, and

$$\mathbf{Y} = \begin{bmatrix} 1 \\ 1 \\ \cdot \\ \cdot \\ \cdot \\ 1 \end{bmatrix}$$

From (13.5.13) and (13.5.14), we find that

$$\hat{w} = \frac{1}{N + (\sigma^2/\gamma^2)} \sum_{n=1}^{N} x_n$$

and

$$\mathbf{\Gamma}_w = \frac{\sigma^2}{N + (\sigma^2/\gamma^2)}$$

Notice that these results are significantly more difficult to obtain using (13.5.10) and (13.5.11). ◄

14

RECURSIVE ESTIMATION

14.1 INTRODUCTION

In many estimation problems, the observed data is received sequentially in time and up-to-date parameter estimates (i.e., estimates using all the available data) are required. Theoretically, the entire past observed data sequence could be stored and a new estimate computed using the appropriate equations in Chapter 13 as each new data sample is received. Clearly this approach is not practical when the observed data sequence is long, as is often the case. In this chapter, we generalize the linear minimum mean square error estimation problem somewhat and allow the parameter vector to be the state of a dynamical system driven by white noise. We find that each up-to-date estimate can be computed from the last previous estimate and the new received data sample. Thus only the last estimate must be stored. In addition to eliminating the problem of storing the entire past observed data sequence, this recursive estimation algorithm is computationally more efficient than computing the estimates directly from the entire observed data sequence at each step. The recursive solution to the discrete-time estimation problem was first published by Kalman [67]. A solution to the analogous continuous-time problem was published a year later by Kalman and Bucy [68]. These estimation algorithms are now called *Kalman-Bucy* or just *Kalman* filters.

In this chapter, we assume that the parameter vector $\mathbf{x}(n)$ is the state of a K-dimensional dynamical system with a state equation of the form

$$\mathbf{x}(n+1) = \boldsymbol{\varphi}(n+1, n)\mathbf{x}(n) + \mathbf{v}(n) \tag{14.1.1}$$

where $\boldsymbol{\varphi}(n_1, n_2)$ is a $K \times K$ state transition matrix and $\mathbf{v}(n)$ is a zero-mean K-dimensional vector random process with

$$\text{cov}\{\mathbf{v}(n), \mathbf{v}(k)\} = E\{\mathbf{v}(n)\mathbf{v}^\dagger(k)\} = \mathbf{Q}(n)\delta_{nk} \tag{14.1.2}$$

371

The properties of $\varphi(n_1, n_2)$ are discussed in Section 6.3. To simplify the notation used in that section, we will now represent the time t_n simply by the integer n. The observed data $\mathbf{y}(n)$ will be assumed to have the form

$$\mathbf{y}(n) = \mathbf{C}(n)\mathbf{x}(n) + \mathbf{w}(n) \tag{14.1.3}$$

where $\mathbf{C}(n)$ is a known $N \times K$ matrix and $\mathbf{w}(n)$ is a zero-mean N-dimensional vector random process with

$$\text{cov}\{\mathbf{w}(n), \mathbf{w}(k)\} = \mathbf{R}(n)\,\delta_{nk} \tag{14.1.4}$$

and

$$\text{cov}\{\mathbf{v}(n), \mathbf{w}(k)\} = \mathbf{0} \tag{14.1.5}$$

The signal $\mathbf{w}(n)$ represents the noise introduced in the observation process. We will assume that the initial state $\mathbf{x}(n_0)$ is a zero-mean vector random variable with

$$\text{cov}\{\mathbf{x}(n_0)\} = \mathbf{P}(n_0) \tag{14.1.6}$$

and is uncorrelated with $\mathbf{v}(n)$ and $\mathbf{w}(k)$ for $n, k \geq n_0$. This model is illustrated in Fig. 14.1.1.

The problem is to find, for each $n \geq n_0$, the linear minimum mean square error estimates of the components of $\mathbf{x}(i)$ from the data observed from time n_0 up to the present time n, that is, $\mathbf{y}(n_0), \ldots, \mathbf{y}(n)$. This is called the *filtering problem* if $i = n$, the *prediction problem* if $i > n$, and the *smoothing problem* if $n_0 \leq i < n$. The Wiener filter problem discussed in Chapter 7 is included in this formulation as well as the problem of estimating the state of an actual dynamical system since any random process with a rational power spectral density can be modeled as a white noise sequence passed through a time-invariant dynamical system.

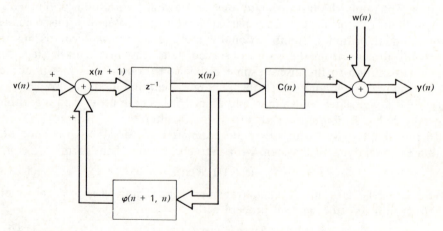

FIGURE 14.1.1.   Model for the estimation problem.

14.2 ESTIMATION USING THE INNOVATIONS PROCESS

Consider the space $Y(n)$ of scalar random variables generated by the components of the observed data $\mathbf{y}(n_0), \ldots, \mathbf{y}(n)$. The random variables in this space have the form

$$\sum_{i=n_0}^{n} \sum_{k=1}^{N} a_{ik} y_k(i)$$

where $y_k(i)$ is the kth component of $\mathbf{y}(i)$ and $\{a_{ik}\}$ is a set of constants. $Y(n)$ is a subspace of the Hilbert space S of finite-variance scalar random variables defined in Section 13.5. Thus the optimum estimates of the components of $\mathbf{x}(i)$ from the observed data are the projections of the components onto $Y(n)$. We will denote these estimates by

$$\hat{\mathbf{x}}(i \mid n) = \mathcal{E}\{\mathbf{x}(i)/Y(n)\} \qquad (14.2.1)$$

We will use the *innovations* approach of Kailath [65,66] to solve the estimation problem. This approach is similar to Kalman's original approach. The innovations process associated with the observed data process is defined to be

$$\boldsymbol{\alpha}(n) = \mathbf{y}(n) - \hat{\mathbf{y}}(n \mid n-1) \qquad (14.2.2)$$

where

$$\hat{\mathbf{y}}(n \mid n-1) = \mathcal{E}\{\mathbf{y}(n)/Y(n-1)\} \qquad (14.2.3)$$

is the optimum prediction of $\mathbf{y}(n)$ from the past observed data. According to the Projection Theorem, $\boldsymbol{\alpha}(n) \perp Y(n-1)$ so that

$$E\{\boldsymbol{\alpha}(n)\mathbf{y}^{\dagger}(k)\} = \mathbf{0} \qquad \text{for} \qquad n_0 \le k < n \qquad (14.2.4)$$

Thus $\boldsymbol{\alpha}(n) \perp Y(k)$ for $n_0 \le k < n$. In a sense, $\boldsymbol{\alpha}(n)$ is the new information contained in the observation $\mathbf{y}(n)$. Since $\boldsymbol{\alpha}(k)$ is in $Y(k)$, it follows that

$$E\{\boldsymbol{\alpha}(k)\boldsymbol{\alpha}^{\dagger}(n)\} = \mathbf{0} \qquad \text{for} \qquad k \ne n \qquad (14.2.5)$$

Therefore, the innovations process is a sequence of uncorrelated vector random variables.

The technique used to form the innovations process is similar to the Gram-Schmidt orthonormalization technique discussed in Section 12.10. Using the same reasoning as in that section, it follows that the sets of vectors $\{\mathbf{y}(n_0), \ldots, \mathbf{y}(n)\}$ and $\{\boldsymbol{\alpha}(n_0), \ldots, \boldsymbol{\alpha}(n)\}$ both generate the space $Y(n)$. Thus the optimum estimates can be expressed as linear combinations of either set. We will find that it is significantly easier to solve the estimation problem using the innovations process.

We will need an expression for $\mathbf{V}_\alpha(n) = \text{cov}\{\boldsymbol{\alpha}(n)\}$. As a first step, the state equation can be recursively solved to yield

$$\mathbf{x}(k) = \boldsymbol{\varphi}(k, n_0)\mathbf{x}(n_0) + \sum_{i=n_0}^{k-1} \boldsymbol{\varphi}(k, i+1)\mathbf{v}(i) \qquad (14.2.6)$$

for $k > n_0$. Thus $\mathbf{x}(k)$ is a linear combination of $\mathbf{v}(n_0), \dots, \mathbf{v}(k-1)$, and $\mathbf{x}(n_0)$. By hypothesis, $\mathbf{w}(n)$ is uncorrelated with $\mathbf{x}(n_0)$ and $\mathbf{v}(k)$ for n, $k \geq n_0$. Therefore

$$E\{\mathbf{x}(k)\mathbf{w}^\dagger(n)\} = \mathbf{0} \qquad \text{for} \qquad n, k \geq n_0 \qquad (14.2.7)$$

Also, $\mathbf{y}(k) = \mathbf{C}(k)\mathbf{x}(k) + \mathbf{w}(k)$ is a linear combination of $\mathbf{x}(n_0)$, $\mathbf{v}(n_0), \dots,$ $\mathbf{v}(k-1)$, and $\mathbf{w}(k)$. Therefore

$$E\{\mathbf{y}(k)\mathbf{w}^\dagger(n)\} = \mathbf{0} \qquad \text{for} \qquad n_0 \leq k \leq n-1 \qquad (14.2.8)$$

so that

$$\mathbf{w}(n) \perp Y(k) \qquad \text{and} \qquad \mathscr{E}\{\mathbf{w}(n)/Y(k)\} = \boldsymbol{\theta} \qquad (14.2.9)$$

for $n_0 \leq k \leq n-1$. Similarly

$$E\{\mathbf{y}(k)\mathbf{v}^\dagger(n)\} = \mathbf{0} \qquad \text{for} \qquad n_0 \leq k \leq n \qquad (14.2.10)$$

so that

$$\mathbf{v}(n) \perp Y(k) \qquad \text{and} \qquad \mathscr{E}\{\mathbf{v}(n)/Y(k)\} = \boldsymbol{\theta} \qquad (14.2.11)$$

for $n_0 \leq k \leq n$. Consequently

$$
\begin{aligned}
\hat{\mathbf{y}}(n \mid n-1) &= \mathscr{E}\{\mathbf{C}(n)\mathbf{x}(n) + \mathbf{w}(n)/Y(n-1)\} \\
&= \mathbf{C}(n)\mathscr{E}\{\mathbf{x}(n)/Y(n-1)\} + \mathscr{E}\{\mathbf{w}(n)/Y(n-1)\} \\
&= \mathbf{C}(n)\hat{\mathbf{x}}(n \mid n-1) \qquad (14.2.12)
\end{aligned}
$$

and

$$\boldsymbol{\alpha}(n) = \mathbf{y}(n) - \mathbf{C}(n)\hat{\mathbf{x}}(n \mid n-1) \qquad (14.2.13)$$

Substituting for $\mathbf{y}(n)$ gives

$$\boldsymbol{\alpha}(n) = \mathbf{C}(n)\tilde{\mathbf{x}}(n \mid n-1) + \mathbf{w}(n) \qquad (14.2.14)$$

where

$$\tilde{\mathbf{x}}(n \mid n-1) = \mathbf{x}(n) - \hat{\mathbf{x}}(n \mid n-1) \qquad (14.2.15)$$

Therefore

$$
\begin{aligned}
\mathbf{V}_\alpha(n) &= E\{\boldsymbol{\alpha}(n)\boldsymbol{\alpha}^\dagger(n)\} \\
&= \mathbf{C}(n)E\{\tilde{\mathbf{x}}(n \mid n-1)\tilde{\mathbf{x}}^\dagger(n \mid n-1)\}\mathbf{C}^\dagger(n) + \mathbf{C}(n)E\{\tilde{\mathbf{x}}(n \mid n-1)\mathbf{w}^\dagger(n)\} \\
&\quad + E\{\mathbf{w}(n)\tilde{\mathbf{x}}^\dagger(n \mid n-1)\}\mathbf{C}^\dagger(n) + E\{\mathbf{w}(n)\mathbf{w}^\dagger(n)\} \qquad (14.2.16)
\end{aligned}
$$

The middle two terms on the right-hand side of (14.2.16) are zero as a result of (14.2.7) and (14.2.9). The quantity

$$\mathbf{P}(n) = E\{\tilde{\mathbf{x}}(n \mid n-1)\tilde{\mathbf{x}}^\dagger(n \mid n-1)\} \qquad (14.2.17)$$

is the error covariance matrix for the optimum prediction of $\mathbf{x}(n)$ from the data observed up through time $n-1$. Thus

$$\mathbf{V}_{\alpha}(n) = \mathbf{C}(n)\mathbf{P}(n)\mathbf{C}^{\dagger}(n) + \mathbf{R}(n) \qquad (14.2.18)$$

We now derive a general formula for the optimum up-to-date estimate of $\mathbf{x}(i)$ from the innovations process. This formula is specialized to the filtering and prediction problem in Section 14.3 and the smoothing problem in Section 14.4. The optimum estimate can be written in the form

$$\hat{\mathbf{x}}(i \mid n) = \sum_{k=n_0}^{n} \mathbf{A}_i(k)\boldsymbol{\alpha}(k)$$

where $\{\mathbf{A}_i(k)\}$ is a set of $K \times N$ constant matrices. According to the Projection Theorem

$$E\left\{\left[\mathbf{x}(i) - \sum_{k=n_0}^{n} \mathbf{A}_i(k)\boldsymbol{\alpha}(k)\right]\boldsymbol{\alpha}^{\dagger}(m)\right\} = \mathbf{0} \qquad (14.2.19)$$

for $m = n_0, \ldots, n$. Using the orthogonality property of the innovations process (14.2.19) reduces to

$$E\{\mathbf{x}(i)\boldsymbol{\alpha}^{\dagger}(m)\} = \mathbf{A}_i(m)E\{\boldsymbol{\alpha}(m)\boldsymbol{\alpha}^{\dagger}(m)\}$$

so that

$$\mathbf{A}_i(m) = E\{\mathbf{x}(i)\boldsymbol{\alpha}^{\dagger}(m)\}\mathbf{V}_{\alpha}^{-1}(m)$$

Therefore

$$\hat{\mathbf{x}}(i \mid n) = \sum_{k=n_0}^{n} E\{\mathbf{x}(i)\boldsymbol{\alpha}^{\dagger}(k)\}\mathbf{V}_{\alpha}^{-1}(k)\boldsymbol{\alpha}(k) \qquad (14.2.20)$$

The kth term in the sum is the projection of $\mathbf{x}(i)$ onto the subspace generated by $\boldsymbol{\alpha}(k)$.

14.3 FILTERING AND PREDICTION

Following Kalman's approach, we first solve the problem of predicting one step into the future and then see that the solutions to the filtering problem and the problem of predicting any number of steps into the future are simply related to the one-step prediction solution. According to (14.2.20), the optimum estimate of $\mathbf{x}(n+1)$ from the data observed through time n is

$$\hat{\mathbf{x}}(n+1 \mid n) = \sum_{k=n_0}^{n} E\{\mathbf{x}(n+1)\boldsymbol{\alpha}^{\dagger}(k)\mathbf{V}_{\alpha}^{-1}(k)\boldsymbol{\alpha}(k) \qquad (14.3.1)$$

which can also be written as

$$\hat{\mathbf{x}}(n+1\mid n) = \sum_{k=n_0}^{n-1} E\{\mathbf{x}(n+1)\boldsymbol{\alpha}^{\dagger}(k)\}\mathbf{V}_{\boldsymbol{\alpha}}^{-1}(k)\boldsymbol{\alpha}(k)$$

$$+ E\{\mathbf{x}(n+1)\boldsymbol{\alpha}^{\dagger}(n)\}\mathbf{V}_{\boldsymbol{\alpha}}^{-1}(n)\boldsymbol{\alpha}(n) \tag{14.3.2}$$

Using (14.2.11) and the fact that $\boldsymbol{\alpha}(k)$ is in $Y(k)$, it follows that for $n_0 \le k \le n$

$$E\{\mathbf{x}(n+1)\boldsymbol{\alpha}^{\dagger}(k)\} = E\{[\boldsymbol{\varphi}(n+1, n)\mathbf{x}(n) + \mathbf{v}(n)]\boldsymbol{\alpha}^{\dagger}(k)\}$$

$$= \boldsymbol{\varphi}(n+1, n)E\{\mathbf{x}(n)\boldsymbol{\alpha}^{\dagger}(k)\} \tag{14.3.3}$$

Therefore

$$\sum_{k=n_0}^{n-1} E\{\mathbf{x}(n+1)\boldsymbol{\alpha}^{\dagger}(k)\}\mathbf{V}_{\boldsymbol{\alpha}}^{-1}(k)\boldsymbol{\alpha}(k) = \boldsymbol{\varphi}(n+1, n) \sum_{k=n_0}^{n-1} E\{\mathbf{x}(n)\boldsymbol{\alpha}^{\dagger}(k)\}\mathbf{V}_{\boldsymbol{\alpha}}^{-1}(k)\boldsymbol{\alpha}(k)$$

$$= \boldsymbol{\varphi}(n+1, n)\hat{\mathbf{x}}(n\mid n-1) \tag{14.3.4}$$

Using (14.2.13) for $\boldsymbol{\alpha}(n)$ and letting

$$\mathbf{G}(n) = E\{\mathbf{x}(n+1)\boldsymbol{\alpha}^{\dagger}(n)\}\mathbf{V}_{\boldsymbol{\alpha}}^{-1}(n) \tag{14.3.5}$$

it follows from (14.3.2) and (14.3.4) that

$$\hat{\mathbf{x}}(n+1\mid n) = \boldsymbol{\varphi}(n+1, n)\hat{\mathbf{x}}(n\mid n-1) + \mathbf{G}(n)[\mathbf{y}(n) - \mathbf{C}(n)\hat{\mathbf{x}}(n\mid n-1)] \tag{14.3.6}$$

Thus the optimum up-to-date estimate can be computed from the last previous estimate and the new observed data vector.

The optimum gain matrix $\mathbf{G}(n)$ can be expressed in terms of the one-step prediction error covariance matrix $\mathbf{P}(n)$ defined by (14.2.17). According to (14.3.3), (14.2.14), and (14.2.7)

$$E\{\mathbf{x}(n+1)\boldsymbol{\alpha}^{\dagger}(n)\} = \boldsymbol{\varphi}(n+1, n)E\{\mathbf{x}(n)[\mathbf{C}(n)\tilde{\mathbf{x}}(n\mid n-1) + \mathbf{w}(n)]^{\dagger}\}$$

$$= \boldsymbol{\varphi}(n+1, n)E\{\mathbf{x}(n)\tilde{\mathbf{x}}^{\dagger}(n\mid n-1)\}\mathbf{C}^{\dagger}(n) \tag{14.3.7}$$

Since $\tilde{\mathbf{x}}(n\mid n-1) = \mathbf{x}(n) - \hat{\mathbf{x}}(n\mid n-1)$ is orthogonal to $Y(n-1)$ and $\hat{\mathbf{x}}(n\mid n-1)$ is in $Y(n-1)$, $\mathbf{x}(n)$ can be replaced by $\tilde{\mathbf{x}}(n\mid n-1)$ in (14.3.7) so that

$$E\{\mathbf{x}(n+1)\boldsymbol{\alpha}^{\dagger}(n)\} = \boldsymbol{\varphi}(n+1, n)E\{\tilde{\mathbf{x}}(n\mid n-1)\tilde{\mathbf{x}}^{\dagger}(n\mid n-1)\}\mathbf{C}^{\dagger}(n)$$

$$= \boldsymbol{\varphi}(n+1, n)\mathbf{P}(n)\mathbf{C}^{\dagger}(n) \tag{14.3.8}$$

Therefore, according to (14.3.5), (14.2.18), and (14.3.8)

$$\mathbf{G}(n) = \boldsymbol{\varphi}(n+1, n)\mathbf{P}(n)\mathbf{C}^{\dagger}(n)[\mathbf{C}(n)\mathbf{P}(n)\mathbf{C}^{\dagger}(n) + \mathbf{R}(n)]^{-1} \tag{14.3.9}$$

As it stands, the formula for $\mathbf{G}(n)$ is not particularly useful, since $\mathbf{P}(n)$ must be known. To solve this problem, we will now derive a formula for recursively computing $\mathbf{P}(n)$. The error in the one-step prediction of $\mathbf{x}(n+1)$ can be written as

$$\tilde{\mathbf{x}}(n+1 \mid n) = \mathbf{x}(n+1) - \hat{\mathbf{x}}(n+1 \mid n)$$
$$= \boldsymbol{\varphi}(n+1, n)\mathbf{x}(n) + \mathbf{v}(n) - \boldsymbol{\varphi}(n+1, n)\hat{\mathbf{x}}(n \mid n-1)$$
$$- \mathbf{G}(n)[\mathbf{y}(n) - \mathbf{C}(n)\hat{\mathbf{x}}(n \mid n-1)]$$

Substituting (14.1.3) for $\mathbf{y}(n)$ and rearranging terms yields

$$\tilde{\mathbf{x}}(n+1 \mid n) = [\boldsymbol{\varphi}(n+1, n) - \mathbf{G}(n)\mathbf{C}(n)]\tilde{\mathbf{x}}(n \mid n-1) + \mathbf{v}(n) - \mathbf{G}(n)\mathbf{w}(n) \quad (14.3.10)$$

Since $\tilde{\mathbf{x}}(n \mid n-1)$, $\mathbf{v}(n)$ and $\mathbf{w}(n)$ are mutually uncorrelated, we find that

$$\mathbf{P}(n+1) = \text{cov} \{\tilde{\mathbf{x}}(n+1 \mid n)\}$$
$$= [\boldsymbol{\varphi}(n+1, n) - \mathbf{G}(n)\mathbf{C}(n)]\mathbf{P}(n)[\boldsymbol{\varphi}(n+1, n) - \mathbf{G}(n)\mathbf{C}(n)]^{\dagger}$$
$$+ \mathbf{Q}(n) + \mathbf{G}(n)\mathbf{R}(n)\mathbf{G}^{\dagger}(n) \quad (14.3.11)$$

By expanding the first term on the right-hand side, using (14.3.9) for $\mathbf{G}(n)$, and combining terms (14.3.11) can be reduced to

$$\mathbf{P}(n+1) = \boldsymbol{\varphi}(n+1, n)\boldsymbol{\Gamma}(n)\boldsymbol{\varphi}^{\dagger}(n+1, n) + \mathbf{Q}(n) \quad (14.3.12)$$

where

$$\boldsymbol{\Gamma}(n) = \mathbf{P}(n) - \mathbf{P}(n)\mathbf{C}^{\dagger}(n)[\mathbf{C}(n)\mathbf{P}(n)\mathbf{C}^{\dagger}(n) + \mathbf{R}(n)]^{-1}\mathbf{C}(n)\mathbf{P}(n) \quad (14.3.13)$$

We will see shortly that $\boldsymbol{\Gamma}(n)$ is the error covariance matrix for the filtered estimate $\hat{\mathbf{x}}(n \mid n)$.

We now have all the equations required to recursively generate the solution to the one-step prediction problem. In summary, these equations are

1. $\hat{\mathbf{x}}(n+1 \mid n) = \boldsymbol{\varphi}(n+1, n)\hat{\mathbf{x}}(n \mid n-1) + \mathbf{G}(n)[\mathbf{y}(n) - \mathbf{C}(n)\hat{\mathbf{x}}(n \mid n-1)]$

$$(14.3.6)$$

2. $\mathbf{G}(n) = \boldsymbol{\varphi}(n+1, n)\mathbf{P}(n)\mathbf{C}^{\dagger}(n)[\mathbf{C}(n)\mathbf{P}(n)\mathbf{C}^{\dagger}(n) + \mathbf{R}(n)]^{-1}$ (14.3.9)

3. $\mathbf{P}(n+1) = \boldsymbol{\varphi}(n+1, n)\boldsymbol{\Gamma}(n)\boldsymbol{\varphi}^{\dagger}(n+1, n) + \mathbf{Q}(n)$ (14.3.12)

4. $\boldsymbol{\Gamma}(n) = \mathbf{P}(n) - \mathbf{P}(n)\mathbf{C}^{\dagger}(n)[\mathbf{C}(n)\mathbf{P}(n)\mathbf{C}^{\dagger}(n) + \mathbf{R}(n)]^{-1}\mathbf{C}(n)\mathbf{P}(n)$ (14.3.13)

Assuming that $\hat{\mathbf{x}}(n \mid n-1)$ and $\mathbf{P}(n)$ are known, these equations can be used in the following sequence:

1. Compute $\mathbf{G}(n)$ using (14.3.9).
2. Compute $\hat{\mathbf{x}}(n+1 \mid n)$ using (14.3.6).
3. Compute $\boldsymbol{\Gamma}(n)$ using (14.3.13).
4. Compute $\mathbf{P}(n+1)$ using (14.3.12).

The sequence can then be repeated with n replaced by $n+1$, etc. The initial conditions can be obtained by observing that

$$\hat{\mathbf{x}}(n+1 \mid n) = \mathscr{E}\{\mathbf{x}(n+1)/Y(n)\}$$
$$= \mathscr{E}\{\boldsymbol{\varphi}(n+1, n)\mathbf{x}(n) + \mathbf{v}(n)/Y(n)\}$$
$$= \boldsymbol{\varphi}(n+1, n)\hat{\mathbf{x}}(n \mid n) \tag{14.3.14}$$

Thus the initial estimate should be

$$\hat{\mathbf{x}}(n_0 \mid n_0 - 1) = \boldsymbol{\varphi}(n_0, n_0 - 1)\hat{\mathbf{x}}(n_0 - 1 \mid n_0 - 1)$$

However, $\hat{\mathbf{x}}(n_0 - 1 \mid n_0 - 1)$ is the optimum estimate of $\mathbf{x}(n_0 - 1)$ given no observed data since the observations were assumed to start at time n_0. This estimate is simply the mean of $\mathbf{x}(n_0 - 1)$ which we will take to be $\boldsymbol{0}$. Therefore

$$\hat{\mathbf{x}}(n_0 \mid n_0 - 1) = \boldsymbol{0} \qquad \text{and} \qquad \mathbf{P}(n_0) = \text{cov}\{\mathbf{x}(n_0)\} \tag{14.3.15}$$

are the required initial conditions.

The recursive solution to the estimation problem appears formidable at first. However, we should remember that the filter will be implemented by machine and not by hand. The recursive solution is ideally suited for implementation with a digital computer.

We find immediately from (14.3.14) that the filtered estimate $\hat{\mathbf{x}}(n \mid n)$ can be computed as

$$\hat{\mathbf{x}}(n \mid n) = \boldsymbol{\varphi}^{-1}(n+1, n)\hat{\mathbf{x}}(n+1 \mid n) = \boldsymbol{\varphi}(n, n+1)\hat{\mathbf{x}}(n+1 \mid n) \tag{14.3.16}$$

To solve the problem of predicting more than one step into the future, we can first recursively solve the state equation starting at time $n+1$ to obtain

$$\mathbf{x}(n+r) = \boldsymbol{\varphi}(n+r, n+1)\mathbf{x}(n+1) + \sum_{i=n+1}^{n+r-1} \boldsymbol{\varphi}(n+r, i+1)\mathbf{v}(i) \tag{14.3.17}$$

for $r \geq 2$. Since $\mathbf{v}(i)$ is orthogonal to $Y(n)$ for $n \leq i$, it follows that

$$\hat{\mathbf{x}}(n+r \mid n) = \mathscr{E}\{\mathbf{x}(n+r)/Y(n)\} = \boldsymbol{\varphi}(n+r, n+1)\hat{\mathbf{x}}(n+1 \mid n) \tag{14.3.18}$$

for $r \geq 2$. Notice that (14.3.16) is identical to (14.3.18) with $r = 0$.

A block diagram of the optimum filtering and prediction filter based on the one-step prediction algorithm is shown in Fig. 14.3.1. It is interesting to observe that a model of the dynamical system for generating the message $\mathbf{C}(n)\mathbf{x}(n)$ is imbedded in the filter. The model is driven by a linear transformation of the error between the observed data $\mathbf{y}(n)$ and its predicted value $\hat{\mathbf{y}}(n \mid n-1)$. This error is the innovations process $\boldsymbol{\alpha}(n)$. Since the optimum filter is a negative feedback system, we would expect the state of the imbedded model to track the state of the actual system generating $\mathbf{y}(n)$.

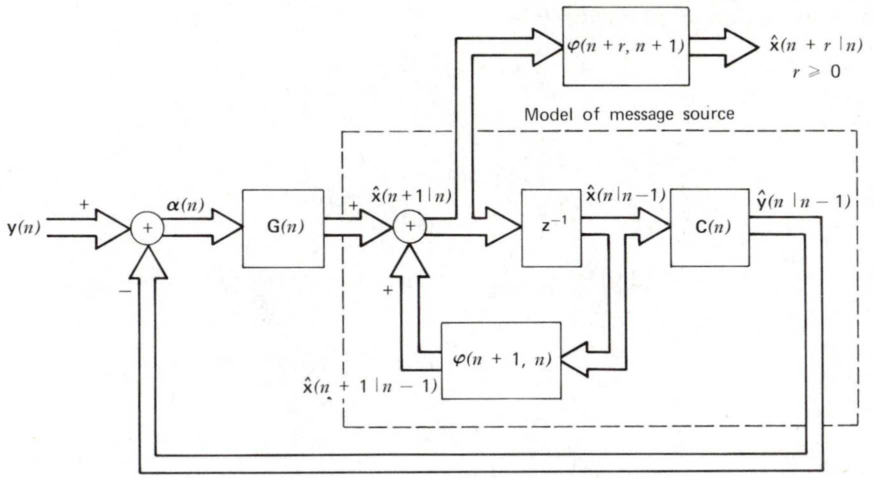

FIGURE 14.3.1.   Optimum filter based on the one-step prediction algorithm.

Example 14.3.1

Suppose that we wish to estimate the real scalar random variable x from the real scalar observations $y(n) = x + w(n)$ for $n = 0, 1, \ldots$ where $E\{x\} = E\{w(n)\} = 0$, $E\{xw(n)\} = 0$, $E\{w(n)w(k)\} = R\delta_{nk}$, and var $\{x\} = \sigma^2$. The random variable x can be considered to be the state of the degenerate dynamical system

$$x(n+1) = \varphi(n+1, n)x(n) + v(n)$$

where $\varphi(n+1, n) = 1$ and $v(n) = 0$. Thus for this problem, $Q(n) = $ var $\{v(n)\} = 0$ and $C(n) = 1$. According to (14.3.16), $\hat{x}(n \mid n) = \hat{x}(n+1 \mid n)$. The filter equations reduce to

$$G(n) = \frac{P(n)}{P(n) + R}$$

$$P(n+1) = \Gamma(n) = \text{var}\{x - \hat{x}(n \mid n)\} = \frac{P(n)R}{P(n) + R}$$

and

$$\hat{x}(n+1 \mid n) = \hat{x}(n \mid n-1) + G(n)[y(n) - \hat{x}(n \mid n-1)]$$

$$= \frac{R}{P(n) + R}\,\hat{x}(n \mid n-1) + \frac{P(n)}{P(n) + R}\,y(n)$$

with the initial conditions

$$P(0) = \sigma^2 \quad \text{and} \quad \hat{x}(0 \mid -1) = 0$$

Thus the estimate when $n = 0$ is

$$\hat{x}(1 \mid 0) = \frac{\sigma^2}{\sigma^2 + R} y(0)$$

and

$$P(1) = \Gamma(0) = \frac{R\sigma^2}{\sigma^2 + R}$$

For $n = 1$

$$\hat{x}(2 \mid 1) = \frac{R}{P(1) + R} \hat{x}(1 \mid 0) + \frac{P(1)}{P(1) + R} y(1)$$

$$= \frac{\sigma^2}{2\sigma^2 + R} [y(0) + y(1)]$$

and

$$P(2) = \frac{P(1)R}{P(1) + R} = \frac{R\sigma^2}{2\sigma^2 + R}$$

Continuing the recursion, we find that

$$\hat{x}(n + 1 \mid n) = \frac{\sigma^2}{(n+1)\sigma^2 + R} \sum_{k=0}^{n} y(k)$$

and

$$P(n + 1) = \frac{R\sigma^2}{(n+1)\sigma^2 + R}$$

Compare the results of this example with Example 13.5.2.   ◄

We will now derive a more convenient set of equations for recursively calculating the filtered estimate $\hat{x}(n \mid n)$. According to (14.3.6) and (14.3.16)

$$\hat{\mathbf{x}}(n \mid n) = \boldsymbol{\varphi}(n, n+1)\hat{\mathbf{x}}(n+1 \mid n)$$

$$= \hat{\mathbf{x}}(n \mid n-1) + \mathbf{H}(n)[\mathbf{y}(n) - \mathbf{C}(n)\hat{\mathbf{x}}(n \mid n-1)] \qquad (14.3.19)$$

where

$$\mathbf{H}(n) = \boldsymbol{\varphi}(n, n+1)\mathbf{G}(n) = \mathbf{P}(n)\mathbf{C}^\dagger(n)[\mathbf{C}(n)\mathbf{P}(n)\mathbf{C}^\dagger(n) + \mathbf{R}(n)]^{-1}$$

$$(14.3.20)$$

It follows from (14.3.14) that

$$\hat{\mathbf{x}}(n \mid n-1) = \boldsymbol{\varphi}(n, n-1)\hat{\mathbf{x}}(n-1 \mid n-1)$$

Therefore, (14.3.19) can be written as

$$\hat{\mathbf{x}}(n \mid n) = \boldsymbol{\varphi}(n, n-1)\hat{\mathbf{x}}(n-1 \mid n-1)$$

$$+ \mathbf{H}(n)[\mathbf{y}(n) - \mathbf{C}(n)\boldsymbol{\varphi}(n, n-1)\hat{\mathbf{x}}(n-1 \mid n-1)] \qquad (14.3.21)$$

This is the desired recursion formula for $\hat{\mathbf{x}}(n \mid n)$. Solving the state equation recursively from time n to $n+r$ and projecting the result onto $Y(n)$ gives

$$\hat{\mathbf{x}}(n+r \mid n) = \boldsymbol{\varphi}(n+r, n)\hat{\mathbf{x}}(n \mid n) \qquad (14.3.22)$$

for $r \geq 0$. A block diagram of the filter specified by (14.3.21) and (14.3.22) is shown in Fig. 14.3.2. Once again, the filter is a feedback system with a model of the state equation imbedded in it.

Using (14.3.19) for $\hat{\mathbf{x}}(n \mid n)$, the estimation error can be written as

$$\tilde{\mathbf{x}}(n \mid n) = \mathbf{x}(n) - \hat{\mathbf{x}}(n \mid n) = \tilde{\mathbf{x}}(n \mid n-1) - \mathbf{H}(n)\boldsymbol{\alpha}(n)$$

Therefore, the error covariance matrix is

$$\boldsymbol{\Gamma}(n) = E\{\tilde{\mathbf{x}}(n \mid n)\tilde{\mathbf{x}}^\dagger(n \mid n)\} = \mathbf{P}(n) - E\{\tilde{\mathbf{x}}(n \mid n-1)\boldsymbol{\alpha}^\dagger(n)\}\mathbf{H}^\dagger(n)$$
$$- \mathbf{H}(n)E\{\boldsymbol{\alpha}(n)\tilde{\mathbf{x}}^\dagger(n \mid n-1)\} + \mathbf{H}(n)\mathbf{V}_\alpha(n)\mathbf{H}^\dagger(n) \quad (14.3.23)$$

Using (14.3.3), (14.3.8), and the fact that $\hat{\mathbf{x}}(n \mid n-1)$ is orthogonal to $\boldsymbol{\alpha}(n)$, it follows that

$$E\{\tilde{\mathbf{x}}(n \mid n-1)\boldsymbol{\alpha}^\dagger(n)\} = E\{\mathbf{x}(n)\boldsymbol{\alpha}^\dagger(n)\}$$
$$= \boldsymbol{\varphi}(n, n+1)E\{\mathbf{x}(n+1)\boldsymbol{\alpha}^\dagger(n)\} = \mathbf{P}(n)\mathbf{C}^\dagger(n) \quad (14.3.24)$$

Substituting (14.2.18), (14.3.20), and (14.3.24) into (14.3.23), we find that the filtering error covariance matrix can be written as

$$\boldsymbol{\Gamma}(n) = \mathbf{P}(n) - \mathbf{P}(n)\mathbf{C}^\dagger(n)[\mathbf{C}(n)\mathbf{P}(n)\mathbf{C}^\dagger(n) + \mathbf{R}(n)]^{-1}\mathbf{C}(n)\mathbf{P}(n) \quad (14.3.25)$$

Notice that (14.3.25) is identical to (14.3.13).

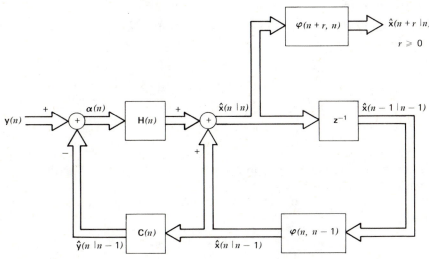

FIGURE 14.3.2. Optimum filter for directly computing $\hat{\mathbf{x}}(n \mid n)$.

In summary, the filtering equations are

1. $\hat{\mathbf{x}}(n \mid n) = \boldsymbol{\varphi}(n, n-1)\hat{\mathbf{x}}(n-1 \mid n-1)$
$$+ \mathbf{H}(n)[\mathbf{y}(n) - \mathbf{C}(n)\boldsymbol{\varphi}(n, n-1)\hat{\mathbf{x}}(n-1 \mid n-1)] \qquad (14.3.21)$$

2. $\mathbf{H}(n) = \mathbf{P}(n)\mathbf{C}^{\dagger}(n)[\mathbf{C}(n)\mathbf{P}(n)\mathbf{C}^{\dagger}(n) + \mathbf{R}(n)]^{-1}$ $\qquad (14.3.20)$

3. $\mathbf{P}(n+1) = \boldsymbol{\varphi}(n+1, n)\boldsymbol{\Gamma}(n)\boldsymbol{\varphi}^{\dagger}(n+1, n) + \mathbf{Q}(n)$ $\qquad (14.3.12)$

4. $\boldsymbol{\Gamma}(n) = \mathbf{P}(n) - \mathbf{P}(n)\mathbf{C}^{\dagger}(n)[\mathbf{C}(n)\mathbf{P}(n)\mathbf{C}^{\dagger}(n) + \mathbf{R}(n)]^{-1}\mathbf{C}(n)\mathbf{P}(n)$ $\qquad \left\{ \begin{array}{l} (14.3.13) \\ \text{and} \\ (14.3.25) \end{array} \right.$

with the initial conditions

$$\hat{\mathbf{x}}(n_0-1 \mid n_0-1) = E\{\mathbf{x}(n_0-1)\} = \mathbf{0} \qquad \text{and} \qquad \mathbf{P}(n_0) = \text{cov}\,\{\mathbf{x}(n_0)\} \quad (14.3.26)$$

These equations can be used in the following sequence:
1. Compute $\mathbf{H}(n)$ using (14.3.20).
2. Compute $\hat{\mathbf{x}}(n \mid n)$ using (14.3.21).
3. Compute $\boldsymbol{\Gamma}(n)$ using (14.3.13).
4. Compute $\mathbf{P}(n+1)$ using (14.3.12).
5. Increment n to $n+1$ and repeat the sequence.

$\mathbf{H}(n)$, $\mathbf{P}(n)$, and $\boldsymbol{\Gamma}(n)$ do not depend on the observed data sequence and can be precomputed. However, when the observed data sequence is long, these quantities are usually computed recursively as the data is received to eliminate large storage requirements.

Example 14.3.2

Suppose that $x(n)$ and $w(n)$ are real discrete-time random processes with the power spectral densities

$$S_{xx}(z) = \frac{0.36}{(1 - 0.8z^{-1})(1 - 0.8z)}$$

$$S_{ww}(z) = 1$$

and

$$S_{xw}(z) = 0$$

The signal $y(n) = x(n) + w(n)$ is observed starting at time $n = 0$ and up-to-date estimates of $x(n)$ are desired.

The signal $x(n)$ can be considered to be the solution of the difference equation

$$x(n+1) = 0.8x(n) + v(n)$$

where $v(n)$ is a white noise sequence with variance $Q(n) = 0.36$. Thus

$\varphi(n+1, n) = 0.8$, $C(n) = 1$, and $R(n) = \text{var}\{w(n)\} = 1$ in this problem. The filtering equations reduce to

$$\hat{x}(n \mid n) = 0.8\hat{x}(n-1 \mid n-1) + H(n)[y(n) - 0.8\hat{x}(n-1 \mid n-1)]$$
$$H(n) = P(n)[P(n)+1]^{-1}$$
$$\Gamma(n) = P(n) - P^2(n)[P(n)+1]^{-1} = P(n)[P(n)+1]^{-1} = H(n) \quad (14.3.27)$$

and

$$P(n+1) = 0.64\Gamma(n) + 0.36 \quad (14.3.28)$$

with the initial conditions

$$\hat{x}(-1 \mid -1) = 0 \quad \text{and} \quad P(0) = \text{var}\{x(0)\} = 1$$

It is interesting to observe the behavior of the filter for large n. Replacing n by $n+1$ in (14.3.27) and using (14.3.28) gives

$$\Gamma(n+1) = \frac{0.64\Gamma(n) + 0.36}{0.64\Gamma(n) + 1.36} \quad (14.3.29)$$

Thus $0 \le \Gamma(n+1) < \Gamma(n)$ so that $\Gamma(\infty) = \lim_{n \to \infty} \Gamma(n)$ must exist. Replacing $\Gamma(n+1)$ and $\Gamma(n)$ by $\Gamma(\infty)$ in (14.3.29) and rearranging terms, we obtain the quadratic equation

$$0.64\Gamma^2(\infty) + 0.72\Gamma(\infty) - 0.36 = 0$$

The positive solution to this equation is

$$\Gamma(\infty) = \tfrac{3}{8}$$

Thus the estimation equation converges to

$$\hat{x}(n \mid n) = 0.8\hat{x}(n-1 \mid n-1) + 0.375[y(n) - 0.8\hat{x}(n-1 \mid n-1)]$$
$$= 0.5\hat{x}(n-1 \mid n-1) + 0.375y(n)$$

Notice that these same results were derived in Example 7.5.2 using the physically realizable discrete-time Wiener filter formula. ◀

14.4 SMOOTHING

We will now investigate the smoothing problem, that is, the problem of estimating $\mathbf{x}(i)$ from $\mathbf{y}(n_0), \ldots, \mathbf{y}(n)$ for $n_0 \le i < n$. Equation 14.2.20, the general estimation. equation, can be written as

$$\hat{\mathbf{x}}(i \mid n) = \sum_{k=n_0}^{i} E\{\mathbf{x}(i)\boldsymbol{\alpha}^{\dagger}(k)\}\mathbf{V}_{\alpha}^{-1}(k)\boldsymbol{\alpha}(k)$$

$$+ \sum_{k=i+1}^{n} E\{\mathbf{x}(i)\boldsymbol{\alpha}^{\dagger}(k)\}\mathbf{V}_{\alpha}^{-1}(k)\boldsymbol{\alpha}(k)$$

$$= \hat{\mathbf{x}}(i \mid i) + \sum_{k=i+1}^{n} E\{\mathbf{x}(i)\boldsymbol{\alpha}^{\dagger}(k)\}\mathbf{V}_{\alpha}^{-1}(k)\boldsymbol{\alpha}(k) \qquad (14.4.1)$$

Thus the smoothed estimates can be computed by modifying the filtered estimates.

We will derive an explicit computational formula by converting (14.4.1) into an alternate form from which we will derive a recursive formula for computing the smoothed estimates backward from time n. As a first step, the innovations sample $\boldsymbol{\alpha}(k)$ can be expressed as

$$\boldsymbol{\alpha}(k) = \mathbf{C}(k)\boldsymbol{\varphi}(k, k-1)\tilde{\mathbf{x}}(k-1 \mid k-1) + \mathbf{C}(k)\mathbf{v}(k-1) + \mathbf{w}(k)$$

where, as usual

$$\tilde{\mathbf{x}}(k-1 \mid k-1) = \mathbf{x}(k-1) - \hat{\mathbf{x}}(k-1 \mid k-1)$$

Then for $i < k$

$$E\{\mathbf{x}(i)\boldsymbol{\alpha}^{\dagger}(k)\} = E\{\tilde{\mathbf{x}}(i \mid i)\boldsymbol{\alpha}^{\dagger}(k)\}$$

$$= E\{\tilde{\mathbf{x}}(i \mid i)\tilde{\mathbf{x}}^{\dagger}(k-1 \mid k-1)\}\boldsymbol{\varphi}^{\dagger}(k, k-1)\mathbf{C}^{\dagger}(k) \qquad (14.4.2)$$

so that

$$\hat{\mathbf{x}}(i \mid n) = \hat{\mathbf{x}}(i \mid i) + \sum_{k=i+1}^{n} E\{\tilde{\mathbf{x}}(i \mid i)\tilde{\mathbf{x}}^{\dagger}(k-1 \mid k-1)\}\boldsymbol{\varphi}^{\dagger}(k, k-1)\mathbf{C}^{\dagger}(k)\mathbf{V}_{\alpha}^{-1}(k)\boldsymbol{\alpha}(k)$$

$$(14.4.3)$$

It follows from (14.3.19) and (14.3.20) that

$$\mathbf{C}^{\dagger}(k)\mathbf{V}_{\alpha}^{-1}(k)\boldsymbol{\alpha}(k) = \mathbf{P}^{-1}(k)[\hat{\mathbf{x}}(k \mid k) - \hat{\mathbf{x}}(k \mid k-1)] \qquad (14.4.4)$$

Therefore, (14.4.3) can be written as

$$\hat{\mathbf{x}}(i \mid n) = \hat{\mathbf{x}}(i \mid i) + \sum_{k=i+1}^{n} E\{\tilde{\mathbf{x}}(i \mid i)\tilde{\mathbf{x}}^{\dagger}(k-1 \mid k-1)\}\boldsymbol{\varphi}^{\dagger}(k, k-1)\mathbf{P}^{-1}(k)$$

$$\times [\hat{\mathbf{x}}(k \mid k) - \hat{\mathbf{x}}(k \mid k-1)] \qquad (14.4.5)$$

For $i = n-1$ (14.4.5) reduces to

$$\hat{\mathbf{x}}(n-1 \mid n) = \hat{\mathbf{x}}(n-1 \mid n-1) + \boldsymbol{\Gamma}(n-1)\boldsymbol{\varphi}^{\dagger}(n, n-1)\mathbf{P}^{-1}(n)[\hat{\mathbf{x}}(n \mid n) - \hat{\mathbf{x}}(n \mid n-1)]$$

$$(14.4.6)$$

where

$$\boldsymbol{\Gamma}(n-1) = E\{\tilde{\mathbf{x}}(n-1 \mid n-1)\tilde{\mathbf{x}}^{\dagger}(n-1 \mid n-1)\}$$

For $i = n - 2$, (14.4.5) becomes

$$\hat{\mathbf{x}}(n-2 \mid n) = \hat{\mathbf{x}}(n-2 \mid n-2) + \mathbf{\Gamma}(n-2)\boldsymbol{\varphi}^\dagger(n-1, n-2)\mathbf{P}^{-1}(n-1)$$
$$\times [\hat{\mathbf{x}}(n-1 \mid n-1) - \hat{\mathbf{x}}(n-1 \mid n-2)]$$
$$+ E\{\tilde{\mathbf{x}}(n-2 \mid n-2)\tilde{\mathbf{x}}^\dagger(n-1 \mid n-1)\}\boldsymbol{\varphi}^\dagger(n, n-1)\mathbf{P}^{-1}(n)$$
$$\times [\hat{\mathbf{x}}(n \mid n) - \hat{\mathbf{x}}(n \mid n-1)] \qquad (14.4.7)$$

Since

$$\hat{\mathbf{x}}(n-1 \mid n-1) = \boldsymbol{\varphi}(n-1, n-2)\hat{\mathbf{x}}(n-2 \mid n-2) + \mathbf{H}(n-1)\boldsymbol{\alpha}(n-1)$$

and

$$\mathbf{x}(n-1) = \boldsymbol{\varphi}(n-1, n-2)\mathbf{x}(n-2) + \mathbf{v}(n-2)$$

we find that

$$\tilde{\mathbf{x}}(n-1 \mid n-1) = \mathbf{x}(n-1) - \hat{\mathbf{x}}(n-1 \mid n-1)$$
$$= \boldsymbol{\varphi}(n-1, n-2)\tilde{\mathbf{x}}(n-2 \mid n-2) + \mathbf{v}(n-2) - \mathbf{H}(n-1)\boldsymbol{\alpha}(n-1)$$
$$(14.4.8)$$

Therefore

$$E\{\tilde{\mathbf{x}}(n-2 \mid n-2)\tilde{\mathbf{x}}^\dagger(n-1 \mid n-1)\} = \mathbf{\Gamma}(n-2)\boldsymbol{\varphi}^\dagger(n-1 \mid n-2)$$
$$- E\{\tilde{\mathbf{x}}(n-2 \mid n-2)\boldsymbol{\alpha}^\dagger(n-1)\}\mathbf{H}^\dagger(n-1)$$
$$(14.4.9)$$

The expectation on the right-hand side of (14.4.9) can be evaluated by letting $i = n - 2$ and $k = n - 1$ in (14.4.2). Using this result gives

$$E\{\tilde{\mathbf{x}}(n-2 \mid n-2)\tilde{\mathbf{x}}^\dagger(n-1 \mid n-1)\}$$
$$= \mathbf{\Gamma}(n-2)\boldsymbol{\varphi}^\dagger(n-1, n-2)[\mathbf{I} - \mathbf{C}^\dagger(n-1)\mathbf{H}^\dagger(n-1)]$$
$$= \mathbf{\Gamma}(n-2)\boldsymbol{\varphi}^\dagger(n-1, n-2)\mathbf{P}^{-1}(n-1)$$
$$\times [\mathbf{P}(n-1) - \mathbf{P}(n-1)\mathbf{C}^\dagger(n-1)\mathbf{H}^\dagger(n-1)] \quad (14.4.10)$$

From (14.3.20) and (14.3.25), it follows that the expression in the rectangular brackets on the right-hand side of (14.4.10) is $\mathbf{\Gamma}(n-1)$. Therefore

$$E\{\tilde{\mathbf{x}}(n-2 \mid n-2)\tilde{\mathbf{x}}^\dagger(n-1 \mid n-1)\} = \mathbf{\Gamma}(n-2)\boldsymbol{\varphi}^\dagger(n-1, n-2)\mathbf{P}^{-1}(n-1)\mathbf{\Gamma}(n-1)$$
$$(14.4.11)$$

Substituting (14.4.11) into (14.4.7) gives

$$\hat{\mathbf{x}}(n-2 \mid n) = \hat{\mathbf{x}}(n-2 \mid n-2) + \mathbf{\Gamma}(n-2)\boldsymbol{\varphi}^\dagger(n-1, n-2)\mathbf{P}^{-1}(n-1)\{\hat{\mathbf{x}}(n-1 \mid n-1)$$
$$+ \mathbf{\Gamma}(n-1)\boldsymbol{\varphi}^\dagger(n, n-1)\mathbf{P}^{-1}(n)[\hat{\mathbf{x}}(n \mid n) - \hat{\mathbf{x}}(n \mid n-1)] - \hat{\mathbf{x}}(n-1 \mid n-2)\} \quad (14.4.12)$$

According to (14.4.6), the first two terms in the braces on the right-hand side of (14.4.12) sum to $\hat{\mathbf{x}}(n-1 \mid n)$. Therefore

$$\hat{\mathbf{x}}(n-2 \mid n) = \hat{\mathbf{x}}(n-2 \mid n-2) + \mathbf{\Gamma}(n-2)\boldsymbol{\varphi}^\dagger(n-1, n-2)\mathbf{P}^{-1}(n-1)$$
$$\times [\hat{\mathbf{x}}(n-1 \mid n) - \hat{\mathbf{x}}(n-1 \mid n-2)] \quad (14.4.13)$$

Continuing in the same manner, it follows that for $n_0 \leq i < n$

$$\hat{\mathbf{x}}(i \mid n) = \hat{\mathbf{x}}(i \mid i) + \mathbf{A}(i)[\hat{\mathbf{x}}(i+1 \mid n) - \hat{\mathbf{x}}(i+1 \mid i)] \qquad (14.4.14)$$

where

$$\mathbf{A}(i) = \boldsymbol{\Gamma}(i)\boldsymbol{\varphi}^\dagger(i+1, i)\mathbf{P}^{-1}(i+1) \qquad (14.4.15)$$

As an additional step, $\hat{\mathbf{x}}(i+1 \mid i)$ can be replaced by $\boldsymbol{\varphi}(i+1, i)\hat{\mathbf{x}}(i \mid i)$ in (14.4.14). Equations 14.4.14 and 14.4.15 provide a means for recursively computing the smoothed estimates as far back from time n as desired after the solution to the filtering problem has been computed and stored.

The error in the smoothed estimate is

$$\tilde{\mathbf{x}}(i \mid n) = \mathbf{x}(i) - \hat{\mathbf{x}}(i \mid n) = \tilde{\mathbf{x}}(i \mid i) - \mathbf{A}(i)[\hat{\mathbf{x}}(i+1 \mid n) - \hat{\mathbf{x}}(i+1 \mid i)]$$

which can be rearranged into the form

$$\tilde{\mathbf{x}}(i \mid n) + \mathbf{A}(i)\hat{\mathbf{x}}(i+1 \mid n) = \tilde{\mathbf{x}}(i \mid i) + \mathbf{A}(i)\hat{\mathbf{x}}(i+1 \mid i) \qquad (14.4.16)$$

We will denote the smoothing error covariance matrix by

$$\boldsymbol{\Lambda}(i \mid n) = \mathrm{cov}\,\{\tilde{\mathbf{x}}(i \mid n)\} \qquad (14.4.17)$$

From the orthogonality condition of the Projection Theorem, it follows that

$$\mathrm{cov}\,\{\tilde{\mathbf{x}}(i \mid n), \hat{\mathbf{x}}(i+1 \mid n)\} = \mathrm{cov}\,\{\tilde{\mathbf{x}}(i \mid i), \hat{\mathbf{x}}(i+1 \mid i)\} = \mathbf{0} \qquad (14.4.18)$$

$$\mathrm{cov}\,\{\hat{\mathbf{x}}(i+1 \mid n)\} = \mathrm{cov}\,\{\mathbf{x}(i+1)\} - \boldsymbol{\Lambda}(i+1 \mid n) \qquad (14.4.19)$$

and

$$\mathrm{cov}\,\{\hat{\mathbf{x}}(i+1 \mid i)\} = \mathrm{cov}\,\{\mathbf{x}(i+1)\} - \mathbf{P}(i+1) \qquad (14.4.20)$$

Taking the covariance of both sides of (14.4.16) and using (14.4.18)–(14.4.20), it follows that

$$\boldsymbol{\Lambda}(i \mid n) = \boldsymbol{\Gamma}(i) + \mathbf{A}(i)[\boldsymbol{\Lambda}(i+1 \mid n) - \mathbf{P}(i+1)]\mathbf{A}^\dagger(i) \qquad (14.4.21)$$

This equation can be solved backward in time starting with $i = n-1$ and the terminal condition $\boldsymbol{\Lambda}(n \mid n) = \boldsymbol{\Gamma}(n)$.

The problem solved in this section is known as the *fixed-interval* smoothing problem. A second smoothing problem that has been posed is known as the *fixed-point* smoothing problem. In the fixed-point problem, i is fixed while n increases. A convenient recursive algorithm for updating the fixed-point estimates can be derived from the fixed-interval equations. A third smoothing problem that has been posed is known as the *fixed-lag* smoothing problem. In the fixed-lag problem, n increases and $i = n - J$ where J is a fixed integer. An efficient recursive algorithm for updating the fixed-lag estimates can be found by combining the fixed-interval and fixed-point equations. A clear derivation of the fixed-point and fixed-lag smoothing algorithms can be found in Reference 118.

14.5 COMMENTS

When the dynamical system is time-invariant and the system input $\mathbf{v}(n)$ and observation noise $\mathbf{w}(n)$ are stationary random processes, the Kalman filter converges to a time-invariant filter in several time constants of the dynamical system. If the signal to be estimated and the observed signal are scalars, then the Kalman filter converges to the realizable Wiener filter presented in Section 7.5. The Wiener filter problem also can be formulated for vector signals and the solution involves matrix spectral factorization. Finding the steady-state Kalman filter equations bypasses the matrix spectral factorization problem. When the observed data sequence is long and initial transients can be ignored, the time-invariant steady-state filter is often used to process the entire sequence because it is more economical to implement. The complete time-varying Kalman filter is particularly useful when the dynamical system and/or signal statistics are time varying and also when optimum estimates are required from the very beginning.

APPENDIX
A

SUMMARY OF THE
THEORY OF RESIDUES

A.1 INTRODUCTION

The theory of residues provides an important tool for evaluating integrals. A brief summary of the theory is presented here for reference. The reader should consult references on complex variables such as Ahlfors [1], Churchill [21], and Kaplan [71] for a detailed presentation of the theory.

A.2 LAURENT SERIES AND THE CLASSIFICATION OF SINGULARITIES

The following theorem is one of the important results in complex variable theory. It states that any function analytic in a ring centered about z_0 can be expanded into a series of positive and negative powers of $z - z_0$. This series is called a *Laurent series*. Much of the theory of residues is based on Laurent series.

THEOREM A.2.1. Laurent's Theorem
Let $F(z)$ be analytic in the ring $R_1 < |z - z_0| < R_2$. Then

$$F(z) = \sum_{n=-\infty}^{\infty} f_n(z - z_0)^{-n} \qquad (A.2.1)$$

where

$$f_n = \frac{1}{2\pi j} \oint_C F(z)(z - z_0)^{n-1} \, dz \qquad (A.2.2)$$

and C is any simple closed curve separating $|z| = R_1$ from $|z| = R_2$. The series converges uniformly for $R_1 < R_1' \leq |z - z_0| \leq R_2' < R_2$.

An isolated singularity of $F(z)$ at z_0 is classified in terms of the Laurent expansion for $F(z)$ about z_0. If

$$F(z) = \sum_{n=-\infty}^{N} f_n(z - z_0)^{-n} \qquad \text{for} \qquad 0 < |z - z_0| < R_2 \qquad (A.2.3)$$

with $0 < N < \infty$ and $f_N \neq 0$, then we say that $F(z)$ has a *pole of order* N at z_0. If $N = 1$, the singularity is also called a *simple pole*. We can write (A.2.3) as

$$F(z) = G(z)/(z - z_0)^N \qquad (A.2.4)$$

where

$$G(z) = f_N + f_{N-1}(z - z_0) + \cdots$$

so that $G(z)$ is analytic in the neighborhood of z_0 and nonzero at z_0. Conversely, it can be shown that any function having the form (A.2.4) has a pole of order N at z_0. If infinitely many negative powers of $z - z_0$ appear in the Laurent series, we say that $F(z)$ has an *essential* singularity at z_0. For example

$$e^{z^{-1}} = \sum_{n=0}^{\infty} z^{-n}/n!$$

has an essential singularity at $z = 0$.

A.3 RESIDUES AT SINGULARITIES IN THE FINITE COMPLEX PLANE

If $F(z)$ has an isolated singularity at z_0, the *residue* of $F(z)$ at z_0 is defined to be

$$\text{Res}\,[F(z), z_0] = \frac{1}{2\pi j} \oint_C F(z) \, dz \qquad (A.3.1)$$

where C is any simple closed path along which $F(z)$ is analytic and which encircles only the singularity of $F(z)$ at z_0.

THEOREM A.3.1
The residue of $F(z)$ at z_0 is

$$\text{Res}\,[F(z), z_0] = f_1 \qquad (A.3.2)$$

where f_1 is the coefficient of $(z - z_0)^{-1}$ in the Laurent expansion of $F(z)$ about z_0.

Proof Let $R_1 = 0$ and $n = 1$ in (A.2.2).

Q.E.D.

In discrete-time system analysis, the singularities of $F(z)$ are normally poles of finite order. The following simple rules are usually sufficient for finding the residues.

RULE 1 Pole of Order 1 at z_0

$$\text{Res}\,[F(z), z_0] = \lim_{z \to z_0} (z - z_0)F(z) \qquad (A.3.3)$$

Proof For a simple pole

$$F(z) = \frac{f_1}{z - z_0} + f_0 + \cdots$$

so that

$$\lim_{z \to z_0} (z - z_0)F(z) = f_1$$

Q.E.D.

RULE 2 Pole of Order 1 at z_0

If $A(z)$ and $B(z)$ are analytic in the neighborhood of z_0, $A(z_0) \neq 0$, and $B(z)$ has a zero of order 1 at z_0, then

$$F(z) = A(z)/B(z)$$

has a pole of order 1 at z_0 and

$$\text{Res}\,[F(z), z_0] = A(z_0)/B'(z_0) \qquad (A.3.4)$$

Proof From (A.3.3) and the fact that $B(z_0) = 0$

$$f_1 = \lim_{z \to z_0} (z - z_0)\frac{A(z)}{B(z) - B(z_0)}$$

$$= A(z_0)/\lim_{z \to z_0} \frac{B(z) - B(z_0)}{z - z_0} = A(z_0)/B'(z_0)$$

Q.E.D.

RULE 3 Pole of Order N at z_0

$$\text{Res}\,[F(z), z_0] = \lim_{z \to z_0} \frac{1}{(N-1)!} \frac{d^{N-1}}{dz^{N-1}}[(z - z_0)^N F(z)] \qquad (A.3.5)$$

Proof If $F(z)$ has a pole of order N at z_0, then

$$F(z) = \frac{f_N}{(z-z_0)^N} + \cdots + \frac{f_1}{z-z_0} + f_0 + \cdots$$

so that

$$(z-z_0)^N F(z) = f_N + \cdots + f_1(z-z_0)^{N-1} + f_0(z-z_0)^N + \cdots$$

This is a Taylor series in $z - z_0$ and f_1 is the coefficient of $(z-z_0)^{N-1}$. Equation A.3.5 is just the Taylor series formula for calculating this coefficient.

Q.E.D.

A.4 RESIDUE AT INFINITY

If $F(z)$ is analytic for $|z| > R$ except possibly at $z = \infty$, then the residue of $F(z)$ at ∞ is defined as

$$\text{Res}\,[F(z), \infty] = \frac{1}{2\pi j} \oint_C F(z)\, dz \qquad (A.4.1)$$

where C is any simple closed path encircling the origin in the negative direction and enclosing all the singularities of $F(z)$ in the finite complex plane.

THEOREM A.4.1

$$\text{Res}\,[F(z), \infty] = -f_1 \qquad (A.4.2)$$

where f_1 is the coefficient of z^{-1} in the Laurent expansion of $F(z)$ about $z = 0$ in the domain $R < |z| < \infty$.

Proof The proof is the same as for Theorem A.2.1.

Q.E.D.

The following rules are usually sufficient for calculating the residue of $F(z)$ at infinity.

RULE 4 First-Order Zero at Infinity

$$\text{Res}\,[F(z), \infty] = -\lim_{z \to \infty} zF(z) \qquad (A.4.3)$$

Proof If $F(z)$ has a first-order zero at infinity, then

$$F(z) = \frac{f_1}{z} + \frac{f_2}{z^2} + \cdots$$

so that

$$\lim_{z \to \infty} zF(z) = f_1$$

Q.E.D.

RULE 5   Zero of Order Greater than One at Infinity

$$\text{Res}\,[F(z), \infty] = 0 \tag{A.4.4}$$

Proof   In this case

$$F(z) = \frac{f_2}{z^2} + \frac{f_3}{z^3} + \cdots$$

and $f_1 = 0$.

Q.E.D.

RULE 6

$$\text{Res}\,[F(z), \infty] = -\text{Res}\,[F(z^{-1})z^{-2}, 0] \tag{A.4.5}$$

Proof

$$F(z) = \cdots + f_{-1}z + f_0 + f_1 z^{-1} + \cdots \qquad \text{for} \qquad R < |z| < \infty$$

so that for $0 < |z| < R$

$$F(z^{-1}) = \cdots + f_{-1}z^{-1} + f_0 + f_1 z + \cdots$$

and

$$F(z^{-1})z^{-2} = \cdots + f_{-1}z^{-3} + f_0 z^{-2} + f_1 z^{-1} + \cdots$$

Therefore

$$f_1 = \text{Res}\,[F(z^{-1})z^{-2}, 0]$$

This reduces the problem to finding a residue at 0 to which Rules 1, 2, and 3 can be applied.

Q.E.D.

A.5 EVALUATION OF INTEGRALS BY CAUCHY'S RESIDUE THEOREM

An important application of the theory of residues is the evaluation of integrals around closed contours. The following theorem presents this remarkable result.

THEOREM A.5.1. Cauchy's Residue Theorem

If $F(z)$ is analytic in a domain D which includes a deleted neighborhood of ∞ and C is any simple closed path in D, then

$$\frac{1}{2\pi j} \oint_C F(z)\,dz = \sum_{k=1}^{N} \text{Res}\,[F(z), a_k] \tag{A.5.1}$$

if $F(z)$ is analytic within C except for isolated singularities at $a_1, \ldots, a_N$. Also

$$\frac{1}{2\pi j} \oint_C F(z)\,dz = -\sum_{k=1}^{M} \text{Res}\,[F(z), b_k] - \text{Res}\,[F(z), \infty] \tag{A.5.2}$$

if $F(z)$ is analytic in the finite complex plane outside C except for isolated singularities at $b_1, \ldots, b_M$.

B

TABLE OF Z-TRANSFORMS OF CAUSAL SIGNALS

The Z-transforms of some causal signals are listed in the following table. The entries under the column labeled $\mathscr{F}(s)$ are the Laplace transforms of $f(t)u(t)$ where $u(t)$ is the unit step function defined as

$$u(t) = \begin{cases} 1 & \text{for} & t \geq 0 \\ 0 & \text{for} & t < 0 \end{cases}$$

The entries under the column labeled $F(z)$ are the Z-transforms of $f(t)u(t)$ so that

$$F(z) = \sum_{n=0}^{\infty} f(nT)z^{-n} \qquad \text{for} \qquad |z| > R$$

The convention of assigning the value $f(0^+)$ to $f(0)$ has been followed. In each case

$$R = \max_k |z_k|$$

where $\{z_k\}$ is the set of points at which $F(z)$ has singularities. The variables b and c in the table designate real numbers.

No.	$f(t)$ or $f(nT)$	$\mathscr{F}(s)$	$F(z)$
1	1	$\dfrac{1}{s}$	$\dfrac{1}{1-z^{-1}}$
2	t	$\dfrac{1}{s^2}$	$\dfrac{Tz^{-1}}{(1-z^{-1})^2}$

No.	$f(t)$ or $f(nT)$	$\mathscr{F}(s)$	$F(z)$
3	t^2	$\dfrac{2}{s^3}$	$\dfrac{T^2 z^{-1}(1+z^{-1})}{(1-z^{-1})^3}$
4	t^3	$\dfrac{3!}{s^4}$	$\dfrac{T^3 z^{-1}(1+4z^{-1}+z^{-2})}{(1-z^{-1})^4}$
5	t^4	$\dfrac{4!}{s^5}$	$\dfrac{T^4 z^{-1}(1+11z^{-1}+11z^{-2}+z^{-3})}{(1-z^{-1})^5}$
6	e^{-at}	$\dfrac{1}{s+a}$	$\dfrac{1}{1-e^{-aT}z^{-1}}$
7	te^{-at}	$\dfrac{1}{(s+a)^2}$	$\dfrac{Te^{-aT}z^{-1}}{(1-e^{-aT}z^{-1})^2}$
8	$t^2 e^{-at}$	$\dfrac{2}{(s+a)^3}$	$\dfrac{T^2 e^{-aT}z^{-1}(1+e^{-aT}z^{-1})}{(1-e^{-aT}z^{-1})^3}$
9	$t^L e^{at}$	$\dfrac{L!}{(s-a)^{L+1}}$	$\dfrac{\partial^L}{\partial a^L}\dfrac{1}{1-e^{aT}z^{-1}}$
10	$1-e^{-at}$	$\dfrac{a}{s(s+a)}$	$\dfrac{(1-e^{-aT})z^{-1}}{(1-z^{-1})(1-e^{-aT}z^{-1})}$
11	$e^{-at}-e^{-dt}$	$\dfrac{d-a}{(s+a)(s+d)}$	$\dfrac{(e^{-aT}-e^{-dT})z^{-1}}{(1-e^{-aT}z^{-1})(1-e^{-dT}z^{-1})}$
12	$at+e^{-at}-1$	$\dfrac{a^2}{s^2(s+a)}$	$\dfrac{(aT+e^{-aT}-1)z^{-1}+(1-e^{-aT}-aTe^{-aT})z^{-2}}{(1-z^{-1})^2(1-e^{-aT}z^{-1})}$
13	$\cos ct$	$\dfrac{s}{s^2+c^2}$	$\dfrac{1-z^{-1}\cos cT}{1-2z^{-1}\cos cT+z^{-2}}$
14	$\sin ct$	$\dfrac{c}{s^2+c^2}$	$\dfrac{z^{-1}\sin cT}{1-2z^{-1}\cos cT+z^{-2}}$
15	$e^{-bt}\cos ct$	$\dfrac{s+b}{(s+b)^2+c^2}$	$\dfrac{1-e^{-bT}z^{-1}\cos cT}{1-2e^{-bT}z^{-1}\cos cT+e^{-2bT}z^{-2}}$
16	$e^{-bt}\sin ct$	$\dfrac{c}{(s+b)^2+c^2}$	$\dfrac{e^{-bT}z^{-1}\sin cT}{1-2e^{-bT}z^{-1}\cos cT+e^{-2bT}z^{-2}}$
17	$\cosh bt$	$\dfrac{s}{s^2-b^2}$	$\dfrac{1-z^{-1}\cosh bT}{1-2z^{-1}\cosh bT+z^{-2}}$
18	$\sinh bt$	$\dfrac{b}{s^2-b^2}$	$\dfrac{z^{-1}\sinh bT}{1-2z^{-1}\cosh bT+z^{-2}}$

No.	$f(t)$ or $f(nT)$	$\mathscr{F}(s)$	$F(z)$
19	$\dfrac{1}{n}u(nT-T)$		$-\log_e(1-z^{-1})$
20	$\dbinom{K}{n}u(KT-nT)$		$(1+z^{-1})^K$
21	$n(n-1)\cdots(n-L+1)$		$\dfrac{L!\,z^{-L}}{(1-z^{-1})^L}$
22	$\dfrac{1}{n!}$		$e^{z^{-1}}$

PROBLEMS

CHAPTER 2

2.1. Let the sampling pulse train be $s(t) = \sum\limits_{n=-\infty}^{\infty} p(t-nT)$ with

$$p(t) = \begin{cases} 1 & \text{for} \quad |t| < a \text{ and } a < T/2 \\ 0 & \text{elsewhere} \end{cases}$$

and

$$F(\omega) = \begin{cases} 1 - \dfrac{|\omega|}{B} & \text{for} \quad |\omega| < B \\ 0 & \text{elsewhere} \end{cases}$$

(a) Find the Fourier series for $s(t)$.
(b) Sketch $F^{\#}(\omega)$ for $\omega_s < 2B$ and $\omega_s > 2B$.

2.2. If $f(t) = e^{-a|t|}$ for $-\infty < t < \infty$ with $a > 0$, find $F^*(\omega)$ and sketch $|F^*(\omega)|$.

2.3. If $f(t) = \cos \omega_0 t$ for $-\infty < t < \infty$, find $F^*(\omega)$.

2.4. If $f(t) = te^{-at}$ for $t > 0$ and 0 elsewhere with $a > 0$, find $F^*(\omega)$.

2.5. Find a closed form expression for $\sum\limits_{n=-\infty}^{\infty} \dfrac{1}{n^2 + a^2}$. Hint: Use the Poisson Sum Formula 2.3.11.

2.6. Let

$$f_1(t) = \begin{cases} f(nT) & \text{for} \quad nT \le t < nT + a \text{ with } 0 < a < T \\ 0 & \text{elsewhere} \end{cases}$$

Notice that $f_1(t)$ represents the samples of $f(t)$ with finite duration flat top pulses. Find $F_1(\omega)$ in terms of $F^*(\omega)$. Hint: Consider applying $f^*(t)$ to a filter with impulse response

$$h(t) = \begin{cases} 1 & \text{for} \quad 0 \le t < a \\ 0 & \text{elsewhere} \end{cases}$$

2.7. In a pulse-amplitude modulation system, the sequence $\{f(nT)\}$ is transmitted by applying the signal $f^*(t) = \sum_{n=-\infty}^{\infty} f(nT)\,\delta(t-nT)$ to a channel with impulse response $c(t)$. The received signal is

$$r(t) = \sum_{n=-\infty}^{\infty} f(nT)c(t-nT).$$

The symbol $f(kT)$ is estimated at the receiver as

$$r(kT) = c_0 f(kT) + \sum_{n \neq k} f(nT)c(kT-nT)$$

The summation on the right is called intersymbol interference. Show that if $\sum_{n=-\infty}^{\infty} C(\omega - n\omega_s) = A$ where A is a real constant, then there is no intersymbol interference. This is known as *Nyquist's Criterion* [97].

2.8. Let

$$C(\omega) = \begin{cases} T & \text{for} & |\omega| \leq \dfrac{\pi}{T}(1-a) \\[2mm] \dfrac{T}{2}\left\{1 - \sin\left[\dfrac{T}{2a}\left(|\omega| - \dfrac{\pi}{T}\right)\right]\right\} & \text{for} & \dfrac{\pi}{T}(1-a) \leq |\omega| \leq \dfrac{\pi}{T}(1+a) \\[2mm] 0 & \text{elsewhere} \end{cases}$$

with $0 < a < 1$. This is known as a *raised cosine characteristic* [87].
(a) Find $C^*(\omega)$ and $c(nT)$.
(b) According to Problem 2.7, does this channel have intersymbol interference?
(c) Show that

$$c(t) = \frac{\sin \pi t/T}{\pi t/T}\,\frac{\cos a\pi t/T}{1 - 4a^2 t^2/T^2}$$

2.9. Find the Hilbert transforms of the following signals:
(a) $\sin \omega_0 t$ for all t
(b) $\cos \omega_0 t$ for all t
(c) $\dfrac{\sin \pi t/T}{\pi t/T}$ for all t
(d) $f(t) = \begin{cases} 1 & \text{for} & |t| < T \\ 0 & \text{elsewhere} \end{cases}$

2.10. Given a band-pass signal $f(t)$ with pass-band $\omega_1 < |\omega| < \omega_2$ and $\omega_0 = \omega_2$:
(a) Find $A(\omega)$ and $B(\omega)$ using (2.5.10) and (2.5.11). Assume a shape for $F(\omega)$ and sketch $A(\omega)$ and $B(\omega)$.
(b) Show that $f(t) = a(t) \cos \omega_0 t + \breve{a}(t) \sin \omega_0 t$

(c) Show that $f(t)$ can be generated by passing $2a(t)\cos\omega_0 t$ through an ideal low-pass filter with cutoff frequency ω_0.

Note: $f(t)$ is the single-sideband, suppressed carrier, amplitude-modulated signal associated with $a(t)$. $F(\omega)$ is called the *lower sideband*.

2.11. A signal $f(t)$ is band-limited to the interval $|\omega| < \omega_s/2$. If $g(t) = f(t)\cos N\omega_s t$ where N is an integer, find the relationship between $F^*(\omega)$ and $G^*(\omega)$. Give an interpolation formula for reconstructing $f(t)$ from $\{g(nT)\}$.

CHAPTER 3

3.1. Find and sketch the impulse responses for the $m = 2$ quadratic reconstruction filters with $r = 0$, 1, and 2. Calculate the corresponding transfer functions. Give equations for the reconstructed signals for $nT \le t < (n+1)T$.

3.2. Find and sketch the impulse responses for the $m = 3$ reconstruction filters with $r = 0$, 1, 2, and 3. Five equations for the reconstructed signals for $nT \le t < (n+1)T$.

3.3. The signal $f(t)$ is known to be band-limited with cutoff frequency π/T. To reconstruct $f(t)$ the sampled signal $f^*(t)$ is first passed through a zero-order hold and then $H(\omega)$. Find $H(\omega)$ so that $y(t)$ is an exact replica of $f(t)$ except for a possible delay.

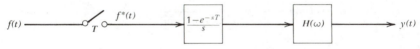

FIGURE P3.3.

3.4. The backward-difference operator ∇ is defined by the equation

$$\nabla f(t) = f(t) - f(t - T)$$

(a) Find the transfer function for ∇, that is $\mathcal{L}\{\nabla f(t)\}/\mathcal{L}\{f(t)\}$.

(b) If $p(t)$ is a polynomial of degree m, prove that $\nabla^k p(t) \equiv 0$ for $k > m$.

3.5. Conceptually we can find $f(t + a)$ by passing $f(t)$ through the nonrealizable filter $\mathcal{H}(s) = e^{sa}$ where $a > 0$.

(a) Show that

$$e^{sa} = \sum_{k=0}^{\infty} \frac{1}{k!}\left(\frac{a}{T}\right)^{[k]}(1 - e^{-sT})^k$$

where

$$\left(\frac{a}{T}\right)^{[k]} = \frac{a}{T}\left(\frac{a}{T}+1\right)\left(\frac{a}{T}+2\right)\cdots\left(\frac{a}{T}+k-1\right) \qquad \text{for} \qquad k=1,2,\ldots,\infty$$

and

$$\left(\frac{a}{T}\right)^{[0]} = 1$$

(b) Show that

$$f(nT+a) = \sum_{k=0}^{\infty} \frac{1}{k!}\left(\frac{a}{T}\right)^{[k]} \nabla^k f(nT)$$

where ∇ is defined in Problem 3.4. This is called the *Newton-Gregory backward-difference extrapolation formula.*

(c) If $f(t)$ is a polynomial of degree m, what does the Newton-Gregory formula reduce to? This provides an alternate approach to polynomial extrapolation.

Hint: $e^{sa} = [1-(1-e^{-sT})]^{-a/T}$.

3.6. Prove that the system shown in Fig. P3.6 is a first-order hold by calculating its impulse response. The parameter d is an infinitesimally small positive number.

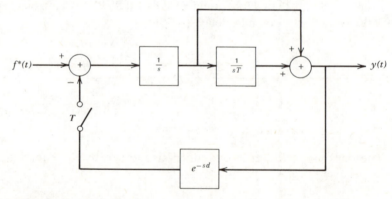

FIGURE P3.6.

CHAPTER 4

4.1. Find the Z-transform of each of the following signals and state the region of convergence. Use the convention of assigning the value $f(0^+)$ to $f(0)$.

(a) $e^{-2t}u(t)\cos t$

(b) $e^t u(t)\sin t$

(c) $te^{-at}u(t)$

(d) $f(t)=\begin{cases}e^{-t} & \text{for} & t\le0\\e^{-2t} & \text{for} & t>0\end{cases}$

(e) $e^{-|t|}\cos2t$

(f) $1/n!\ u(nT)$

(g) $f(nT)=\begin{cases}\dfrac{(-1)^{(n-1)/2}}{n!} & \text{for} & n>0 \text{ and odd}\\0 & & \text{elsewhere}\end{cases}$

4.2.   Each of the following Laplace transforms corresponds to a causal signal. Find these signals. Find the Z-transform of each of these signals using (4.4.22) and state the region of convergence.

(a) $\dfrac{s+1}{(s+2)(s+3)}$

(b) $\dfrac{1}{s^2(s+1)(s+2)}$

(c) $\dfrac{s}{s^2+1}$

(d) $\dfrac{1}{(s-1)^3}$

4.3.   The following Z-transforms correspond to causal signals. Find their inverse Z-transforms by the method of partial fractions.

(a) $\dfrac{1}{(1-2z^{-1})(1-0.5z^{-1})}$

(b) $\dfrac{1+z^{-1}}{1-2z^{-1}\cos\omega_0T+z^{-2}}$

(c) $\dfrac{1+z^{-1}+z^{-2}}{1-1.5z^{-1}+0.5z^{-2}}$

4.4.   Find $f(nT)$ if $F(z)$ is given by:

(a) $\dfrac{1}{(1-z^{-1})(1-2z^{-1})}$ for $1<|z|<2$

(b) $\dfrac{r!\,z^{-r}}{(1-z^{-1})^{r+1}}$ for $r\ge0$ and $|z|>1$

(c) $\dfrac{z^{-5}}{(1-0.5z^{-1})(1-0.5z)}$ for $0.5<|z|<2$

(d) $\dfrac{1}{(1-z^{-1})(1+z^{-1})}$ for $|z|<1$

(e) $\dfrac{z^{-1}}{(1-e^{-T}z^{-1})^2}$ for $|z|>e^{-T}$

4.5.   Find $f(nT)$ if

$$F(z) = \frac{z}{(z-2)(z^2+1)}$$

for

(a) $2 < |z|$
(b) $1 < |z| < 2$
(c) $|z| < 1$

4.6.   Find $f(nT)$ if $F(z) = \dfrac{z^{-2} \sin T}{1 - 2z^{-2} \cos T + z^{-4}}$ for $|z| > 1$.

4.7.   If $F(z) = \dfrac{z^{-2}}{1+z^{-2}}$ for $|z| > 1$, find $f(nT)$ and $\mathscr{L}\{f[(n+1)T]/(nT)\}$.

4.8.   Find $f(nT)$ if $F(z) = \dfrac{1+j}{1-(1+j)z^{-1}}$ for $|z| > \sqrt{2}$. Find $\mathscr{L}\{f(nT)\}$ directly and by (4.6.12).

4.9.   Find $f(0)$ and $\lim\limits_{n\to\infty} f(nT)$, if it exists, for the causal signals with the following Z-transforms:

(a) $\dfrac{1+2z^{-1}}{1-0.7z^{-1}-0.3z^{-2}}$

(b) $\dfrac{z^{-1}}{1-1.5z^{-1}+0.5z^{-3}}$

(c) $\dfrac{1+z^{-1}+z^{-2}}{(1-z^{-1})(1-2z^{-1})}$

4.10.   Let $f(t)$ be a causal signal with Z-transform $F(z)$ for $|z| > R$. Use Theorem 4.7.1 to find $\mathscr{L}\{tf(t)\}$. State the region of convergence. Compare your answer with (4.6.4).

4.11.   Let $f(t) = e^{-at}u(t)$ and $g(t) = \sin t$. Use Theorem 4.7.1 to find $\mathscr{L}\{f(t)g(t)\}$.

4.12.   If $F(z) = \dfrac{0.99}{(1-0.1z^{-1})(1-0.1z)}$ for $\quad 0.1 < |z| < 10$

and

$$G(z) = \frac{1}{1-10z} \quad \text{for} \quad |z| < 0.1$$

find $\mathscr{L}\{f(nT)g(nT)\}$ using Theorem 4.7.1 and by direct calculation using 4.2.1.

4.13.   Let

$$F_k(z) = \frac{\sqrt{1-a^2}}{1-az^{-1}}\left[\frac{1-az}{1-az^{-1}}\right]^k z^{-k}$$

for $|z| > a$ with $0 \le a < 1$ and $k = 0, 1, \dots, \infty$.

Prove that $\displaystyle\sum_{n=0}^{\infty} f_r(nT)f_s(nT) = \delta_{rs}$

Hint: Use Parseval's Theorem.

4.14. Let $f(t) = tu(t)$ and $g(t) = e^{-t}u(t) \sin t$. Find $\displaystyle\sum_{n=0}^{\infty} f(nT)g(nT)$ using Theorem 4.8.1.

4.15. Let $\mathscr{L}\{a(t)\} = A(z)$ for $R_{1a} < |z| < R_{2a}$ and $\mathscr{L}\{b(t)\} = B(z)$ for $R_{1b} < |z| < R_{2b}$. Show that

$$\mathscr{L}\left\{\sum_{k=-\infty}^{\infty} a(kT)b[n-k)T]\right\} = A(z)B(z)$$

for $\max(R_{1a}, R_{1b}) < |z| < \min(R_{2a}, R_{2b})$ by using (4.8.4) and (4.8.5).

4.16. Let $\mathscr{L}\{f(kT)\} = F(z)$ for $|z| > R$ with $R < 1$. Assume that $f(nT)$ is real. If $\operatorname{Re} F^*(\omega) = \cos \omega T$,

(a) Find $\operatorname{Im} F^*(\omega)$ using (4.10.4). Hint: Use the identity

$$\cos yT = \cos \omega T \cos(\omega - y)T + 2 \sin \omega T \cos \frac{(\omega - y)T}{2} \sin \frac{(\omega - y)T}{2}$$

(b) Find $F(z)$ and $f(nT)$.

4.17. Let $f(t)$ be a real causal signal with

$$\operatorname{Re} F^*(\omega) = \frac{0.9 + 0.9 \cos \omega T}{1.01 - 0.2 \cos \omega T}$$

Find $\operatorname{Im} F^*(\omega)$ and $F(z)$.

4.18. Find the minimum phase sequence $f(nT)$ corresponding to $\log|F^*(\omega)| = \cos \omega T$.

4.19. Let $F(z) = A(z)(1 - z_0 z^{-1})$ and $G(z) = A(z)(z^{-1} - z_0)$ where z_0 is real and $z_0 > 1$. Assume $f(t)$ is real.

(a) Show that $|F^*(\omega)| = |G^*(\omega)|$.

(b) Sketch $\arg(1 - z_0 e^{-j\omega T})$ and $\arg(e^{-j\omega T} - z_0)$ for $0 < \omega < \omega_s/2$. Which has less phase lag, $F^*(\omega)$ or $G^*(\omega)$?

(c) Extend the argument to the case where $F(z)$ has a pair of complex conjugate zeros outside the unit circle and conclude that the Z-transform of the minimum phase sequence corresponding to $|F^*(\omega)|$ is obtained by reflecting the zeros of $F(z)$ into the unit circle.

4.20. Prove the converse to Theorem 4.10.1. That is, prove that if $A^*(\omega)$ is real, and $A(z)$ has no singularities on or in the neighborhood of the unit circle, and $B^*(\omega)$ is related to $A^*(\omega)$ by (4.10.4), then $F(z) = A(z) + jB(z)$ has no singularities on or outside the unit circle and $f(nT)$ is a causal signal.

4.21. (a) Show that

$$\sum_{n=-\infty}^{\infty} (nT)^2 |f(nT)|^2 = \frac{1}{\omega_s} \int_{-\omega_s/2}^{\omega_s/2} \left| \frac{d}{d\omega} F^*(\omega) \right|^2 d\omega$$

This gives a measure of the time dispersal of the signal $f(nT)$.

(b) Let $F^*(\omega) = A(\omega)e^{j\theta(\omega)}$ where $A(\omega) = |F^*(\omega)|$ and $\theta(\omega) = \arg F^*(\omega)$. Assuming that $A(\omega)$ is fixed, find $\theta(\omega)$ to minimize the expression in (a).

4.22. Find the solutions to the following difference equations for $n \geq 0$.
(a) $y(n+2) - 1.1y(n+1) + 0.1y(n) = 0$; $y(0) = 0$, $y(1) = 1$
(b) $y(n+1) - 0.5y(n) = \sin n$; $y(0) = -1$
(c) $y(n+3) - 2y(n+2) + 1.25y(n+1) - 0.25y(n) = 1$; $y(2) = y(1) = 0$, $y(0) = 1$

4.23. Consider the infinite resistive ladder network shown in Fig. P4.23.
(a) Find a difference equation relating i_{n-1}, i_n, and i_{n+1}.
(b) Solve this difference equation.

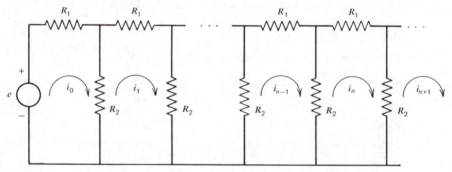

FIGURE P4.23.

4.24. Prove that the N solutions given by (4.11.7) and (4.11.9) to an Nth order, linear, constant coefficient, homogeneous, difference equation are linearly independent.

4.25. Find the discrete-time convolution of the following pairs of signals
(a) $f(nT) = \cos nTu(nT)$ and $g(nT) = 0.2^n u(nT)$
(b) $f(nT) = g(nT) = \begin{cases} 1 & \text{for} \quad 0 \leq n \leq N \\ 0 & \text{otherwise} \end{cases}$

4.26. Let the causal signal $f(nT)$ have a Z-transform $F(z)$ with a region of convergence that includes the unit circle. Under what conditions is there a bounded causal signal $g(nT)$ such that $f(nT)*g(nT) = \delta_{n0}$? How are $F(z)$ and $G(z)$ related?

CHAPTER 5

5.1. In Fig. 5.2.1 let $\mathcal{G}_h(s)$ correspond to the impulse response

$$g_h(t) = \begin{cases} t/T & \text{for} \quad 0 \le t \le T \\ 2 - (t/T) & \text{for} \quad T \le t < 2T \\ 0 & \text{elsewhere} \end{cases}$$

and let $\mathcal{G}_1(s) = 1/s$.

(a) Find the pulse transfer function $Y(z)/F(z)$.

(b) Find the discrete-time signal $y(nT)$ when $f(t)$ is the unit step function $u(t)$ by Z-transform methods. Check your answer by finding the continuous-time output of $g_h(t)$ and then $y(t)$ for all t.

(c) Repeat (b) for $f(t) = tu(t)$.

5.2. The pulse response of a discrete-time system is

$$g(n) = [1 + (0.1)^n + (0.2)^n]u(n)$$

Find a finite-order difference equation relating the input and output of the system.

5.3. Sketch the amplitude and phase responses for the systems with the following pulse transfer functions:

(a) $\dfrac{1}{1 + z^{-1}}$

(b) $\dfrac{z^{-2}}{(1 - 0.8z^{-1})(1 - 0.9z^{-1})}$

(c) $\dfrac{(1 + z^{-1})^2}{1 + 0.16z^{-2}}$

5.4. Sketch the block diagrams for the type 1 and type 2 direct form realizations of the pulse transfer function

$$H(z) = \frac{1 + z^{-1} + z^{-3}}{1 - 0.2z^{-1} + z^{-2}}$$

5.5. Sketch the block diagram for the parallel form realization of the pulse transfer function

$$H(z) = \frac{1 + z^{-1}}{(1 - 0.6z^{-1} + 0.25z^{-2})(1 - z^{-1})}$$

Use the type 1 direct form realization for each branch.

5.6. Sketch the block diagram for a cascade form realization of the pulse transfer function in Problem 5.5.

5.7. In Fig. 5.6.1 let $f(t)$ be a unit step function and let $\mathcal{G}(s)$ consist of a zero-order hold followed by an ideal mass so that

$$\mathcal{G}(s) = \frac{1 - z^{-1}}{s} \cdot \frac{1}{s^2}$$

Assume that $y(0) = \dot{y}(0) = 0$.

(a) Find $D(z)$ so that the closed-loop pulse transfer function is $Y(z)/F(z) = z^{-1}$ and give the difference equation relating the signals $v(nT)$ and $e(nT)$.

(b) Find $E(z)$, $e(nT)$, $V(z)$, and $v(nT)$.

(c) Sketch the continuous-time output of the zero-order hold. Calculate $y(t)$ for all t and sketch.

(d) Repeat (a), (b), and (c) if

$$Y(z)/F(z) = z^{-1}(1 + z^{-1})(\tfrac{5}{4} - \tfrac{3}{4}z^{-1})$$

(e) Which system tracks a step input best?

5.8. In Fig. 5.6.3 let $D(z) = 1$, $\mathcal{G}(s) = 1/s$, $\mathcal{H}(s) = 1/(s+1)$, and $f(t) = u(t)$. For zero initial conditions find $Y(z)$, $E(z)$, $y(nT)$, $e(nT)$, and $\lim_{n \to \infty} y(nT)$.

5.9. Determine which of the following polynomials have all their roots inside the unit circle.

(a) $z^3 - 1.3z^2 + 0.32z - 0.02$

(b) $z^3 - 0.9z^2 - 0.23z + 0.015$

(c) $z^3 + 9z^2 + 26z + 24$

(d) $z^4 + 1.09z^2 + 0.09$

(e) $z^4 - 0.2z^3 - 0.79z^2 + 0.162z - 0.0162$

5.10. The open-loop gain for a discrete-time, negative feedback system is

$$\frac{K}{(1 - 0.5z^{-1})(1 - 2z^{-1})}$$

Determine the positive values of K for which the closed-loop system is stable by using the Nyquist criterion.

5.11. The open-loop gain for a discrete-time, negative feedback system is

$$\frac{K}{(1 - z^{-1})^2}$$

Find the positive values of K for which the closed-loop system is stable by using the Nyquist criterion.

5.12. Sketch the root loci for the systems described in Problems 5.10 and 5.11.

5.13. In Fig. 5.6.1 let $D(z) = 1$, $f(t) = u(t)$, and

$$\mathcal{G}(s) = \frac{1 - z^{-1}}{s} \cdot \frac{1}{s+1}$$

(a) Find $E(z)$, $Y(z)$, $y(nT)$, and $\lim_{n \to \infty} e(nT)$.

(b) Find $Y(z, \delta)$ and $y(nT + \delta)$ for $0 \le \delta < T$.

5.14. Repeat Problem 5.13 for $D(z) = 1/(1 - z^{-1})$.

5.15. In Fig. 5.6.3 let $D(z) = 1$, $\mathcal{G}(s) = 1/(s+1)$, $\mathcal{H}(s) = 1/s$, and $f(t) = u(t)$. Find $Y(z, \delta)$, $y(nT + \delta)$, and $y(\infty)$. Sketch $y(t)$ for all t.

5.16. In Fig. 5.6.1 let $f(t) = u(t)$, $D(z) = 1$, and

$$\mathcal{G}(s) = \frac{1}{s(s+a)}$$

(a) Find $E(z)$.

(b) Assume that $y(t)$ is observed through a sampler operating with period $T/2$. Find $Y(\lambda : T/2)$ and $y(nT/2)$.

5.17. Generalize the result in Example 5.13.1 to the case where N is an arbitrary positive integer. That is, from $Y(\lambda : T/N)$ show that

$$y\left(\frac{kT}{N}\right) = \frac{1 - e^{-a(n+1)T}}{1 - e^{-aT}} e^{-arT/N} \qquad \text{for} \qquad k = nN + r,\ 0 \le r < N,$$

and $n = 0, 1, \ldots$.

5.18. In Fig. 5.14.1 let $F(\lambda : T/N) = 1$, $N = 2$, and

$$\mathcal{G}(s) = \frac{1}{s(s+1)}$$

Find $Y(z)$ three different ways by using (5.14.4), (5.14.10), and (5.14.17).

5.19. A discrete-time signal $f(nT)$ is obtained by skip sampling the signal $f(nT/3)$. If in the interval $|\omega| < 3\omega_s/2$

$$F^\$(\omega) = \begin{cases} 1 & \text{for} & |\omega| < \omega_s/2 \\ 2 - 2|\omega/\omega_s| & \text{for} & \omega_s/2 \le |\omega| < \omega_s \\ 0 & \text{elsewhere} \end{cases}$$

find and sketch $F^*(\omega)$.

CHAPTER 6

6.1. A discrete-time system has the state equation

$$\mathbf{x}(nT+T) = \begin{bmatrix} 0.1 & 0 \\ 0 & 0.9 \end{bmatrix} \mathbf{x}(nT) + \begin{bmatrix} 1 \\ 1 \end{bmatrix} v(nT)$$

and output equation

$$y(nT) = [1 \quad 2]\mathbf{x}(nT)$$

(a) Find the characteristic polynomial and eigenvalues for the system.
(b) Find a closed form for the state transition matrix $\boldsymbol{\varphi}(nT)$ for $0 \leq n$.
(c) Find its pulse transfer function.
(d) Assume that $\mathbf{x}(0) = [0 \quad 1]'$ and $v(nT)$ is a unit step function. Find $\mathbf{x}(nT)$ and $y(nT)$ for $0 \leq n$.

6.2. Repeat Problem 6.1 if

$$\mathbf{x}(nT+T) = \begin{bmatrix} 0 & 1 \\ -0.09 & 1 \end{bmatrix} \mathbf{x}(nT) + \begin{bmatrix} 0 \\ 1 \end{bmatrix} v(nT)$$

and

$$y(nT) = [-1.09 \quad 1]\mathbf{x}(nT) + v(nT)$$

6.3. Repeat Problem 6.1 if

$$\mathbf{x}(nT+T) = \begin{bmatrix} 1 & 0 & 0 \\ 0 & -0.9 & 1 \\ 0 & 0 & -0.9 \end{bmatrix} \mathbf{x}(nT) + \begin{bmatrix} 1 \\ 1 \\ 0 \end{bmatrix} v(nT)$$

$$y(nT) = [1 \quad 0 \quad 1]\mathbf{x}(nT)$$

and

$$\mathbf{x}(0) = [1 \quad 0 \quad 0]^t$$

6.4. Suppose that a system has the state equation

$$\mathbf{x}(nT+T) = \begin{bmatrix} 0 & 1 \\ -0.72 & 1.7 \end{bmatrix} \mathbf{x}(nT) + \begin{bmatrix} 0 \\ 1 \end{bmatrix} v(nT)$$

and output equation

$$y(nT) = [-0.72 \quad 1.7]\mathbf{x}(nT) + v(nT)$$

(a) Show that its eigenvalues are $z_1 = 0.8$ and $z_2 = 0.9$.
(b) Find $\mathbf{x}(nT)$ and $y(nT)$ for $0 \leq n$ if $v(nT)$ is identically zero and $\mathbf{x}(0) = \mathbf{x}_1 = [1 \quad 0.8]^t$.
(c) Find $\mathbf{x}(nT)$ and $y(nT)$ for $0 \leq n$ if $v(nT)$ is identically zero and $\mathbf{x}(0) = \mathbf{x}_2 = [1 \quad 0.9]^t$.

Comment: Notice that $\mathbf{Ax}_1 = z_1\mathbf{x}_1$ and $\mathbf{Ax}_2 = z_2\mathbf{x}_2$. The initial states $\mathbf{x}_1$ and $\mathbf{x}_2$ are called eigenvectors of $\mathbf{A}$ [47]. For an arbitrary initial state, the elements of the state vector and the output will be a linear combination of the "modes" of behavior $(z_1)^n$ and $(z_2)^n$. An initial state corresponding to an eigenvector of $\mathbf{A}$ excites only the mode corresponding to the associated eigenvalue.

6.5. The state equation for a system is

$$\mathbf{x}(nT+T) = \begin{bmatrix} 0.2 & -0.1 \\ -0.1 & 0.2 \end{bmatrix} \mathbf{x}(nT) + \begin{bmatrix} 0.5 \\ 0.5 \end{bmatrix} v(nT)$$

and the output equation is

$$y(nT) = [2 \quad 0]\mathbf{x}(nT)$$

(a) Show that its eigenvalues are 0.1 and 0.3.
(b) Find its pulse transfer function.
(c) Find $\mathbf{X}(z)$ and $Y(z)$ using (6.4.8). Notice that if the initial state is zero, then $\mathbf{x}(nT)$ and $y(nT)$ cannot have components of the form $(0.3)^n$ unless $v(nT)$ does.
(d) A new state vector $\mathbf{x}'(nT)$ is defined by the transformation

$$\mathbf{x}'(nT) = \begin{bmatrix} 1 & -1 \\ 1 & 1 \end{bmatrix} \mathbf{x}(nT)$$

Find the new state equation, i.e., an equation of the form

$$\mathbf{x}'(nT+T) = \mathbf{A}'\mathbf{x}'(nT) + \mathbf{B}'v(nT)$$

Explain the result in (c) using this new state equation.
Comment: This system is said to have an uncontrollable state [144].

6.6. A system has the state equation

$$\mathbf{x}(nT+T) = \begin{bmatrix} 0.4 & -0.3 \\ -0.3 & 0.4 \end{bmatrix} \mathbf{x}(nT) + \begin{bmatrix} 1 \\ 0 \end{bmatrix} v(nT)$$

and output equation

$$y(nT) = [1 \quad 1]\mathbf{x}(nT)$$

(a) Show that its eigenvalues are 0.7 and 0.1.
(b) Find $\mathbf{X}(z)$ using (6.4.8). Observe that $v(nT)$ can excite both modes $(0.7)^n$ and $(0.1)^n$.
(c) Find $Y(z)$ and observe that $y(nT)$ cannot have a component of the form $(0.7)^n$ unless $v(nT)$ does.
(d) A new state vector is defined as in Problem 6.5(d). Find the new

state and output equations. Explain why the output has no component of the form $(0.7)^n$ unless $v(nT)$ does.

Comment: This system is said to have an unobservable state [144].

6.7. Sketch the block diagrams and find the state and output equations for the type 1 direct form realizations of systems with the following pulse transfer functions:

(a) $\dfrac{1}{1 - 0.7z^{-1} + 0.06z^{-2}}$

(b) $\dfrac{1}{1 - z^{-1} + z^{-2} - z^{-3}}$

(c) $\dfrac{1}{1 - 2.8z^{-1} + 2.61z^{-2} - 0.81z^{-3}}$

6.8. Repeat Problem 6.7 for the type 2 direct form realization.

6.9. A filter with a pulse transfer function of the form

$$a_0 + a_1 z^{-1} + \cdots + a_N z^{-N}$$

is called a transversal filter. Sketch the block diagrams and find the state and output equations for the type 1 and 2 direct form realizations of the transversal filter.

6.10. Repeat Problem 6.7 for the standard form realization.

6.11. Sketch the block diagrams and find the state and output equations for the parallel form realizations of the following pulse transfer functions:

(a) $\dfrac{1 - z^{-1}}{(1 - 0.1z^{-1})(1 - 0.8z^{-1})}$

(b) $\dfrac{z^{-2}}{(1 - z^{-1})(1 - 0.9z^{-1})^3}$

6.12. Find a parallel form realization for the pulse transfer function

$$\frac{z^{-1}}{(1 - z^{-1})(1 - 0.9e^{j\pi/4}z^{-1})(1 - 0.9e^{-j\pi/4}z^{-1})}$$

Combine the complex conjugate terms into a single second-order section realized by a type 1 direct form structure. Sketch the block diagram of the resulting system and find the corresponding state and output equations.

6.13. Repeat Problem 6.12 for a cascade form realization.

6.14. Repeat Problem 6.11 using a cascade form realization.

6.15. Show that the pulse transfer function of the system in Fig. P6.15 can be

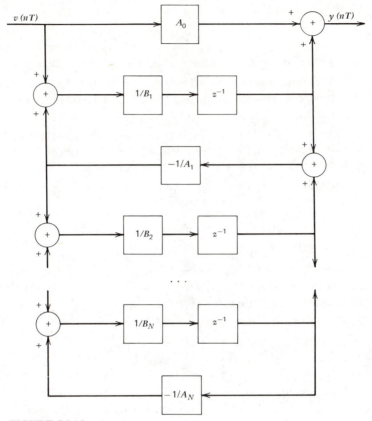

FIGURE P6.15.

written as the continued fraction

$$H(z) = A_0 + \cfrac{1}{B_1 z + \cfrac{1}{A_1 + \cfrac{\cdots}{+\cfrac{1}{A_N}}}}$$

This is a type of discrete-time ladder network [94]. Assign state variables to the outputs of the delay elements and find the corresponding state and output equations.

6.16. Show that the characteristic polynomial for a system with an **A** matrix of the form in (6.5.5) is

$$Q(z) = z^N + b_1 z^{N-1} + \cdots + b_{N-1} z + b_N$$

Hint: Expand the determinant using the bottom row.

6.17. Find the state and output equations and sketch the block diagram for the transpose configuration of the standard form realization shown in Fig. 6.5.3.

6.18. Sketch the block diagram for the transpose configuration of the system shown in Fig. P6.15. Is the structure of the transpose configuration essentially different from the original structure?

6.19. For positive t, the input and output of a time varying linear system are related by the differential equation

$$\ddot{y}(t) + (3t + 2t^{-1})\dot{y}(t) + (2 + 3t^{-1})y(t) = t^{-1}v(t)$$

Choose $x_1(t) = y(t)$ and $x_2(t) = \dot{y}(t)$ as state variables and find the corresponding state and output equations. Find the state transition matrix $\varphi(t, \tau)$. Hint: Substitute $y(t) = t^{-1}w(t)$ into the original differential equation and solve for $w(t)$.

6.20. A system has the state equation

$$\dot{x}(t) = \begin{bmatrix} 0 & 1 \\ -2 & -3 \end{bmatrix} x(t) + \begin{bmatrix} 0 \\ 1 \end{bmatrix} v(t)$$

and output equation

$$y(t) = \begin{bmatrix} 1 & 0 \end{bmatrix} x(t)$$

(a) Find its characteristic polynomial and eigenvalues.
(b) Find the state transition matrix $\varphi(t)$.
(c) Find its transfer function.
(d) If $x(0) = \begin{bmatrix} 1 & 1 \end{bmatrix}'$, find $x(t)$ and $y(t)$ for $0 \le t$ when $v(t)$ is a unit step function.

6.21. Repeat Problem 6.20 if

$$\dot{x}(t) = \begin{bmatrix} -1 & 0 \\ -2 & -2 \end{bmatrix} x(t) + \begin{bmatrix} 1 \\ 0 \end{bmatrix} v(t)$$

and

$$y(t) = \begin{bmatrix} 1 & 1 \end{bmatrix} x(t)$$

6.22. A system has the state equation

$$\dot{x}(t) = \begin{bmatrix} -1+j & 0 \\ 0 & -1-j \end{bmatrix} x(t) + \begin{bmatrix} 1 \\ 1 \end{bmatrix} v(t)$$

and output equation

$$y(t) = [1+j \quad 1-j]\mathbf{x}(t)$$

(a) Find its transfer function.

(b) Assume $\mathbf{x}(10) = [0 \quad 0]'$ and $v(t) = u(t-10)$. Find $\mathbf{x}(t)$ and $y(t)$ for $10 \le t$. Express $y(t)$ as a real function of t.

6.23. Sketch the block diagrams and find the corresponding state and output equations for the type 1 and 2 direct form realizations of the transfer function

$$\frac{s^2+1}{s^3+5s^2+7s+3}$$

6.24. Sketch the block diagram and find the state and output equations for the parallel form realization of the transfer function

$$\frac{s^2+1}{(s+1)(s+2)(s+3)}$$

6.25. Consider the system shown in Fig. P6.25. Assume that the filter $H(s)$ is described by the state equation

$$\dot{\mathbf{x}}(t) = \begin{bmatrix} 0 & 1 \\ -10 & -7 \end{bmatrix}\mathbf{x}(t) + \begin{bmatrix} 0 \\ 1 \end{bmatrix}v(t)$$

and output equation

$$y(t) = [3 \quad 1]\mathbf{x}(t)$$

(a) Find equations relating $\mathbf{x}(t)$ and $y(t)$ for $nT \le t \le nT + T$ to $\mathbf{x}(nT)$ and $f(nT)$.

(b) If $\mathbf{x}(0) = \mathbf{0}$ and $f(nT) = \delta_{n0}$, find $\mathbf{x}(t)$ and $y(t)$ for $0 \le t$.

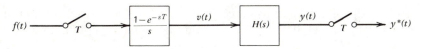

FIGURE P6.25.

CHAPTER 7

7.1. Find $S_{xx}(z)$ if $R_{xx}(\tau) = e^{-|\tau|}\cos \omega_0 \tau$.

7.2. Find $S_{xx}^*(\omega)$ if $R_{xx}(\tau) = 1 + \cos 2\tau$.

7.3. Find $S_{xx}(z)$ if $R_{xx}(nT) = 2\delta_{n0} - (2/3)(0.5)^{|n|}$. Verify that $R_{xx}(nT)$ is actually an autocorrelation function.

7.4. Find $R_{xx}(nT)$ and $E\{x^2(nT)\}$ if $S_{xx}^*(\omega) = 1 - |2\omega/\omega_s|$ for $|\omega| \le \omega_s/2$.

7.5. Find $R_{xx}(nT)$ if

$$S_{xx}(z) = \frac{1}{(1-0.1z^{-1})(1-0.5z^{-1})(1-0.1z)(1-0.5z)}$$

7.6. Represent

$$S_{xx}(z) = \frac{1}{(1-0.5z^{-1})(1-0.5z)} + 1$$

in the factored form indicated in Theorem 7.2.1.

7.7. Find $S_{xx}(z)$ using (7.3.8) if

$$S_{xx}(\omega) = \frac{1}{(\omega^2 + 1)(\omega^2 + 4)}$$

Find $R_{xx}(nT)$ from $S_{xx}(z)$.

7.8. Let $a(nT)$ be a sequence of independent random variables that can take on only the values 0 and 1 each with probability 1/2. This is known as a binary random sequence. A new sequence $b(nT)$ that can take on only the values 1, 0, and -1 is defined as

$$b(nT) = a(nT)(-1)^{x(nT)}$$

where

$$x(nT) = \sum_{k=-\infty}^{n} a(kT)$$

Notice that the nonzero samples in $b(nT)$ alternate between 1 and -1. Find $E\{b(nT)\}$, $R_{bb}(nT)$, and $S_{bb}(z)$. Sketch $S_{bb}^*(\omega)$.

7.9. A discrete-time signal $a(nT)$ with $S_{aa}(z) = 1$ is passed through the filter $H(z) = 1 - z^{-2}$ resulting in the signal $b(nT)$. Find $S_{bb}(z)$, $R_{bb}(nT)$, and $E\{b^2(nT)\}$. Sketch $S_{bb}^*(\omega)$.

7.10. A signal $x(nT)$ with $S_{xx}(z) = (1-z^{-1})(1-z)$ is passed through the filter $H(z) = 1/(1-0.3z^{-1})$. If the output signal is $y(nT)$, find $S_{yy}(z)$ and $R_{yy}(nT)$.

7.11. A signal $b(nT)$ is corrupted by additive noise $v(nT)$. An attempt is made to eliminate the noise by passing the observed signal $c(nT) = b(nT) + v(nT)$ through the low-pass filter $1/(1-0.8z^{-1})$. Let

$$S_{bb}(z) = \frac{1}{(1-0.9z^{-1})(1-0.9z)}$$

$$S_{vv}(z) = \sigma^2$$

and

$$S_{bv}(z) = 0$$

(a) Find the signal to noise ratio at the filter input, i.e., $\rho_i = E\{b^2(nT)\}/E\{v^2(nT)\}$

(b) Let the component of the filter output resulting from the signal $b(nT)$ be $y_b(nT)$ and the component resulting from the noise $v(nT)$ be $y_v(nT)$. Find the output signal to noise ratio, i.e.,

$$\rho_0 = E\{y_b^2(nT)\}/E\{y_v^2(nT)\}$$

(c) Find the improvement $I = \rho_0/\rho_i$.

7.12. The input to a filter with the pulse transfer function $H(z)$ is $a(nT)$ and its output is $b(nT)$. If $a(nT)$ is a white noise sequence with $S_{aa}(z) = \sigma^2$, find $S_{ba}(z)$ and $R_{ba}(nT)$. Describe a method based on this result for estimating the impulse response samples of an unknown system.

7.13. Let $x(nT)$ and $v(nT)$ have the sampled power spectral densities shown in Fig. P7.13 and assume that $R_{xv}(nT) = 0$. Find and sketch $H^*(\omega)$, the pulse transfer function of the optimum nonrealizable filter for estimating $x(nT)$ from $y(nT) = x(nT) + v(nT)$. Calculate the corresponding mean square error.

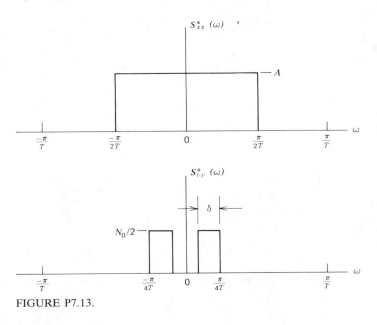

FIGURE P7.13.

7.14. Find the pulse response of the optimum nonrealizable filter for estimating $x(nT)$ from $y(nT)$ for the situation described in Example 7.5.2. Calculate the mean square error.

7.15. Let $y(nT) = x(nT) + v(nT)$ with $S_{vv}(z) = 1$, $S_{xv}(z) = 0$, and

$$S_{xx}(z) = \frac{0.75}{(1 - 0.5z^{-1})(1 - 0.5z)}$$

Find the pulse transfer function of the optimum filter for estimating $x(nT - 2T)$ from the data $\{y(kT)\}_{k=-\infty}^{n}$. Give the difference equation relating the input and output of this filter. Calculate the mean square error.

7.16. Let

$$S_{xx}(z) = \frac{(1 - 0.1z^{-1})(1 - 0.1z)}{(1 - 0.9z^{-1})(1 - 0.9z)}$$

Find the pulse transfer function of the optimum realizable filter for estimating $x(nT + 2T)$, i.e., for predicting two samples into the future. State the difference equation relating its input and output. Calculate the mean square error.

7.17. Repeat Problem 7.16 for

$$S_{xx}(z) = \frac{1}{(1 - 0.1z^{-1})(1 - 0.5z^{-1})(1 - 0.1z)(1 - 0.5z)}$$

7.18. Let

$$S_{xx}(z) = \frac{\sigma^2}{D(z)D(z^{-1})}$$

where σ^2 is a positive constant and

$$D(z) = 1 + \sum_{k=1}^{N} d_k z^{-k}$$

has all its roots inside the unit circle. Show that the pulse transfer function of the optimum realizable filter for predicting one sample into the future is

$$H(z) = -\sum_{k=0}^{N-1} d_{k+1} z^{-k}$$

and that the mean square prediction error is σ^2. State the difference equation relating the input and output of the optimum predictor.

7.19. A discrete-time random process $x(nT)$ with a sampled power spectral density $S_{xx}(z)$ of the type in Problem 7.18 is known as an Nth order autoregressive process. Show that in this case it is only necessary to know the N samples $x(nT - NT + T), \ldots, x(nT - T), x(nT)$ to optimally predict $x(nT + LT)$ for any finite, positive integer L.

7.20. Let $R_{xx}(nT) = a^{|n|}$ with $-1 < a < 1$. Using the recursive equations (7.6.30)–(7.6.36) show that the optimum analysis filter is $D_n(z) = 1 - az^{-1}$ and the corresponding mean square prediction error is $p_n = 1 - a^2$ for $n > 0$. Give the difference equation for optimally predicting one sample into the future.

7.21. If $R_{xx}(nT) = e^{-|nT|} \cos \omega_1 nT$, recursively calculate $D_2(z)$, p_2, and k_2 using (7.6.30)–(7.6.36).

7.22. Find $D_4(z)$ if the partial correlation coefficients are $k_n = 0.5$ for $n = 1, \ldots, 4$.

7.23. Find $D_n(z)$ and k_n for $n \leq 3$ using (7.6.41) and (7.6.42) if

$$D_3(z) = 1 - 0.6z^{-1} + 0.11z^{-2} - 0.006z^{-3}$$

Is the synthesis filter $1/D_3(z)$ stable?

7.24. Let $R_{xx}(nT) = \cos \omega_1 NT$. Find $D_1(z)$, $D_2(z)$, k_1, k_2, p_1, and p_2. Is $1/D_2(z)$ stable? Give the difference equation for optimum one-step prediction using the last two samples.

7.25. The samples $x(0), \ldots, x(MT)$ are observed. Let

$$X(z) = \sum_{n=0}^{M} x(nT)z^{-n}$$

$$C(z) = X(z)X(z^{-1})$$

and

$$E(z) = X(z)D_N(z)$$

where $D_N(z)$ is given by (7.6.3). Show that all the results of Section 7.6 hold if the autocorrelation function $R_{xx}(nT)$ is replaced by $c(nT)$ corresponding to $C(z)$, the mean square prediction error p_N is replaced by the sum squared error

$$p_N' = \sum_{k=-\infty}^{\infty} e^2(kT)$$

and similarly for v_N and q_N.

7.26. Find state and output equations for the analysis and synthesis filters shown in Figs. 7.6.1 and 7.6.2.

7.27. A digital baseband pulse amplitude modulation transmitter can be modeled as the upper branch in Fig. 7.7.2. The input samples $a(nT)$ normally can take on a finite set of amplitudes. $G(\omega)$ is frequently chosen so that $g(nT) = \delta_{n0}$ resulting in no intersymbol interference. The continuous-time signal $c(t)$ is transmitted over an analog channel. If

$S_{aa}(z) = A^2$ and

$$G(\omega) = \begin{cases} T & \text{for} & |\omega| < \omega_s/2 \\ 0 & \text{elsewhere} \end{cases}$$

find $S_{cc}(\omega)$ and the transmitted power $E\{c^2(t)\}$ using the results in Section 7.7.

7.28. Repeat Problem 7.27 if $G(\omega)$ has the raised cosine characteristics described in Problem 2.8.

7.29. Sometimes controlled intersymbol interference is introduced for spectral shaping. Repeat Problem 7.27 if $a(nT)$ is replaced by $a'(nT) = a(nT) - a(nT - 2T)$. This is known as class IV partial response shaping [87].

7.30. Repeat Problem 7.27 if $a(nT)$ is replaced by $b(nT)$ from Problem 7.8 and $g(t) = u(t) - u(t - T/2)$ where $u(t)$ is the unit step function. This is the T-carrier signal format in the telephone network.

7.31. A simplified model for a pulse amplitude modulation communication system is shown in Fig. P7.31. The information to be transmitted is the sequence $a(nT)$. $G_T(\omega)$ is the transmitter filter and $c(t)$ is the continuous-time transmitted signal. The analog channel simply adds noise $v(t)$ with power spectral density $S_{vv}(\omega)$. $G_R(\omega)$ is the receiver filter. Normally $G(\omega) = G_T(\omega)G_R(\omega)$ is chosen so that $r(nT) = a(nT)$ when the additive noise $v(t)$ is not present. Assume that $S_{aa}(z) = A$ and that $v(t)$ and $a(nT)$ are independent.

(a) Calculate the transmitted power $E\{c^2(t)\} = S$.

(b) Calculate the output noise power N, i.e., the power in $r(t)$ resulting from $v(t)$, in terms of $S_{vv}(\omega)$ and $G_R(\omega)$.

(c) Find the receiver filter $G_R(\omega)$ that minimizes the output noise power subject to the constraints that $G(\omega) = G_T(\omega)G_R(\omega)$ and the transmitted power S is fixed.

(d) Show that when $S_{vv}(\omega)$ is a constant and $G(\omega)$ is real, the optimum transmitter and receiver filters can be chosen so that

$$|G_T(\omega)| = |G_R(\omega)| = |G(\omega)|^{1/2}$$

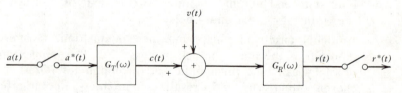

FIGURE P7.31.

(e) If $G(\omega)$ has the raised cosine characteristic described in Problem 2.8 and $S_{vv}(\omega) = N_0/2$, find the theoretically maximum output signal to noise ratio $E\{a^2(nT)\}/N$ with the constraints of (c).

7.32. Let $S_{ff}(\omega) = 2a/(a^2 + \omega^2)$. Find the transfer function of the optimum nonrealizable filter for reconstructing $f(t)$ from its samples. Sketch its impulse response. Calculate the mean square error and compare with Example 7.8.2.

7.33. Let the desired output of the reconstruction filter be $f(t-d)$. Show that the mean square error for the optimum realizable reconstruction filter is nonincreasing with d and converges to that of the nonrealizable filter when d becomes infinite.

7.34. Let the noise $v(t)$ be zero, $L(\omega) = e^{-j\omega d}$, and $S_{ff}(\omega) = 2a/(a^2 + \omega^2)$. Find the impulse response of the optimum realizable reconstruction filter and the corresponding mean square error. Sketch the impulse response for $d = T/2$, T, and $d > T$.

7.35. Let $R_{ff}(\tau) = \frac{20}{9}(0.8)^{|\tau/T|} - \frac{2}{3}(0.5)^{|\tau/T|}$, $L(\omega) = 1$, and $v(t) = 0$. Find the transfer function, impulse response, and mean square error for the optimum nonrealizable reconstruction filter.

7.36. Repeat Problem 7.35 for the optimum realizable reconstruction filter.

7.37. Let $R_{ff}(\tau) = e^{-a|\tau|}$. Find the mean square errors for reconstructing $f(t)$ from noiseless samples using the zero-order hold, first-order hold, and linear point connector. Compare your results with Example 7.8.2 and Problem 7.32.

CHAPTER 8

8.1. Suppose that the receiver in a particular communication system is implemented digitally using a sampling rate of $f_s = \omega_s/(2\pi) = 8$ kHz. A band-pass filter of the form $G(z) = K/(1 + a_1 z^{-1} + a_2 z^{-2})$ is desired to recover a pilot tone at 3.2 kHz. The amplitude response, i.e., $20 \log_{10} |G^*(2\pi f)|$, is required to be 0 dB at 3.2 kHz and no greater than -40 dB at 2.8 kHz. Find suitable values for K, a_1, and a_2, and give a difference equation for realizing $G(z)$. Sketch the phase and amplitude responses of your filter.

8.2. A simple digital band-pass filter with zeros at $\omega = 0$ and $\omega_s/2$ and a peak near $\omega_s/4$ is required. The amplitude response must meet the specifications $|G^*(\omega_s/4)| = 1$ and $|G^*(\omega_s/8)| = \frac{1}{100}$. Find a pulse transfer function with a second-degree numerator and denominator that meets these requirements. State the difference equations for a type 1 direct form realization of your filter.

8.3. A digital differentiation filter is to be designed using the guard filter and Z-transform method. An ideal differentiator has the transfer function s. A linear point connector with the transfer function given by (3.8.3) is to be used as the guard filter. Find $G_1(z)$ and give a difference equation realization. Sketch the amplitude and phase responses of $G_1(z)$ and compare with the desired responses.

8.4. Repeat Problem 8.3 if the desired transfer function is $1/s$ which corresponds to ideal integration. Show that the resulting digital integrator is equivalent to the classical trapezoidal integration rule.

8.5. Find the pulse transfer function of a low-pass digital filter with a 3 dB cutoff frequency of $\omega_s/6$ using the bilinear transformation and a third-order Butterworth filter.

8.6. Using a fourth-order Butterworth filter as the prototype analog filter and the bilinear transformation, design a digital low-pass filter with a 3 dB cutoff frequency of $\omega_s/6$. Show how to realize the digital filter as the cascade of two second-order type 1 direct form sections and give the required difference equations.

8.7. Using a second-order Butterworth filter as the analog prototype low-pass filter and the transformation specified by (8.5.21), (8.5.22), and (8.5.24), find the pulse transfer function of a digital band-pass filter with upper and lower 3 dB cutoff frequencies of $\omega_s/6$ and $\omega_s/12$. Sketch the amplitude response of the digital filter.

8.8. The envelope delay for a channel is

$$\Delta(\omega) = 16\,T\left(\omega T - \frac{\pi}{2}\right)^2 /\pi^2 \qquad \text{for} \qquad 0 \le \omega < \omega_s/2$$

Using trial and error, attempt to make this delay flat over the band $\omega_s/8 \le \omega \le 3\omega_s/8$ by a cascade of two second-order digital all-pass networks.

8.9. Let $G_1(z)$, $G_2(z)$, and $G_3(z)$ be second-order all-pass filters in Fig. P8.9. By trial and error choose these filters so that the phase responses of $H_1(z) = G_1(z)G_2(z)$ and $H_2(z) = G_1(z)G_3(z)$ are approximately linear and separated by 90° over the band $\omega_s/10 \le \omega \le \omega_s/4$. This type of structure is known as a phase splitter and is used to generate output signals that are Hilbert transform pairs.

8.10. Consider the cascade of first-order filters

$$\prod_{k=0}^{\infty} G_k(z)$$

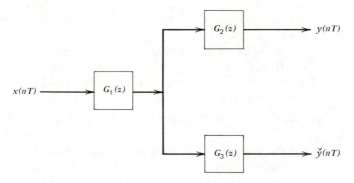

FIGURE P8.9.   A digital phase splitter.

where

$$G_0(z) = \frac{(1-a^2)^{1/2}}{1-az^{-1}} \quad \text{and} \quad G_k(z) = \frac{z^{-1}-a}{1-az^{-1}} \quad \text{for} \quad 1 \le k$$

with $-1 < a < 1$. Notice that the pulse transfer function from the input of the chain to the output of $G_k(z)$ is the function $F_k(z)$ defined in Problem 4.13. The corresponding time sequences $f_k(nT)$ are known as discrete-Laguerre sequences.

(a) Find the amplitude response of $F_k(z)$ for $k \ge 0$.

(b) A desired causal pulse response $g(nT)$ is to be approximated by the sequence

$$\hat{g}(nT) = \sum_{k=0}^{N} b_k f_k(nT)$$

Find the set of coefficients $b_0, \dots, b_N$ that minimize

$$e^2 = \sum_{n=0}^{\infty} [g(nT) - \hat{g}(nT)]^2$$

Hint: Use the result of Problem 4.13.

(c) Do the coefficients found in (b) minimize the frequency domain error

$$\int_{-\omega_s/2}^{\omega_s/2} |G^*(\omega) - \hat{G}^*(\omega)|^2 \, d\omega$$

8.11. A nonrecursive, linear phase, low-pass, digital filter approximating the ideal response

$$G^*(\omega) = \begin{cases} 1 & \text{for} \quad |\omega| \le \omega_s/4 \\ 0 & \text{for} \quad \omega_s/4 < |\omega| < \omega_s/2 \end{cases}$$

is desired.

(a) Find the coefficients for a 21 tap filter using the Fourier series and Hamming window.

(b) Plot the amplitude response in dB and the phase response for the filter in (a).

(c) Repeat (b) for the 21 tap filter obtained by using the rectangular truncation window.

8.12. An ideal full-band differentiator has the frequency response $G^*(\omega) = j\omega$ for $|\omega| \le \omega_s/2$.

(a) Find the Fourier coefficients of $G^*(\omega)$ if $\omega_s = 2$.

(b) Calculate the coefficients for the 19 tap nonrecursive filter obtained using the Hamming window.

(c) Plot the amplitude and phase responses of the filter in (b).

(d) Approximately how many taps are required to obtain the same usable bandwidth when the Blackman window is used?

8.13. Find the coefficients of a 15 tap nonrecursive filter that approximates the Hilbert transform filter $G^*(\omega) = -j$ sign ω for $|\omega| < \omega_s/2$ using the weighted least integral square error method with the weighting function

$$Q(\omega) = \begin{cases} 1 & \text{for} \quad \frac{2}{15} < |\omega/\omega_s| < 5.5/15 \\ 0 & \text{elsewhere} \end{cases}$$

Plot the resulting frequency response and compare with Fig. 8.8.2.

8.14. Another method suggested for reducing the Gibbs' ripple is to specify the desired frequency response with smooth transitions between constant levels and then use the Fourier coefficients with rectangular truncation. Suppose that the desired frequency response of a partial band Hilbert transform filter is specified as

$$G^*(\omega) = \begin{cases} -j \sin \dfrac{\omega\pi}{a\omega_s} & \text{for} \quad 0 < \omega < a\omega_s/2 \\ -j & \text{for} \quad a\omega_s/2 < \omega < (1-a)\omega_s/2 \\ j \sin \dfrac{(\omega - \omega_s/2)\pi}{a\omega_s} & \text{for} \quad (1-a)\omega_s/2 < \omega < \omega_s/2 \end{cases}$$

with $G^*(-\omega) = -G^*(\omega)$ for $|\omega| \le \omega_s/2$ and $0 \le a \le \frac{1}{2}$.

(a) Calculate the Fourier coefficients of $G^*(\omega)$.

(b) Plot the amplitude response, $|G_7^*(\omega)|$, and phase response for the 15 tap, nonrecursive filter using the coefficients g_n for $n = -7, \ldots, 7$ when $a = \frac{4}{15}$. Compare the results with Fig. 8.8.2.

8.15. A linear phase, low-pass, 31 tap, nonrecursive filter with a cutoff frequency of $7\omega_s/31$ desired.

(a) Plot the amplitude response of the frequency sampling filter with $N = 31$ and

$$G'_k = \begin{cases} 1 & \text{for} & k = 0, \ldots, 7 \\ 0 & \text{for} & k = 8, \ldots, 15 \end{cases}$$

(b) Calculate the pulse response g_n.

(c) By trial and error, vary G'_k for $k = 6, 7, 8, 23, 24,$ and 25 to reduce the ripple in the amplitude response.

8.16. A 31 tap, linear phase, nonrecursive filter that approximates the normalized full-band differentiator $G^*(\omega) = j\omega/\omega_s$ for $|\omega| < \omega_s/2$ is desired.

   (a) Plot the amplitude and phase responses of the required frequency sampling filter with $jG'_k = G^*(k\omega_s/31)$ for $k = 0, \ldots, 15$.

   (b) By trial and error, vary G'_k for $k = 14, 15, 16,$ and 17 to minimize the ripple in the amplitude response.

   (c) Calculate the pulse response of the resulting filter.

8.17. Derive an interpolation formula similar to (8.10.7) when the sampling points are the N zeros of $1 + z^{-N}$. Find a formula similar to (8.10.12) for calculating the pulse response.

8.18. Transform the low-pass filter given by (8.5.15) with the 3 dB cutoff frequency $\omega_s/4$ into a low-pass filter with the cutoff frequency $\omega_s/8$. Give the difference equations for a type 1 direct form realization.

8.19. Transform the low-pass filter given by (8.5.15) into a high-pass filter with a 3 dB cutoff frequency of $3\omega_s/8$.

8.20. Transform the low-pass filter given by (8.5.15) into a band-pass filter with upper and lower 3 dB cutoff frequencies of $\omega_s/4$ and $\omega_s/8$.

8.21. Transform the low-pass filter given by (8.5.15) into a band-stop filter with upper and lower 3 dB cutoff frequencies of $5\omega_s/16$ and $\omega_s/4$.

CHAPTER 9

9.1. Find the decimal equivalents of the following normalized two's complement numbers: (a) 01111 (b) 0011 (c) 1111 (d) 1011.

9.2. Find the two's complement representations for the following decimal numbers: (a) $\frac{3}{4}$ (b) $-\frac{3}{4}$ (c) $\frac{7}{8}$ (d) $-\frac{7}{16}$.

9.3. Let X be a random variable with the probability density function

$$f(x) = \begin{cases} \dfrac{2}{11}\left(1 - \dfrac{|x|}{5.5}\right) & \text{for} & |x| \le 5.5 \\ 0 & \text{elsewhere} \end{cases}$$

If X is rounded to the nearest integer:
(a) Find the probability density function for the roundoff error.
(b) Calculate the mean and variance of the roundoff error.
(c) Calculate the signal to quantization noise ratio.

9.4. The input to an analog to digital converter is equally likely to fall anywhere between 10 and -10 volts. Find the number of bits required to achieve an 80 dB signal to quantization noise ratio.

9.5. Calculate and plot the amplitude response in dB of the filter in Problem 8.11(a) if the coefficients are rounded to multiples of (a) $q = 2^{-5}$ and (b) $q = 2^{-10}$.

9.6. The denominator of a pulse transfer function is

$$B(z) = (1 - 0.8z^{-1})(1 - 0.9z^{-1})(1 - 0.99z^{-1})$$
$$= 1 + b_1 z^{-1} + b_2 z^{-2} + b_3 z^{-3}$$

Evaluate (9.4.10) at each of the zeros of $B(z)$.

9.7. Find the zeros of $B(z) = 1 + b_1 z^{-1} + b_2 z^{-2}$ in terms of b_1 and b_2. Assume that the zeros are complex and calculate their magnitude r and arguments $\pm\theta$. Calculate the partial derivatives of r and θ with respect to b_1 and b_2. When is θ most sensitive to changes in b_1 and b_2?

9.8. Find the pulse transfer function of a low-pass digital filter with a 3 dB cutoff frequency of $\omega_s/16$ using the bilinear transformation and a third-order Butterworth prototype analog filter. Estimate the coefficient accuracy required for direct and cascade form realizations.

9.9. Compare the output noise power due to fixed point finite word length arithmetic for parallel and type 0 direct form realizations of

$$G(z) = \frac{0.1}{(1 - 0.9z^{-1})(1 - 0.95z^{-1})}$$

9.10. Find the section ordering that minimizes the output noise power caused by fixed point finite word length arithmetic for cascade form realization of

$$G(z) = \frac{1}{(1 - 0.5z^{-1})(1 - 0.9z^{-1})}$$

9.11. Find the output noise power caused by fixed point finite word length arithmetic if $G(z)$ in Problem 9.10 is realized in the form of the lattice network discussed at the end of Section 7.6.

9.12. Show that the roundoff error sequences at different taps in a nonrecursive filter with even symmetry are not mutually uncorrelated random sequences.

9.13. Suppose $G(z)$ in Problem 9.10 is implemented as a type 1 direct form realization. Assume that each output is calculated by summing full accuracy products and rounding the total sum to the nearest integer. Let the input be zero.
   (a) Find the deadband.
   (b) If the initial conditions are $y_2(-T) = 0$ and $y_2(-2T) = 1$, calculate the future outputs until a steady-state pattern emerges.
   (c) Repeat (b) if $y_2(-T) = 0$ and $y_2(-2T) = 10$.
   (d) Repeat (b) if $y_2(-T) = 10$ and $y_2(-2T) = 10$.

9.14. Repeat parts (a) and (b) of Problem 9.13 if $G(z) = 1/(1 + 1.5z^{-1} + 0.56z^{-2})$. Compare the result with the appropriate bound in (9.8.20).

9.15. Using (9.8.7), find bounds on limit cycles of period $L = 1, 2, 3,$ and 4 in a direct form realization of $G(z) = 1/(1 - 0.9z^{-1})^2$. Compare these results with the bound given by (9.8.20).

9.16. For the filter in Example 9.9.1, calculate the variance σ_y^2 of the ideal output $y(nT)$ when the input $x(nT)$ is a white noise sequence with variance σ_x^2. Using (9.9.19) calculate the output noise to signal ratio $R_{ww}(0)/\sigma_y^2$ for floating point finite word length arithmetic. Calculate the output noise to signal ratio when fixed point finite word length arithmetic is used with $q = 2^{-K}$. Plot the two noise to signal ratios as a function of β for $K = 10$.

9.17. Find the output noise to signal ratio when

$$G(z) = \frac{1}{(1 - 0.8z^{-1})(1 - 0.9z^{-1})}$$

is implemented in type 1 direct form using floating point finite word length arithmetic. Assume that the input sampled power spectral density is $S_{xx}(z) = a^2$.

9.18. This problem shows that limit cycles can also exist with floating point arithmetic [119]. Consider the difference equation

$$y(n) = (1 - 2^{-K})y(n-1) + (1 - 2^{-K})2^{-K}y(n-2)$$

Assume that all machine numbers have the form of (9.9.1) with c limited to K bits. If the initial conditions are $y(0) = y(1) = -1$, calculate the future output values of $y(n)$ using floating point finite word length arithmetic.

CHAPTER 10

10.1. Find a closed form expression for DFT$\{f_n\}$ if f_n for $n = 0, \ldots, N-1$ is
   (a) e^{-anT}

(b) $\cos \omega_0 nT$, ω_0 arbitrary

(c) $\sin \omega_0 nT$, ω_0 arbitrary

(d) 1

(e) n

10.2. Let m be an integer in the interval $(0, N/2)$. Express IDFT$\{F_k\}$ as a real function with a real argument if for $k = 0, \ldots, N-1$

(a) $F_k = \begin{cases} e^{j\theta} N/2 & \text{for} \quad k = m \\ e^{-j\theta} N/2 & \text{for} \quad k = N - m \\ 0 & \text{elsewhere} \end{cases}$

(b) $F_k = \begin{cases} -je^{j\theta} N/2 & \text{for} \quad k = m \\ je^{-j\theta} N/2 & \text{for} \quad k = N - m \\ 0 & \text{elsewhere} \end{cases}$

10.3. The Z-transform inversion formula can be written as

$$f(nT) = \frac{1}{\omega_s} \int_0^{\omega_s} F^*(\omega) e^{j\omega nT} \, d\omega$$

if the region of convergence includes the unit circle. Show that the result $\hat{f}(nT)$, obtained when this integral is computed numerically by evaluating the integrand at the frequencies $k\omega_s/N$ for $k = 0, \ldots, N-1$ and using the rectangular integration rule, is $\hat{f}(nT) =$ IDFT$\{F^*(k\omega_s/N)\}$.

10.4. Let $f(n)$ be a sequence with the two-sided Z-transform $F(z)$ and the unit circle included in the region of convergence. By applying the IDFT formula to the infinite series expansion for $F^*(k\omega_s/N)$, show that

$$\text{IDFT}\{F^*(k\omega_s/N)\} = \sum_{m=-\infty}^{\infty} f(n + mN)$$

Discuss how this result relates to the error introduced by evaluating the Z-transform inversion formula numerically as described in Problem 10.3. Notice that (9.8.12) is identical with this result and was derived by a different method.

10.5. Let $f(nT) = e^{-anT} u(n)$ with $a > 0$. Find $F(z)$ and $F^*(\omega)$. Compute $f_n = \text{IDFT}\{F^*(k\omega_s/N)\}$ using the result of Problem 10.4 and compare with $f(nT)$.

10.6. A signal $f(nT)$ with period N is the input to a discrete-time system with the frequency response $H^*(\omega)$. Show that the output can be written as

$$y(nT) = \frac{1}{N} \sum_{k=0}^{N-1} H^*(k\omega_s/N) F_k e^{jk\omega_s nT/N}$$

where F_k is the DFT of the N-point sequence $f(0), \ldots, f(NT-T)$. Hint: Express $f(nT)$ as a sum of N sampled complex exponentials using the IDFT formula. Then use sinusoidal steady-state analysis.

10.7. The periodic sequence

$$f(nT) = \begin{cases} 1 & \text{for} \quad n \text{ a multiple of } N \\ 0 & \text{elsewhere} \end{cases}$$

is transmitted over a channel with the unknown frequency response $H^*(\omega)$. At the receiver, the N output samples $y(0), \ldots, y(NT-T)$ are measured and Y_k, the corresponding DFT, is computed. How is Y_k related to $H^*(k\omega_s/N)$? Hint: See Problem 10.6.

10.8. If $F_k = \text{DFT}\{f_n\}$, show that $\text{DFT}\{F_n\} = Nf_{-k}$.

10.9. The first row of an $N \times N$ matrix $\mathbf{A}$ is $a_0, a_1, \ldots, a_{N-1}$. Each successive row is the previous row cyclically shifted one position to the right. Show that the eigenvalues of $\mathbf{A}$ are $A_k = \text{DFT}\{a_n\}$ and the corresponding eigenvectors are $\mathbf{X}_k^t = [1, W^k, W^{2k}, \ldots, W^{(N-1)k}]$ where $W = e^{-j2\pi/N}$. In other words, show that $\mathbf{AX}_k = A_k \mathbf{X}_k$. Hint: Use (10.4.6).

10.10. We will say that a real N-point sequence f_n is DFT band-limited if $F_k = \text{DFT}\{f_n\} = 0$ for $m \le k \le N-m$. Suppose that $(r-1)N$ zeros are inserted in the middle of F_k to form the rN-point DFT

$$F_k' = \begin{cases} F_k & \text{for} \quad k = 0, \ldots, m-1 \\ 0 & \text{for} \quad k = m, \ldots, rN-m \\ F_{k-(r-1)N} & \text{for} \quad k = rN-m+1, \ldots, rN-1 \end{cases}$$

Show that the rN point sequence $f_n' = \text{IDFT}\{F_k'\}$ has the property that $rf_{nr}' = f_n$ for $n = 0, \ldots, N-1$. This is a technique for DFT band-limited interpolation.

10.11. Digital data is commonly transmitted over an HF channel by differentially phase modulating a set of parallel tones. Suppose that the transmitted signal has the form

$$f(t) = \sum_{k=k_1}^{k_2} \sqrt{2} A \cos(k\omega_0 t + \theta_{i,k}) \qquad \text{for} \qquad iT_0 \le t \le (i+1)T_0$$

where $\omega_0 = 2\pi/T_0$ and $\{\theta_{i,k}\}$ are independent random variables, each equally likely to have one of the $M = 2^m$ values $r2\pi/M$ for $r = 0, \ldots, M-1$. Assume that the signal $y(t) = f(t) + v(t)$ is received where $v(t)$ is additive noise. The received signal is sampled to form the N-point sequence $y_{i,n} = y(iT_0 + nT_0/N)$ for $n = 0, \ldots, N-1$ and $Y_{i,k} = \text{DFT}\{y_{i,n}\} = F_{i,k} + V_{i,k}$ is computed.

(a) Neglecting the noise, how large must N be to avoid aliasing?

(b) Assuming that $v(t) \equiv 0$, show that $\{\theta_{i,k}\}$ can be determined from $Y_{i,k}$.

(c) Assume that the noise samples are uncorrelated and have variance σ^2 and zero mean. Find the output signal to noise ratio $E\{|F_{i,k}|^2\}/E\{|V_{i,k}|^2\}$.

10.12. Let

$$f_n = \begin{cases} 1 & \text{for} & 0 \le n \le 10 \\ 0 & \text{for} & 10 < n < 99 \end{cases}$$

and

$$g_n = \begin{cases} 1 & \text{for} & n = 0 \\ 0 & \text{for} & 1 \le n \le 89 \\ 1 & \text{for} & 90 \le n \le 99 \end{cases}$$

Find and sketch the sequence h_n corresponding to $H_k = \text{DFT}\{f_n\}\text{DFT}\{g_n\}$.

10.13. Let $f_n = \cos(2\pi n/N)$ and $g_n = \sin(2\pi n/N)$ for $n = 0, \ldots, N-1$. Find closed form expressions for

(a) the cyclic convolution of f_n and g_n;

(b) the cyclic correlation of f_n and g_n;

(c) the cyclic correlation of f_n with itself;

(d) the cyclic correlation of g_n with itself.

10.14. Let

$$f(n) = \begin{cases} 1 & \text{for} & 0 \le n \le 3 \\ 0 & \text{elsewhere} \end{cases} \quad \text{and} \quad g(n) = \begin{cases} n & \text{for} & 0 \le n \le 7 \\ 0 & \text{elsewhere} \end{cases}$$

(a) Find and sketch the discrete-time convolution of $f(n)$ and $g(n)$.

(b) Let $f_n = f(n)$ and $g_n = g(n)$ for $n = 0, \ldots, 7$. Find and sketch the cyclic convolution of f_n and g_n.

10.15. Using Theorem 10.6.1, find a closed form expression for $\sum_{n=0}^{N-1} f_n \bar{g}_n$ if

(a) $f_n = g_n = \cos(2\pi n/N)$

(b) $f_n = \cos(2\pi n/N)$, $g_n = \sin(2\pi n/N)$

(c) $f_n = g_n = \text{IDFT}\{e^{j2\pi nk^2/N}\}$

10.16. The following FORTRAN subroutine computes an in-place FFT of the complex array X.

(a) Show that the "DO 3" loop rearranges X into bit reversed order. Hint: See the flow chart on p. 198 of Gold and Rader [39].

(b) Is it a decimation in time or decimation in frequency algorithm?

(c) Notice that the required powers of W are computed recursively according to the rule $W^{J+1} = WW^J$.

```
SUBROUTINE FFT(X,M)
COMPLEX X(1024),U,W,T
N=2**M
N2=N/2
N1=N-1
J=1
DO 3 I=1,N1
IF(I.GE.J) GO TO 1
T=X(J)
X(J)=X(I)
X(I)=T
1   K=N2
2   IF(K.GE.J) GO TO 3
J=J-K
K=K/2
GO TO 2
3   J=J+K
PI=3.141592653589793
DO 5 L=1,M
LE=2**L
LE1=LE/2
U=(1.0,0.0)
W=CMPLX(COS(PI/LE1),SIN(PI/LE1))
DO 5 J=1, LE1
DO 4 I=J,N,LE
ID=I+LE1
T=X(ID)*U
X(ID)=X(I)-T
4   X(I)=X(I)+T
5   U=U*W
RETURN
END
```

10.17. The decimation in time FFT butterfly shown in Fig. 10.7.2 does not explicitly indicate that $H_k W_N^k$ needs to be computed only once and can then be used in the computation of both F_k and $F_{k+(N/2)}$.

   (a) Draw an equivalent flow graph with one additional branch that explicitly indicates this fact.

   (b) Redraw the flow graph in Fig. 10.7.5 using the result of (a).

10.18. Redraw the flow graph in Fig. 10.7.6 to explicitly indicate the more efficient computational structure suggested by (10.7.15).

10.19. Let f_n be an $N = qp$-point sequence. By generalizing the first step of

the radix 2 decimation in time algorithm, derivè a formula for computing DFT$\{f_n\}$ by combining p DFTs of q-point subsequences. Assuming
that the q-point DFTs are known, estimate the number of complex
additions and multiplications required for combining them into the
complete N-point DFT.

10.20. Specialize the result of Problem 10.19 to the case where $p = 4$ and
$q = 2$. In the radix 2 decimation in time algorithm, the fact that
$W_N^{k+(N/2)} = -W_N^k$ was used to minimize the number of multiplications.

(a) Show how to use the fact that $W_N^{k+(rN/4)} = j^r W_N^k$ to minimize
multiplications in this case and draw the corresponding flow
graph.

(b) Compare the number of complex multiplications and additions
required by this $2 + 4$ algorithm with those required by the radix
2 decimation in time algorithm. Do not count multiplication by j
as a complex multiplication.

10.21. Let $f(n)$ be a signal which can be nonzero only for $0 \le n \le N - 1$. Then
$F(z)$ evaluated along the spiral $z_k = AU^{-k}$ for $k = 0, \ldots, M - 1$,
where A and U are arbitrary complex numbers, is

$$F(z_k) = \sum_{n=0}^{N-1} f(n) z_k^{-n} \qquad \text{for} \qquad k = 0, \ldots, M-1$$

(a) Using Bleustein's [11] trick of letting $nk = [n^2 + k^2 - (k-n)^2]/2$,
show that

$$F(z_k) = U^{k^2/2} \sum_{n=0}^{N-1} g(n) h(k-n)$$

where $g(n) = f(n) A^{-n} U^{n^2/2}$ and $h(n) = U^{-n^2/2}$. This is known as
the chirp Z-transform algorithm [112].

(b) Discuss how to use the FFT to implement the chirp Z-transform
algorithm.

(c) How must A and U be chosen to evaluate $F(z)$ at M points
uniformly distributed along the arc of the unit circle $e^{j\theta}$ for
$\theta_1 \le \theta \le \theta_2$?

10.22. The pulse response of a desired filter is

$$g(n) = \begin{cases} 1 & \text{for} \quad 0 \le n \le 4 \\ 0 & \text{elsewhere} \end{cases}$$

The signal to be filtered is the semi-infinite square-wave

$$f(n) = \begin{cases} 0 & \text{for} \quad n < 0 \\ 1 & \text{for} \quad 0 \le n \le 4 \\ -1 & \text{for} \quad 5 \le n \le 9 \end{cases}$$

and

$$f(n+10) = f(n) \qquad \text{for} \qquad n \geq 0$$

For tutorial purposes, assume that 16-point DFTs must be used.

(a)   For both the overlap-add and overlap-save methods, sketch the 16-point input sequences $f_{0,n}$ and $f_{1,n}$ and corresponding output sequences $y_{0,n}$ and $y_{1,n}$. Piece the two sections together appropriately and sketch the result.

(b)   Compute the filtered output sequence directly by discrete-time convolution and compare with the result obtained in (a). Explain the importance of any discrepancies.

10.23.   Show how to modify the overlap-save method so that the last N_2-1 points of each output section are deleted rather than the first N_2-1 points.

10.24.   Suppose that in a particular situation, the input data is available only in blocks of N_1 points, the pulse response of the desired filter has duration N_2, and an FFT algorithm with $N > N_1 + N_2 - 1$ must be used. In addition, assume that data from the next future block to be received cannot be used to fill out the present block. Discuss the modifications necessary in the overlap-add and overlap-save methods.

CHAPTER 11

11.1.   Most versions of FORTRAN have a library function for generating a sequence of uncorrelated random numbers uniformly distributed over $(0, 1)$.

(a)   Write a FORTRAN program to:
   1. generate a sequence $v(n)$ for $n = 1, \ldots, N + 100$ of uncorrelated random variables with zero mean and unity variance;
   2. filter $v(n)$ by $H(z) = 1/(1 - 0.9z^{-1})$ to obtain a correlated sequence $x(n)$;
   3. directly compute for $x(n)$ the sample autocorrelation functions $R_N(m)$ for $N = 10, 100, 1000$, and $0 \leq m \leq 10$, neglecting the first 100 values of $x(n)$ to avoid transient effects.

(b)   Find the theoretical autocorrelation function for $x(n)$ assuming that it is a stationary random process. Compare the three sample autocorrelation functions computed in (b) with the theoretical one by plotting the results.

11.2.   Compute var$\{R_N(m)\}$ using (11.2.8) if $f(n)$ is a Gaussian, discrete-time, random process with the autocorrelation function

(a) $\sigma^2 \delta_{m,0}$

(b) $a^{|n|}, -1 < a < 1$

11.3. Derive (11.5.16) from (11.5.15) and derive (11.5.19) from (11.5.18).

11.4. Let $f(n)$ be a white Gaussian noise sequence. Find $\mathrm{var}\{\hat{S}^*(\omega)\}$ for a rectangular data window and $M/2 \le K \le M-1$ using the results in Section 11.7. Compare $\mathrm{var}\{\hat{S}^*(\omega)\}$ for $K = M/2$ and $K = M$.

11.5. Let $f(n)$ have the sampled power spectral density

$$S(z) = \frac{1}{D(z)D(z^{-1})}$$

where

$$D(z) = (1 - j0.95z^{-1})(1 + j0.95z^{-1})$$

(a) Plot $10 \log_{10} S^*(\omega)$.

(b) Plot $10 \log_{10} E\{I_m^*(\omega)\}$ for a rectangular data window and $M = 16$, 32, and 64.

(c) Simulate $f(n)$ by filtering a sequence of random numbers by $1/D(z)$. Discard the first 100 or so filtered points to avoid transients and store $N = 256$ additional points. Compute and plot $10 \log_{10} \hat{S}^*(\omega)$ for $M = 16$, 32, and 64 using a rectangular data window and observe the trade-off between bias and variance.

11.6. Repeat Problem 11.5 using a Hamming data window.

11.7. Evaluate Λ as defined by (11.8.23) for the Hanning, Hamming, and Bartlett lag windows. Compare the results assuming that M is relatively large.

11.8. Suppose that a continuous-time random process $f(t)$ has a power spectral density $S(\omega)$ that is symmetric about a positive frequency ω_0, $S(\omega) \equiv 0$ for $|\omega - \omega_0| > B$ on the positive frequency axis, and B/ω_0 is small. The signal $f(t)$ is sampled at the rate $\omega_s = \omega_0/K > 2B$ where K is an integer. Show that we can estimate $S(\omega)$ for $|\omega - \omega_0| < B$ by estimating $S^*(\omega)$ for $|\omega| < B$.

11.9. Estimate the parameters for first- and second-order autoregressive models using the 256 points generated in Problem 11.5. Plot the two estimated power spectral densities and compare with those computed for Problem 11.5.

CHAPTER 12

12.1. Let F be the field of real numbers and V be the set of n tuples $\mathbf{x} = (x_1, \ldots, x_n)$ where $x_1, \ldots, x_n$ are complex numbers. Let vector addition and scalar multiplication be defined as in Example 12.2.1. Prove that V is a vector space.

12.2. Let F be the field of real numbers and V be the set of all pairs $[x, y]$ of real numbers. Let

$$[x, y] + [x_1, y_1] = [y + y_1, x + x_1]$$

and

$$c[x, y] = [cy, -cx]$$

Show that V is not a vector space.

12.3. Using the vector space axioms, prove that

$$(\mathbf{x}_1 + \mathbf{x}_2) + (\mathbf{x}_3 + \mathbf{x}_4) = [\mathbf{x}_2 + (\mathbf{x}_3 + \mathbf{x}_1)] + \mathbf{x}_4$$

12.4. (a) Find the only two subspaces of R^1.
  (b) Describe the subspaces of R^2 geometrically.
  (c) Describe the subspaces of R^3 geometrically.

12.5. Let $\mathbf{x}_1, \ldots, \mathbf{x}_n$ be linearly independent vectors in a vector space V. Let $\mathbf{x} = a_1\mathbf{x}_1 + \cdots + a_n\mathbf{x}_n$ and $\mathbf{y} = c_1\mathbf{x}_1 + \cdots + c_n\mathbf{x}_n$ where the a's and c's are scalars. Prove that $\mathbf{x} = \mathbf{y}$ if and only if $a_i = c_i$ for $i = 1, \ldots, n$.

12.6. Let M be a linearly independent subset of a vector space V. Let $\mathbf{x}$ be any vector in V which is not in M. Prove that the set obtained by adjoining $\mathbf{x}$ to M is linearly independent.

12.7. Let F be the field of real numbers and V be the set of real functions on $[0, \infty]$. Vector addition and scalar multiplication are defined in the ordinary sense for real functions. Are the vectors e^{-nt} for $n = 0, \ldots, N$ linearly independent?

12.8. Let F be the field of real numbers and V be the set of real continuous functions on the interval $[a, b]$. Let scalar multiplication and vector addition be defined in the usual sense for functions. Prove that

$$\max_{a \leq t \leq b} |x(t)|, \qquad x(t) \text{ in } V$$

is a norm.

12.9. Show that

$$\int_a^b |x(t)| \, dt$$

is another norm for the space V defined in Problem 12.8.

12.10. Consider the normed vector space with F the field of real numbers, V the set of real continuous functions on $[0, 2]$, and

$$\|\mathbf{x}\| = \int_0^2 |x(t)| \, dt$$

(a)   Prove that

$$x_n(t) = \begin{cases} 0 & \text{for} & 0 \le t \le 1 - 1/n \\ [(t-1)n + 1]/2 & \text{for} & 1 - 1/n \le t \le 1 + 1/n \\ 1 & \text{for} & 1 + 1/n \le t \le 2 \end{cases}$$

is a Cauchy sequence.

(b)   Is this space complete?

12.11.   Let F be the field of complex numbers and V be the space of $n \times 1$ matrices over F. Show that $(\mathbf{x}, \mathbf{y}) = \mathbf{y}^\dagger \mathbf{x}$ is an inner product. The symbol $\dagger$ denotes taking the conjugate transpose.

12.12.   Let V be the space of $n \times n$ matrices over the field F of complex numbers. The trace of a square matrix $\mathbf{X}$ is defined as the sum of its diagonal elements and will be denoted by tr $(\mathbf{X})$. Prove that $(\mathbf{X}, \mathbf{Y}) = $ tr $(\mathbf{XY}^\dagger) = $ tr $(\mathbf{Y}^\dagger \mathbf{X})$ is an inner product.

12.13.   Show that if $(\mathbf{x}, \mathbf{y}) = 0$ for all $\mathbf{y}$ in an inner produce space V, then $\mathbf{x} = \mathbf{0}$.

12.14.   Consider the space of polynomials on $[0, T]$ over the field of real numbers with

$$(\mathbf{x}, \mathbf{y}) = \int_0^T x(t)y(t)\, dt$$

Find the projection of t^2 onto the subspace of first-degree polynomials. Compute the norm of the error between t^2 and the projection.

12.15.   Let $Q(\omega)$ be a real, nonnegative, even function and F be the field of complex numbers. Let V be the space of complex functions on $[-\omega_s/2, \omega_s/2]$ such that

$$\int_{-\omega_s/2}^{\omega_s/2} |G(\omega)|^2 \, d\omega < \infty$$

(a)   Show that

$$(G, H) = \frac{1}{\omega_s} \int_{-\omega_s/2}^{\omega_s/2} G(\omega)\bar{H}(\omega)Q(\omega) \, d\omega$$

is an inner product.

(b)   Let $G(\omega)$ be a function in V. Find the set of complex numbers $\{h_n\}$ that minimizes $v^2 = \|G - H\|^2$ if

$$H(\omega) = \sum_{n=-r}^{r} h_n e^{-j\omega n T}$$

Find the minimum value for v^2.

(c)   Compare your answer with the results of Section 8.9.

12.16. Consider the vector space R^3 with $(\mathbf{x}, \mathbf{y}) = x_1 y_1 + x_2 y_2 + x_3 y_3$. Let $\mathbf{v} = (1, 1, 0)$ and $\mathbf{w} = (1, 1, 1)$.
   (a) Find the minimum norm vector $\mathbf{x}$ such that $(\mathbf{x}, \mathbf{v}) = 6$.
   (b) Find the minimum norm vector $\mathbf{x}$ such that $(\mathbf{x}, \mathbf{v}) = 2$ and $(\mathbf{x}, \mathbf{w}) = 4$.

12.17. Let V be the space of real continuous functions on $[-1, 1]$ with

$$(\mathbf{x}, \mathbf{y}) = \int_{-1}^{1} x(t) y(t) \, dt$$

   If M is the subspace of even functions, find $M^{\perp}$.

12.18. Consider the inner product space defined in Problem 12.14.
   (a) Convert the vectors 1 and t into an orthonormal pair.
   (b) Using the orthonormal pair, find the projection of t^2 onto the subspace of first-degree polynomials and the norm of the error. Compare your results with Problem 12.14.

12.19. Show that the infinite sequence

$$\frac{1}{\sqrt{2\pi}} e^{-jnt} \qquad \text{for} \qquad n = 0, \pm 1, \pm 2, \ldots$$

   is an orthonormal set in $L^2[-\pi, \pi]$.

12.20. Use the Gram-Schmidt procedure to find an orthonormal basis for the subspace of $L^2[0, \infty]$ spanned by the three functions $x_n(t) = t^{n-1} e^{-t} u(t)$, $n = 1$, 2, and 3.

12.21. Two Hilbert spaces H and H' are said to be isomorphic if there is a one-to-one correspondence between their vectors such that if $\mathbf{x} \leftrightarrow \mathbf{x}'$ and $\mathbf{y} \leftrightarrow \mathbf{y}'$, then
   (1) $\mathbf{x} + \mathbf{y} \leftrightarrow \mathbf{x}' + \mathbf{y}'$
   (2) $c\mathbf{x} \leftrightarrow c\mathbf{x}'$
   (3) $(\mathbf{x}, \mathbf{y})_H = (\mathbf{x}', \mathbf{y}')_{H'}$
   Let $\{\mathbf{u}_i\}_{i=1}^{\infty}$ be a complete orthonormal basis for L^2 and let $\mathbf{x}' = \{x_i\}_{i=1}^{\infty}$ where x_i is the Fourier coefficient of $\mathbf{x}$ with respect to $\mathbf{u}_i$. Show that $\mathbf{x}'$ is in l^2 and that this correspondence between L^2 and l^2 is an isomorphism.

CHAPTER 13

13.1. A signal $x(t)$ is sampled at times $t_1 < t_2 < \cdots < t_N$. Show that the best straight line approximation to $x(t)$ in the sense of minimizing the sum of the squares of the errors at the sampling times is

$$\hat{x}(t) = \bar{x} + c(t - \bar{t})$$

where

$$\bar{t} = \frac{1}{N} \sum_{n=1}^{N} t_n$$

$$\tilde{x} = \frac{1}{N} \sum_{n=1}^{N} x(t_n)$$

and

$$c = \frac{\displaystyle\sum_{n=1}^{N} (t_n - \bar{t})[x(t_n) - \tilde{x}]}{\displaystyle\sum_{n=1}^{N} (t_n - \bar{t})^2}$$

13.2. Suppose that a discrete-time signal $b(n)$ is transmitted over a channel with an unknown pulse transfer function and the distorted signal $x(n)$ is received. In an attempt to undo the effects of the channel, $x(n)$ is passed through a $2K+1$ tap transversal filter to generate the signal

$$r(n) = \sum_{k=-K}^{K} c_k x(n-k)$$

Using measurements of $r(n)$ for $n = 0, \ldots, N-1$ and $x(n)$ for $n = -K, \ldots, N-1+K$, and assuming that $b(n)$ is known, find the set of filter coefficients $\{c_k\}$ that minimizes

$$\sum_{n=0}^{N-1} |r(n) - b(n)|^2$$

Note: Systems with this structure are commonly used in high-speed wireline data communications.

13.3. Find the linear minimum-variance unbiased estimate of the unknown constant c and the variance of the estimate if the observed data is

$$x_n = c + e_n \qquad \text{for} \qquad n = 1, \ldots, N$$

where $\{e_n\}$ is a set of zero-mean random variables with $E\{e_n e_k\} = \sigma^2 \delta_{nk}$.

13.4. Repeat Problem 13.3 if $E\{e_n e_k\} = a^{|n-k|}$ with $-1 < a < 1$. Explain the results in the limiting cases where $a = 1$, 0, and -1.

13.5. The output $f(n)$ of a particular stable system is described by the difference equation

$$f(n) = bf(n-1)$$

where b is a known constant. The initial condition $f(0)$ is an unknown constant. Find the linear minimum-variance unbiased estimate of $f(0)$

if the observed data is

$$x(n) = f(n) + e(n) \qquad \text{for} \qquad n = 0, \ldots, N-1$$

where $e(n)$ is a white noise sequence with variance σ^2. Compute the variance of the estimate and find its limit as N becomes infinite.

13.6. The *maximum likelihood* estimate of a parameter vector $\boldsymbol{\beta}$ from an observed data vector $\mathbf{x}$ is defined to be the parameter vector that maximizes the conditional probability density function $f(\mathbf{x}/\boldsymbol{\beta})$. Suppose that

$$\mathbf{x} = \mathbf{Y}\boldsymbol{\beta} + \mathbf{e}$$

as in (13.4.1) and that all variables are real. If $\mathbf{e}$ is a Gaussian vector random variable, show that maximizing the conditional probabilty density function is equivalent to minimizing the quadratic form

$$(\mathbf{x} - \mathbf{Y}\boldsymbol{\beta})' \mathbf{Q}^{-1} (\mathbf{x} - \mathbf{Y}\boldsymbol{\beta})$$

Show that the linear minimum-variance unbiased estimate $\hat{\boldsymbol{\beta}}$ given by (13.4.20) minimizes this quadratic form.

Hint: The probability density function evaluated at the point $\boldsymbol{\eta}$ for an N-dimensional Gaussian vector random variable $\mathbf{w}$ with mean $\boldsymbol{\mu}$ and covariance matrix $\mathbf{V}$ is

$$f(\boldsymbol{\eta}) = (2\pi)^{-N/2} |\mathbf{V}|^{-1/2} \exp\left[-\tfrac{1}{2}(\boldsymbol{\eta} - \boldsymbol{\mu})' \mathbf{V}^{-1} (\boldsymbol{\eta} - \boldsymbol{\mu})\right]$$

where $|\mathbf{V}|$ is the determinant of $\mathbf{V}$.

13.7. Let w be a real zero-mean random variable with variance γ^2 and $e_1, \ldots, e_n$ be real zero-mean random variables with $E\{e_n e_k\} = a^{|n-k|}$ where $-1 < a < 1$. Also assume that $E\{e_n w\} = 0$ for $n = 1, \ldots, N$. Find the linear minimum mean square error estimate of w and the corresponding mean square error if the observed data is $x_n = w + e_n$ for $n = 1, \ldots, N$.

13.8. Let $x(t)$ be a zero-mean random process with the autocorrelation function $R(\tau) = e^{-a|\tau|}$ where $a > 0$. From the two samples $x(t_1)$ and $x(t_2)$ with $t_1 < t_2$, find the linear minimum mean square error estimate of $x(t)$ and the corresponding mean square error if
(a) $t < t_1$
(b) $t_1 \le t \le t_2$
(c) $t_2 < t$

13.9. Repeat Problem 13.8 if $x(t)$ is a nonstationary random process with the autocorrelation function $R(t, \tau) = \text{minimum } (t, \tau)$.

13.10.   In a particular experiment, the observed data has the form

$$x(t_n) = c + (t_n - \bar{t})b + e(t_n) \quad \text{for} \quad n = 1, \ldots, N$$

where $c, b, e(t_1), \ldots, e(t_N)$ are mutually uncorrelated zero-mean random variables with $\text{var}\{c\} = \sigma_c^2$, $\text{var}\{b\} = \sigma_b^2$, $\text{var}\{e(t_n)\} = \sigma^2$ for all n, and

$$\bar{t} = \frac{1}{N} \sum_{n=1}^{N} t_n$$

Find the linear minimum mean square error estimates of c and b and the corresponding error covariance matrix.

13.11.   Suppose that $\mathbf{x}$ and $\mathbf{w}$ are vector random variables with $E\{\mathbf{x}\} = \boldsymbol{\mu}$, $E\{\mathbf{w}\} = \boldsymbol{\lambda}$, $\text{cov}\{\mathbf{x}\} = \mathbf{V}_{xx}$, $\text{cov}\{\mathbf{w}\} = \mathbf{V}_{ww}$, and $\text{cov}\{\mathbf{w}, \mathbf{x}\} = \mathbf{V}_{wx}$. Find the matrices $\mathbf{b}$ and $\mathbf{A}$ so that the components of

$$\hat{\mathbf{w}} = \mathbf{b} + \mathbf{A}(\mathbf{x} - \boldsymbol{\mu})$$

are the minimum mean square error estimates of the components of $\mathbf{w}$. Find the error covariance matrix for the optimum estimates.

13.12.   Let $\mathbf{x}$ and $\mathbf{w}$ be real N- and K-dimensional vector random variables, respectively, with the means and covariances defined in Problem 13.11. Assume that the components of $\mathbf{x}$ and $\mathbf{w}$ are jointly Gaussian.

(a)   Show that the conditional probability density function for $\mathbf{w}$ given $\mathbf{x}$ evaluated at the points $\boldsymbol{\eta}$ and $\boldsymbol{\zeta}$ is

$$f_{\mathbf{w}/\mathbf{x}}(\boldsymbol{\eta}/\boldsymbol{\zeta}) = (2\pi)^{-K/2} |\mathbf{V}|^{-1/2} \exp\{-\tfrac{1}{2}[\boldsymbol{\eta} - \mathbf{m}(\boldsymbol{\zeta})]' \mathbf{V}^{-1}[\boldsymbol{\eta} - \mathbf{m}(\boldsymbol{\zeta})]\}$$

where

$$\mathbf{m}(\boldsymbol{\zeta}) = \boldsymbol{\lambda} + \mathbf{V}_{wx}\mathbf{V}_{xx}^{-1}(\boldsymbol{\zeta} - \boldsymbol{\mu})$$

$$\mathbf{V} = \mathbf{V}_{ww} - \mathbf{V}_{wx}\mathbf{V}_{xx}^{-1}\mathbf{V}_{xw} = \text{cov}\{\mathbf{w} - \mathbf{m}(\mathbf{x})\}$$

and $|\mathbf{V}|$ is the determinant of $\mathbf{V}$.

(b)   The *maximum a posteriori probability* (MAP) estimate of $\mathbf{w}$ from $\mathbf{x}$ is defined to be the value of $\boldsymbol{\eta}$ that maximizes $f_{\mathbf{w}/\mathbf{x}}(\boldsymbol{\eta}/\mathbf{x})$. Show that the MAP estimate in this case is $\hat{\mathbf{w}} = \mathbf{m}(\mathbf{x})$ and is identical to the linear minimum mean square error estimate derived in Problem 13.11.

13.13.   The conditional expectation of a vector random variable $\mathbf{w}$ given a vector random variable $\mathbf{x}$ when $f_{\mathbf{w}/\mathbf{x}}(\boldsymbol{\eta}/\boldsymbol{\zeta})$ exists is

$$E\{\mathbf{w}/\mathbf{x}\} = \int_{-\infty}^{\infty} \boldsymbol{\eta} f_{\mathbf{w}/\mathbf{x}}(\boldsymbol{\eta}/\mathbf{x}) \, d\boldsymbol{\eta}$$

where the integral of a column vector is the column vector of integrals of the individual components. It can be shown that $E\{\mathbf{w}/\mathbf{x}\}$ is the minimum mean square error estimate of $\mathbf{w}$ from $\mathbf{x}$ when the estimator is not constrained to be linear [30]. If $\mathbf{x}$ and $\mathbf{w}$ are jointly Gaussian as in Problem 13.12, show that

$$E\{\mathbf{w}/\mathbf{x}\} = \mathbf{m}(\mathbf{x})$$

Thus we can conclude that the unconstrained minimum mean square error estimator is linear in the Gaussian case.

13.14. Let $x = b + e$ where e is a zero-mean Gaussian random variable with variance σ^2, and b is a discrete random variable that can have the values 1 and -1 each with probability $1/2$. Assume that b and e are statistically independent. Show that the unconstrained minimum mean square error estimate of b from x defined in Problem 13.13 is $E\{b/x\} = \tanh(x/\sigma^2)$. Find the linear minimum mean square error estimate of b from x and compare with the unconstrained estimate by sketching the two as a function of x.

CHAPTER 14

14.1. A discrete-time random process $x(n)$ is generated by the difference equation

$$x(n+1) = bx(n) + v(n)$$

where $-1 < b < 1$ and $v(n)$ is a white noise sequence with variance Q. The signal

$$y(n) = x(n) + w(n)$$

is observed starting at time n_0. Assume that $w(n)$ is a white noise sequence with variance R and is uncorrelated with $v(n)$.
(a) Write the recursive equations for computing $\hat{x}(n+1 \mid n)$ using the one-step prediction algorithm. Include initial conditions.
(b) Find the limits of $G(n)$, $P(n)$, and $\Gamma(n)$ as n becomes infinite.

14.2. Let

$$\mathbf{x}(n+1) = \begin{bmatrix} 0 & 1 & 0 & 0 & \cdots & 0 & 0 \\ 0 & 0 & 1 & 0 & \cdots & 0 & 0 \\ & \cdot & & & & & \cdot \\ & \cdot & & & & & \cdot \\ & \cdot & & & & & \cdot \\ 0 & 0 & 0 & 0 & \cdots & 0 & 1 \\ -d_N & \cdots & & & -d_2 & -d_1 \end{bmatrix} \mathbf{x}(n) + \begin{bmatrix} 0 \\ 0 \\ \cdot \\ \cdot \\ \cdot \\ 0 \\ v(n) \end{bmatrix}$$

and

$$\mathbf{y}(n) = \mathbf{x}(n)$$

where $v(n)$ is a white noise sequence with variance σ^2. Find $\hat{x}(n+1 \mid n)$, $\mathbf{P}(n)$, and $\mathbf{\Gamma}(n)$. Compare this problem with Problem 7.18.

14.3. Let

$$\mathbf{x}(n+1) = \begin{bmatrix} 0 & 1 \\ -d_2 & -d_1 \end{bmatrix} \mathbf{x}(n) + \begin{bmatrix} 0 \\ v(n) \end{bmatrix}$$

and

$$y(n) = [0 \quad 1]\mathbf{x}(n)$$

where $v(n)$ is a white noise sequence with variance σ^2. Assume that the observations begin at time $n_0 = 0$.

(a) Find $\hat{x}(n+1 \mid n)$, $\mathbf{G}(n)$, $\mathbf{P}(n)$, and $\mathbf{\Gamma}(n)$ for $n = 0$, 1, and 2. Write explicit equations for the components of $\hat{x}(n+1 \mid n)$ and combine terms where possible.

(b) What are the steady-state values of $\mathbf{G}(n)$, $\mathbf{P}(n)$, and $\mathbf{\Gamma}(n)$? For what n are the steady-state values reached?

(c) Compare this problem with Problem 14.2.

14.4. Suppose that the problem model is generalized slightly by changing (14.1.5) to

$$\operatorname{cov}\{\mathbf{v}(n), \mathbf{w}(k)\} = \mathbf{S}(n)\,\delta_{nk}$$

Show that the solution to the one-step prediction problem is given by (14.3.6) and (14.3.13) with (14.3.9) changed to

$$\mathbf{G}(n) = [\boldsymbol{\varphi}(n+1, n)\mathbf{P}(n)\mathbf{C}^{\dagger}(n) + \mathbf{S}(n)][\mathbf{C}(n)\mathbf{P}(n)\mathbf{C}^{\dagger}(n) + \mathbf{R}(n)]^{-1}$$

and (14.3.12) changed to

$$\begin{aligned}
\mathbf{P}(n+1) &= \boldsymbol{\varphi}(n+1, n)\mathbf{\Gamma}(n)\boldsymbol{\varphi}^{\dagger}(n+1, n) - \mathbf{S}(n) \\
&\quad \times [\mathbf{C}(n)\mathbf{P}(n)\mathbf{C}^{\dagger}(n) + \mathbf{R}(n)]^{-1}\mathbf{C}(n)\mathbf{P}(n)\boldsymbol{\varphi}^{\dagger}(n+1, n) \\
&\quad - \boldsymbol{\varphi}(n+1, n)\mathbf{P}(n)\mathbf{C}^{\dagger}(n)[\mathbf{C}(n)\mathbf{P}(n)\mathbf{C}^{\dagger}(n) + \mathbf{R}(n)]^{-1}\mathbf{S}^{\dagger}(n) \\
&\quad - \mathbf{S}(n)[\mathbf{C}(n)\mathbf{P}(n)\mathbf{C}^{\dagger}(n) + \mathbf{R}(n)]^{-1}\mathbf{S}^{\dagger}(n) + \mathbf{Q}(n) \\
&= \boldsymbol{\varphi}(n+1, n)\mathbf{P}(n)\boldsymbol{\varphi}^{\dagger}(n+1, n) - \mathbf{G}(n) \\
&\quad \times [\mathbf{C}(n)\mathbf{P}(n)\mathbf{C}^{\dagger}(n) + \mathbf{R}(n)]\mathbf{G}^{\dagger}(n) + \mathbf{Q}(n)
\end{aligned}$$

14.5. Without noise, the input $v(n)$ and output $s(n)$ of a particular communication channel are related by the equation

$$s(n) = \sum_{k=0}^{N} c_k v(n-k)$$

The noisy observed signal is

$$y(n) = s(n) + e(n)$$

Assume that $E\{v(n)\} = E\{e(n)\} = 0,$ $E\{v(n)v(k)\} = a^2 \delta_{nk},$ $E\{e(n)e(k)\} = b^2 \delta_{nk},$ and $E\{v(n)e(k)\} = 0$ for all n and k.

(a) If the state of the channel is defined to be $\mathbf{x}(n) = [v(n-1),$ $v(n-2), \ldots, v(n-N)]',$ find the state and output equations for the channel. Sketch a block diagram for the channel and label the state variables.

(b) Using the results of Problem 14.4, find the equations for recursively computing $\hat{\mathbf{x}}(n+1 \mid n)$. Sketch a block diagram of the estimation filter and label the points corresponding to the components of $\hat{x}(n+1 \mid n)$.

14.6. A system is described by the state equation

$$\mathbf{x}(nT + T) = \begin{bmatrix} 1 & T \\ 0 & 1 \end{bmatrix} \mathbf{x}(nT)$$

and observation equation

$$y(nT) = \begin{bmatrix} 1 & 0 \end{bmatrix} \mathbf{x}(nT) + w(nT)$$

Assume that $x_1(0) = 0,$ $E\{x_2(0)\} = 0,$ $E\{x_2^2(0)\} = a^2,$ $E\{w(nT)\} = 0,$ $E\{w(nT)w(kT)\} = b^2 \delta_{nk},$ and $E\{x_2(0)w(kT)\} = 0.$

(a) Verify that the system could be an object moving with a constant velocity and that $y(nT)$ would then represent noisy position measurements.

(b) Assume that the observations begin at time zero. Write the equations for recursively computing $\hat{x}(nT \mid nT)$. Verify that for $n \geq 1$

$$\mathbf{P}(nT) = \frac{b^2}{f(n-1)} \begin{bmatrix} (nT)^2 & nT \\ nT & 1 \end{bmatrix}$$

$$\mathbf{\Gamma}(nT) = \frac{b^2}{f(n)} \begin{bmatrix} (nT)^2 & nT \\ nT & 1 \end{bmatrix}$$

and

$$H(nT) = \frac{nT}{f(n)} \begin{bmatrix} nT \\ 1 \end{bmatrix}$$

where

$$f(n) = f(n-1) + (nT)^2$$

with the initial condition

$$f(0) = b^2/a^2$$

(c) Investigate and intuitively justify the filter behavior for large n.

14.7. Suppose that $x(n)$ is generated by the scalar first-order difference equation

$$x(n+1) = b_1 x(n) + v_1(n)$$

and that the observed data is

$$y(n) = x(n) + w(n)$$

where

$$w(n+1) = b_2 w(n) + v_2(n)$$

Assume that $v_1(n)$ and $v_2(n)$ are mutually uncorrelated white noise sequences with variances σ_1^2 and σ_2^2, respectively. Thus the observation noise $w(n)$ is not white in this case. Show that by forming the augmented state vector $\mathbf{x}(n) = [x(n) \quad w(n)]^t$, the problem can be converted into one to which the standard Kalman filter equations apply.

14.8. Consider the problem described in Example 14.3.2. Write the smoothing equations for computing $\hat{x}(n-1 \mid n)$. Find the steady-state smoothing filter and compare the steady-state filtering and one-step smoothing mean-square errors.

BIBLIOGRAPHY
AND
REFERENCES

1. Ahlfors, L. V., *Complex Analysis*, McGraw-Hill, New York, 1953.
2. Atal, B. S., and S. L. Hanauer, "Speech Analysis and Synthesis by Linear Prediction of the Speech Wave," *Journal of the Acoustical Society of America*, Vol. 50, No. 2, August 1971, pp. 637–655.
3. Athans, M., and P. L. Falb, *Optimal Control: An Introduction to the Theory and Its Applications*, McGraw-Hill, New York, 1966.
4. Bartlett, M. S., *An Introduction to Stochastic Processes with Special Reference to Methods and Applications*, Cambridge University Press, New York, 1953.
5. Bennett, W. R., "Spectra of Quantized Signals," *Bell System Technical Journal*, Vol. 27, July 1948, pp. 446–472.
6. Bergland, G. D., "A Fast Fourier Transform Algorithm Using Base 8 Iterations," *Mathematics of Computation*, Vol. 22, April 1968, pp. 275–279.
7. Beutler, F. J., "Sampling Theorems and Bases in Hilbert Space," *Information and Control*, Vol. 4, No. 2–3, September 1961, pp. 97–117.
8. Birkhoff, G., and S. MacLane, *A Survey of Modern Algebra*, Macmillan, New York, 1953.
9. Blackman, R. B., *Linear Data-Smoothing and Prediction in Theory and Practice*, Addison-Wesley, Reading, Massachusetts, 1965.
10. Blackman, R. B., and J. W. Tukey, *The Measurement of Power Spectra from the Point of View of Communications Engineering*, Dover Publications, New York, 1958 (an unabridged and corrected republication of the work originally published in January and March, 1958, in Vol. XXXVII of the *Bell System Technical Journal*).
11. *Bluestein, L. I., "A Linear Filtering Approach to the Computation of Discrete Fourier Transform," *IEEE Transactions on Audio and Electroacoustics*, Vol. AU-18, No. 4, December 1970, pp. 451–455.
12. Bogner, R. E., and A. G. Constantinides (Eds.), *Introduction to Digital Filtering*, Wiley-Interscience, New York, 1975.

* References with a star are reprinted in Reference 111.

13. Brigham, E. O., *The Fast Fourier Transform*, Prentice-Hall, Englewood Cliffs, New Jersey, 1974.
14. Brockett, R. W., *Finite Dimensional Linear Systems*, Wiley, New York, 1970.
15. Burrus, C. S., "Block Realization of Digital Filters," *IEEE Transactions on Audio and Electroacoustics*, Vol. AU-20, No. 4, October 1972, pp. 230–235.
16. Cadzow, J. A., *Discrete-Time Systems*, Prentice-Hall, Englewood Cliffs, New Jersey, 1973.
17. Calahan, D. A., *Modern Network Synthesis*, Hayden, New York, 1964.
18. Chang, S. S. L., *Synthesis of Optimum Control Systems*, McGraw-Hill, New York, 1961.
19. Chu, A., and R. E. Crochiere, "Comments and Experimental Results on Optimal Digital Ladder Structures," *IEEE Transactions on Audio and Electroacoustics*, Vol. AU-20, No. 4, October 1972, pp. 317–318.
20. Chu, Y., *Digital Computer Design Fundamentals*, McGraw-Hill, New York, 1962.
21. Churchill, R. V., *Introduction to Complex Variables and Applications*, 2nd Ed., McGraw-Hill, New York, 1960.
22. Coddington, E. A., and N. Levinson, *Theory of Ordinary Differential Equations*, McGraw-Hill, New York, 1955.
23.*Constantinides, A. G., "Spectral Transformations for Digital Filters," *Proceedings of the Institution of Electrical Engineers*, Vol. 117, August 1970, pp. 1585–1590.
24.*Cooley, J. W., P. A. W. Lewis, and P. D. Welch, "Historical Notes on the Fast Fourier Transform," *IEEE Transactions on Audio and Electroacoustics*, Vol. AU-15, No. 2, June 1967, pp. 76–79.
25. Cooley, J. W., P. A. Lewis, and P. D. Welch, "The Fast Fourier Transform and Its Applications," *IEEE Transactions on Education*, Vol. 12, No. 1, March 1969, pp. 27–34.
26.*Cooley, J. W., and J. W. Tukey, "An Algorithm for the Machine Calculation of Complex Fourier Series," *Mathematics of Computation*, Vol. 19, April 1965, pp. 297–301.
27. Crochiere, R. E., "Digital Ladder Structures and Coefficient Sensitivity," *IEEE Transactions on Audio and Electroacoustics*, Vol. AU-20, No. 4, October 1972, pp. 240–246.
28. Deczky, A. G., "Synthesis of Recursive Digital Filters Using the Minimum p-Error Criterion," *IEEE Transactions on Audio and Electroacoustics*, Vol. AU-20, No. 4, October 1972, pp. 257–263.
29. DeRusso, P. M., R. J. Roy, and C. M. Close, *State Variables for Engineers*, Wiley, New York, 1965.
30. Deutsch, R., *Estimation Theory*, Prentice-Hall, Englewood Cliffs, New Jersey, 1965.
31. Dunford, N., and J. T. Schwartz, *Linear Operators*, Interscience, New York, 1958.
32.*Ebert, P. M., J. E. Mazo, and M. G. Taylor, "Overflow Oscillations in Digital Filters," *Bell System Technical Journal*, Vol. 48, November 1969, pp. 2999–3020.
33. Evans, W. R., "Graphical Analysis of Control Systems," *Transactions of AIEE*, Vol. 67, 1948, pp. 547–551.
34. Freeman, H., *Discrete-Time Systems*, Wiley, New York, 1965.
35. Gauss, K. F., *Theory of the Motion of the Heavenly Bodies Moving About the Sun in Conic Sections*, Dover Publications, New York, 1963, Book II, Section 3 (reprint).
36. Gentleman, W. M., and G. Sande, "Fast Fourier Transforms for Fun and Profit," 1966 Fall Joint Computer Conference, *AFIPS Proceedings*, Vol. 29, Spartan, Washingtron, D.C., 1966, pp. 563–578.
37.*Gibbs, A. J., "The Design of Digital Filters," *Australian Telecommunications Research Journal*, Vol. 4, 1970, pp. 29–34.
38. Gold, B., and K. L. Jordan, "A Note on Digital Filter Synthesis," *Proceedings of the IEEE* (*Letters*), Vol. 56, October 1968, pp. 1717–1718.

* References with a star are reprinted in Reference 111.

39. Gold, B., and C. M. Rader, *Digital Processing of Signals*, McGraw-Hill, New York, 1969.
40. Gray, A. H., Jr., and J. D. Markel, "Digital Lattice and Ladder Filter Synthesis," *IEEE Transactions on Audio and Electroacoustics*, Vol. AU-21, No. 6, December 1973, pp. 491–500.
41. Grenander, U., and M. Rosenblatt, *Statistical Analysis of Stationary Time Series*, Wiley, New York, 1957.
42. Hannan, E. J., *Time Series Analysis*, Methuen, London, 1960.
43. Helms, H. D., "Nonrecursive Digital Filters: Design Methods for Achieving Specifications on Frequency Response," *IEEE Transactions on Audio and Electroacoustics*, Vol. AU-16, September 1968, pp. 336–342.
44. Helms, H. D., and L. R. Rabiner (Eds.), *Literature In Digital Signal Processing: Terminology and Permuted Title Index*, IEEE Press, New York, 1973.
45. Helms, H. D., J. F. Kaiser, and L. R. Rabiner (Eds.), *Literature in Digital Signal Processing: Author and Permuted Title Index*, IEEE Press, New York, 1975.
46.*Herrmann, O., "Design of Nonrecursive Digital Filters with Linear Phase," *Electronics Letters*, Vol. 6, No. 11, May 28, 1970, pp. 328–329.
47. Hoffman, K., and R. Kunze, *Linear Algebra*, Prentice-Hall, Englewood Cliffs, New Jersey, 1961.
48.*Hofstetter, E., A. V. Oppenheim, and J. Siegel, "A New Technique for the Design of Nonrecursive Digital Filters," *Proceedings of the Fifth Annual Princeton Conference on Information Sciences and Systems*, March 1971, pp. 64–72.
49. Huelsman, L. P. (Ed.), *Active Filters: Lumped, Distributed, Integrated, Digital and Parametric*, McGraw-Hill, New York, 1970, Ch. 5.
50. Hwang, S. Y., "Realization of Canonical Digital Networks," *IEEE Transactions on Acoustics, Speech, and Signal Processing*, Vol. ASSP-22, No. 1, February 1974, pp. 27–39.
51. *IEEE Transactions on Audio and Electroacoustics*, Vol. AU-15, No. 2, June 1967.
52. *IEEE Transactions on Audio and Electroacoustics*, Vol. AU-18, No. 2, June 1970.
53. *IEEE Transactions on Audio and Electroacoustics*, Vol. AU-20, No. 4, October 1972.
54. Itakura, F., and S. Saito, "A Statistical Method for Estimation of Speech Spectral Density and Format Frequencies," *Electronics and Communications in Japan*, Vol. 53-A, No. 1, 1970, pp. 36–43.
55. Itakura, F., S. Saito, T. Koike, H. Sawabe, and M. Nishikawa, "An Audio Response Unit Based on Partial Correlation," *IEEE Transactions on Communication Technology*, Vol. COM-20, No. 4, August 1972, pp. 792–797.
56.*Jackson, L. B., "An Analysis of Limit Cycles Due to Multiplication Rounding in Recursive Digital (Sub)Filters," *Proceedings of the Seventh Annual Allerton Conference on Circuit and System Theory*, October 1969, pp. 69–78.
57.*Jackson, L. B., "On the Interaction of Roundoff Noise and Dynamic Range in Digital Filters," *Bell System Technical Journal*, Vol. 49, February 1970, pp. 159–184.
58. Jenkins, G. M., and D. G. Watts, *Spectral Analysis and Its Applications*, Holden-Day, San Fransisco, 1968.
59. Jordan, C., *Calculus of Finite Differences*, 2nd Ed., Chelsea, New York, 1950.
60. Jury, E. I., *Sampled-Data Control Systems*, Wiley, New York, 1958.
61. Jury, E. I., "A Stability Test for Linear Discrete Systems Using a Simple Division," *Proceedings of the IRE*, Vol. 49, No. 12, December 1961, p. 1948.
62. Jury, E. I., *Theory and Application of the Z-Transform Method*, Wiley, New York, 1964.
63. Jury, E. I., and B. H. Bharucha, "Notes on the Stability Criterion for Linear Discrete Systems," *IRE Transactions on Automatic Control*, Vol. AC-6, February 1961, pp. 88–90.

* References with a star are reprinted in Reference 111.

64. Jury, E. I., and J. Blanchard, "A Stability Test for Linear Discrete Systems in Table Form," *Proceedings of the IRE*, Vol. 49, No. 12, December 1961, pp. 1947–1948.
65. Kailath, T., "An Innovations Approach to Least-Squares Estimation—Part I: Linear Filtering in Additive White Noise," *IEEE Transactions on Automatic Control*, Vol. AC-13, No. 6, December 1968, pp. 646–654.
66. Kailath, T., and P. Frost, "An Innovations Approach to Least-Squares Estimation—Part II: Linear Smoothing in Additive White Noise," *IEEE Transactions on Automatic Control*, Vol. AC-13, No. 6, December 1968, pp. 655–660.
67. Kalman, R. E., "A New Approach to Linear Filtering and Prediction Problems," *Transactions of the ASME, Journal of Basic Engineering*, Vol. 82D, March 1960, pp. 34–45.
68. Kalman, R. E., and R. Bucy, "New Results in Linear Filtering and Prediction Theory," *Transactions of the ASME, Journal of Basic Engineering*, Vol. 83D, March 1961, pp. 95–108.
69. Kaneko, T., and B. Liu, "Round-Off Error of Floating-Point Digital Filters," *Proceedings of the Sixth Annual Allerton Conference on Circuit and System Theory*, October 1968, pp. 219–227.
70. Kang, G. S., "Application of Linear Prediction Encoding to a Narrowband Voice Digitizer," NRL Report 7774, Naval Research Laboratory, Washington, D.C., October 31, 1974.
71. Kaplan, W., *Advanced Calculus*, Addison-Wesley, Reading, Massachusetts, 1952.
72. Kaplan, W., *Operational Methods for Linear Systems*, Addison-Wesley, Reading, Massachusetts, 1962.
73. Kelley, J. L., Jr., and C. Lochbaum, "Speech Synthesizer," *Proceedings of the Stockholm Speech Communications Seminar*, R.I.T., Stockholm, September 1962.
74. Knowles, J. B., and R. Edwards, "Effects of a Finite-Word-Length Computer in a Sampled-Data Feedback System," *Proceedings of the Institution of Electrical Engineers*, Vol. 112, No. 6, June 1965, pp. 1197–1207.
75. Kolmogorov, A. N., "Interpolation and Extrapolation von Stationären Zufälligen Folgen," *Bulletin of the Academy of Science* (USSR), Ser. Math. Vol. 5, 1941, pp. 3–14.
76. Kolmogorov, A. N., "Interpolation and Extrapolation of Stationary Random Sequences," translated by W. Doyle and J. Selin, RM-3090-PR, The Rand Corp., Santa Monica, California, 1962.
77. Kolmogorov, A. N., and S. V. Fomin, *Elements of the Theory of Functional Analysis*, Vols. 1 and 2, Graylock Press, Rochester, New York, 1957.
78. Koopmanns, L. H., *Spectral Analysis of Time Series*, Academic Press, New York, 1974.
79. Kuo, B., *Analysis and Synthesis of Sampled-Data Control Systems*, Prentice-Hall, Englewood Cliffs, New Jersey, 1963.
80. Kuo, B. C., *Discrete Data Control Systems*, Prentice-Hall, Englewood Cliffs, New Jersey, 1970.
81. Kuo, F. F., and J. F. Kaiser (Eds), *System Analysis by Digital Compter*, Wiley, New York, 1966, Ch. 7.
82. Lee, R. C. K., *Optimal Estimation, Identification, and Control*, Research Monograph No. 28, MIT Press, Cambridge, Massachusetts, 1964.
83. Legendre, A. M., *Nouvelles méthodes pour la détermination des orbites des cométes*, Paris, 1806.
84. Linden, D. A., and N. M. Abramson, "A Generalization of the Sampling Theorem," *Information and Control*, Vol. 3, No. 1, March 1960, pp. 26–31.
85. Lindorf, D. P., *Theory of Sampled-Data Control Systems*, Wiley, New York, 1965.
86. Long, L. J., and T. N. Trick, "An Absolute Bound on Limit Cycles Due to Roundoff Errors in Digital Filters," *IEEE Transactions on Audio and Electroacoustics*, Vol. AU-21, No. 1, February 1973, pp. 27–30.

87. Lucky, R. W., J. Salz, and E. J. Weldon, Jr., *Principles of Data Communications*, McGraw-Hill, New York, 1968.
88. Luenberger, D. G., *Optimization by Vector Space Methods*, Wiley, New York, 1969.
89. Marden, M., *The Geometry of the Zeros of a Polynomial in the Complex Plane*, American Mathematical Society, New York, 1949, pp. 148–155.
90. McClellan, J. H., and T. W. Parks, "A Unified Approach to the Design of Optimum FIR Linear Phase Digital Filters," *IEEE Transactions on Circuit Theory*, Vol. CT-20, November 1973, pp. 697–701.
91. McClellan, J. H., T. W. Parks, and L. R. Rabiner, "A Computer Program for Designing Optimum FIR Linear Phase Digital Filters," *IEEE Transactions on Audio and Electroacoustics*, Vol. AU-21, No. 6, December 1973, pp. 506–526.
92. Mermelstein, P., "Calculation of Vocal-Tract Transfer Functions for Speech Synthesis Applications," *Proceedings of the Seventh International Congress on Acoustics*, Paper 23C13, 1972, pp. 173–176.
93. Milne-Thomson, L. M., *The Calculus of Finite Differences*, Macmillan, London, 1933.
94. Mitra, S. K., and R. J. Sherwood, "Canonic Realizations of Digital Filters Using Continued Fraction Expansion," *IEEE Transactions on Audio and Electroacoustics*, Vol. AU-20, No. 3, August 1972, pp. 185–194.
95. Mitra, S. K., and R. J. Sherwood, "Digital Ladder Networks," *IEEE Transactions on Audio and Electroacoustics*, Vol. AU-21, No. 1, February 1973, pp. 30–36.
96. Monroe, A. J., *Digital Processes for Sampled Data Systems*, Wiley, New York, 1962.
97. Nyquist, H., "Certain Topics in Telegraph Transmission Theory," *Transactions of the AIEE*, Vol. 47, April 1928, pp. 617–644.
98. Oliver, B. M., J. R. Pierce, and C. E. Shannon, "The Philosophy of PCM," *Proceedings of the IRE*, Vol. 36, No. 11, November 1948, pp. 1324–1331.
99. Oppenheim, A. V. (Ed.), *Papers on Digital Signal Processing*, MIT Press, Cambridge, Massachusetts, 1969.
100. Oppenheim, A. V., and R. W. Schafer, *Digital Signal Processing*, Prentice-Hall, Englewood Cliffs, New Jersey, 1975.
101. Oppenheim, A. V., R. W. Schafer, and T. G. Stockham, "Nonlinear Filtering of Multiplied and Convolved Signals," *Proceedings of the IEEE*, Vol. 56, August 1968, pp. 1264–1291.
102. Otnes, R. K., and L. Enochson, *Digital Time Series Analysis*, Wiley, New York, 1972.
103. Papoulis, A., *The Fourier Integral and Its Applications*, McGraw-Hill, New York, 1962.
104. Papoulis, A., *Probability, Random Variables, and Stochastic Processes*, McGraw-Hill, New York, 1965.
105. Parker, S. R., and S. F. Hess, "Limit-Cycle Oscillations in Digital Filters," *IEEE Transactions on Circuit Theory*, Vol. CT-18, No. 6, November 1971, pp. 687–697.
106. Parks, T. W., and J. H. McClellan, "Chebyshev Approximation for Nonrecursive Digital Filters with Linear Phase," *IEEE Transactions on Circuit Theory*, Vol. CT-19, March 1972, pp. 189–194.
107. Petersen, D. P., and D. Middleton, "Sampling and Reconstruction of Wave-Number-Limited Functions in N-Dimensional Euclidean Space," *Information and Control*, Vol. 5, No. 4, December 1962, pp. 279–323.
108. Porter, A., and F. Stoneman, "A New Approach to the Design of Pulse-Monitored Servo Systems," *Proceedings of the IEE*, Vol. 97, Pt. II, 1950, p. 597.
109. Rabiner, L. R., "Linear Program Design of Finite Impulse Response (FIR) Digital Filters," *IEEE Transactions on Audio and Electroacoustics*, Vol. AU-20, No. 4, October 1972, pp. 280–288.
110. Rabiner, L. R., and B. Gold, *Theory and Application of Digital Signal Processing*, Prentice-Hall, Englewood Cliffs, New Jersey, 1975.

111. Rabiner, L. R., and C. M. Rader (Eds.), *Digital Signal Processing*, IEEE Press, New York, 1972.

112.*Rabiner, L. R., R. W. Schafer, and C. M. Rader, "The Chirp z-Transform Algorithm," *IEEE Transactions on Audio and Electroacoustics*, Vol. AU-17, June 1969, pp. 86–92.

113.*Rader, C. M., "An Improved Algorithm for High-Speed Autocorrelation with Applications to Spectral Estimation," *IEEE Transactions on Audio and Electroacoustics*, Vol. AU-18, No. 4, December 1970, pp. 493–441.

114. Ragazzini, J. R., and G. F. Franklin, *Sampled-Data Control Systems*, McGraw-Hill, New York, 1958.

115. Riesz, F., and B. Sz-Nagy, *Functional Analysis*, Frederick Ungar, New York, 1955.

116. Robinson, E. A., *Statistical Communication and Detection with Special Reference to Digital Processing of Radar and Seismic Signals*, Hafner, New York, 1967, Chapters 7 and 9.

117. Royden, H. L., *Real Analysis*, Macmillan, New York, 1963.

118. Sage, A. P., and J. L. Melsa, *Estimation Theory with Applications to Communications and Control*, McGraw-Hill, New York, 1971.

119. Sandberg, I. W., "Floating-Point-Roundoff Accumulation in Digital Filter Realizations," *Bell System Technical Journal*, Vol. 46, October 1967, pp. 1775–1791.

120. Sandberg, I. W., and J. F. Kaiser, "A Bound on Limit Cycles in Fixed-Point Implementations of Digital Filters," *IEEE Transactions on Audio and Electroacoustics*, Vol. AU-20, No. 2, June 1972, pp. 110–112.

121.*Singleton, R. C., "A Method for Computing the Fast Fourier Transform with Auxiliary Memory and Limited High-Speed Storage," *IEEE Transactions on Audio and Electroacoustics*, Vol. AU-15, No. 2, June 1967, pp. 91–97.

122.*Singleton, R. C., "An Algorithm for Computing the Mixed Radix Fast Fourier Transform," *IEEE Transactions on Audio and Electroacoustics*, Vol. AU-17, No. 2, June 1969, pp. 93–103.

123. Slepian, D. and H. O. Pollak, "Prolate Spheroidal Wave Functions, Fourier Analysis and Uncertainty—I and II," *Bell System Technical Journal*, Vol. 40, No. 1, January 1961, pp. 43–84.

124. Stanley, W. D., *Digital Signal Processing*, Reston Publishing Co., Reston, Virginia, 1975.

125. Steiglitz, K., "Power-Spectrum Identification for Adaptive Systems," *IEEE Transactions on Applications and Industry*, Vol. 83, May 1964, pp. 195–197.

126.*Steiglitz, K., "Computer-Aided Design of Recursive Digital Filters," *IEEE Transactions on Audio and Electroacoustics*, Vol. AU-18, No. 2, June 1970, pp. 123–129.

127. Steiglitz, K., *An Introduction to Discrete Systems*, Wiley, New York, 1974.

128.*Stockham, T. G., Jr., "High-Speed Convolution and Correlation," 1966 Spring Joint Computer Conference, *AFIPS Proceedings*, Vol. 28, Spartan, Washington, D.C., pp. 229–233.

129. Storer, J. E., *Passive Network Synthesis*, McGraw-Hill, New York, 1957.

130. Szego, G., *Orthogonal Polynomials*, American Mathematical Society Colloquium Publications, Vol. 23, 1939, Ch. 11.

131. Tou, J. T., *Digital and Sampled-Data Control Systems*, McGraw-Hill, New York, 1959.

132. Tou, J. T., *Modern Control Theory*, McGraw-Hill, New York, 1964.

133. Tretter, S. A., and K. Steiglitz, "Power-Spectrum Identification in Terms of Rational Models," *IEEE Transactions on Automatic Control*, Vol. AC-12, No. 2, April 1967, pp. 185–188.

134. Tsypkin, Y. Z., *Theory of Impulse Systems* (in Russian), State Publisher for Physical-Mathematical Literature, Moscow, 1958, pp. 423–428.

135.*Voelcker, H. B., and E. E. Hartquist, "Digital Filtering Via Block Recursion," *IEEE*

* References with a star are reprinted in Reference 111.

Transactions on Audio and Electroacoustics, Vol. AU-18, No. 2, June 1970, pp. 169–176.

136. Weinberg, L., *Network Analysis and Synthesis,* McGraw-Hill, New York, 1962.
137. Welch, P. D., "The Use of Fast Fourier Transform for the Estimation of Power Spectra," *IEEE Transactions on Audio and Electroacoustics,* Vol. AU-15, No. 2, June 1967, pp. 70–73.
138. Whittle, P., *Prediction and Regulation by Linear Least-Squares Methods,* English Universities Press, London, 1963.
139. Widrow, B. "A Study of Rough Amplitude Quantization by Means of Nyquist Sampling Theory," *IRE Transactions on Circuit Theory,* Vol. CT-3, No. 4, December 1956, pp. 266–276.
140. Widrow, B., "Statistical Analysis of Amplitude Quantized Sampled-Data Systems," *AIEE Transactions, Applications and Industry,* No. 52, January 1961, pp. 555–568.
141. Wiener, N., *The Extrapolation, Interpolation, and Smoothing of Stationary Time Series,* Wiley, New York, 1949.
142. Wilkinson, J. H., *Rounding Errors in Algebraic Processes,* Prentice-Hall, Englewood Cliffs, New Jersey, 1963.
143. Yen, J. L., "On Nonuniform Sampling of Bandwidth-Limited Signals," *IRE Transactions on Circuit Theory,* Vol. CT-3, December 1956, pp. 251–257.
144. Zadeh, L. A., and C. A. Desoer, *Linear System Theory,* McGraw-Hill, New York, 1963.
145. Zohar, S., "Toeplitz Matrix Inversion: The Algorithm of W. F. Trench," *Journal of the Association for Computing Machinery,* Vol. 16, No. 4, October 1969, pp. 592–601.
146. Zverev, A. I., *Handbook of Filter Synthesis,* Wiley, New York, 1967.

INDEX